S0-CEE-325

Recent Results in Cancer Research 104

Recent Results in Cancer Research

Volume 95: Spheroids in Cancer Research
Edited by H. Acker, J. Carlsson, R. Durand, R. M. Sutherland
1984. 83 figures, 12 tables. IX, 183. ISBN 3-540-13691-6

Volume 96: Adjuvant Chemotherapy of Breast Cancer
Edited by H.-J. Senn
1984. 98 figures, 91 tables. X, 243. ISBN 3-540-13738-6

Volume 97: Small Cell Lung Cancer
Edited by S. Seeber
1985. 44 figures, 47 tables. VII, 166. ISBN 3-540-13798-X

Volume 98: Perioperative Chemotherapy
Edited by U. Metzger, F. Largiadèr, H.-J. Senn
1985. 48 figures, 45 tables. XII, 157. ISBN 3-540-15124-9

Volume 99: Peptide Hormones in Lung Cancer
Edited by K. Havemann, G. Sorenson, C. Gropp
1985. 100 figures, 63 tables. XII, 248. ISBN 3-540-15504-X

Volume 100: Therapeutic Strategies in Primary and Metastatic Liver Cancer
Edited by Ch. Herfarth, P. Schlag, P. Hohenberger
1986. 163 figures, 104 tables. XIV, 327. ISBN 3-540-16011-6

Volume 101: Locoregional High-Frequency Hyperthermia and Temperature Measurement
Edited by G. Bruggmoser, W. Hinkelbein, R. Engelhardt, M. Wannenmacher
1986. 96 figures, 8 tables. IX, 143. ISBN 3-540-15501-5

Volume 102: Epidemiology of Malignant Melanoma
Edited by R. P. Gallagher
1986. 15 figures, 70 tables. IX, 169. ISBN 3-540-16020-5

Volume 103: Preoperative (Neoadjuvant) Chemotherapy
Edited by J. Ragaz, P. R. Band, J. H. Goldie
1986. 58 figures, 49 tables. IX, 162. ISBN 3-540-16129-5

Hyperthermia and the Therapy of Malignant Tumors

Edited by C. Streffer

With 52 Figures and 63 Tables

Springer-Verlag
Berlin Heidelberg New York
London Paris Tokyo

Professor Dr. rer. nat. Christian Streffer

Institut für Medizinische Strahlenbiologie
Universitätsklinikum Essen
Hufelandstrasse 55, 4300 Essen 1, FRG

ISBN 3-540-17250-5 Springer-Verlag Berlin Heidelberg New York
ISBN 0-387-17250-5 Springer-Verlag New York Berlin Heidelberg

Library of Congress Cataloging-in-Publication Data. Hyperthermia and the therapy of malignant tumors. (Recent results in cancer research; 104) Includes bibliographies and index. 1. Thermotherapy. 2. Cancer-Treatment. I. Streffer, Christian, 1934–. II. Series. [DNLM: 1. Hyperthermia, Induced. 2. Neoplasms-therapy. W1 RE106P v. 104/ QZ 266 H9895] RC261.R35 vol. 104 616.99'4 s 86-29792 [RC271.T5] [616.99'406329]

Printed in Germany

Typesetting, printing, and binding: Appl, Wemding
2125/3140-5 4 3 2 1 0

Preface

Tumour therapy depends essentially on being able to destroy the clonogenic activity of tumour cells while keeping the damage to the normal tissue low. Clinical experience shows that tumour response varies greatly even if tumours with the same localisation, clinical, and histopathological staging are compared. Some tumours appear to be resistant to conventional radiotherapy (X-rays, γ-rays or fast electrons) or chemotherapy. In these cases new therapy modalities are necessary. Combined therapy modalities seem to have advantages for some resistant tumours; one possibility of such a treatment is to combine radiotherapy or chemotherapy with hyperthermia. This means that the local tumour, the tumour region or even the whole body of the patient has to be heated to temperatures between 40° to 45° C (in case of whole body hyperthermia to 42° C maximal) for a certain time (usually 30–60 min are adequate).

Hyperthermia has a long tradition in medicine as a treatment modality for various diseases. Inscriptions of the old Egyptians and texts of the Greeks have pointed out its importance. Usually whole body hyperthermia has been used by the induction of fever. Local hyperthermia began around 1900 when Westermark treated unresectable cervix carcinomas with hot water in a metallic coil. By the beginning of this century an increase of radiation effects was hypothesised with hypothermia and later observed. However, only in the 1960s and 1970s were systematic investigations started which showed radiosensitisation and chemosensitisation by hyperthermia in cells and tissues including tumours.

The biological data already reported from intensive studies substantially favour the clinical application of hyperthermia in combined modalities for tumour therapy. However, it has been shown that in order to understand the action and mechanisms of hyperthermia and in order to develop optimal therapy modalities cell biological phenomena as well as molecular, metabolic, and physiological phenomena must be considered. Although these mechanisms are not completely understood far larger problems are presented yet, by the techniques employed for raising temperatures in tumours in a desired homogeneous and local mode, besides measuring and recording temperatures. Nevertheless, more and more clinicians are trying to use hyperthermia in tumour therapy. Several thousand patients have been treated with this modality in various regimes during past few years. It appears therefore reasonable to review the data from the literature in a critical way. Hyperthermia is certainly not a miracle but it may help to improve the therapeutic success for some tumours.

I have to thank the authors of the different chapters for their very helpful cooperation, the editors, Professor Herfarth and Professor Senn, for their stimulating and motivating invitation to publish this book in this series with a high scientific reputation. I also would like to thank the publisher, Springer-Verlag, for helping to make this book successful.

Essen, Dezember 1986 Christian Streffer

Contents

List of Contributors

R. Engelhardt
Medizinische Universitätsklinik
Hugstetter Strasse 55, 7800 Freiburg, FRG

J. W. Hand
MRC Cyclotron Unit, Hammersmith Hospital
Du Cance Road, London W12 OHS, United Kingdom

F. Kallinowski
Abteilung für Angewandte Physiologie, Universität Mainz
Saarstrasse 21, 6500 Mainz, FRG

M. Molls
Strahlenklinik, Universitätsklinikum Essen
Hufelandstrasse 55, 4300 Essen 1, FRG

E. Scherer
Strahlenklinik, Universitätsklinikum Essen
Hufelandstrasse 55, 4300 Essen 1, FRG

C. Streffer
Institut für Medizinische Strahlenbiologie, Universitätsklinikum
Essen, Hufelandstrasse 55, 4300 Essen 1, FRG

D. van Beuningen
Institut für Medizinische Strahlenbiologie, Universitätsklinikum
Essen, Hufelandstrasse 55, 4300 Essen 1, FRG

P. Vaupel
Abteilung für Angewandte Physiologie, Universität Mainz
Saarstrasse 21, 6500 Mainz, FRG

Heat Delivery and Thermometry in Clinical Hyperthermia

J. W. Hand

MRC Cyclotron Unit, Hammersmith Hospital,
Du Cane Road, London W12 OHS, United Kingdom

Introduction

The potential role of hyperthermia in the management of human cancer will be difficult to assess until heating techniques and thermometry systems capable of delivering and monitoring safe, predictable and reproducible treatments are developed. These requirements present major challenges to biomedical engineers and physicists and are currently far from being solved. However, significant improvements to the technology associated with clinical hyperthermia have been achieved during the 1980s and the purpose of this chapter is to present an overview of some of the techniques which are currently available.

In the first section two non-invasive techniques in which electromagnetic power is used to induce hyperthermia in lesions located within 3-4 cm of the body surface are discussed. The much greater problem of treating tumours in deep anatomical sites is the subject of the second section. The techniques described there deposit energy throughout a large volume and differences in energy absorption and blood flow between tumour and normal tissues are relied upon to produce higher temperatures in the tumour than in normal tissue. The third section discusses invasive methods of inducing hyperthermia. Such techniques can be applied not only to superficial lesions but also to those in deep anatomical locations when combined with surgical techniques. Alternative heating methods employing ultrasound are discussed in the fourth section. Selective heating in some deep sites is feasible since the small wavelength associated with ultrasound allows energy to be deposited within relatively small volumes of tissue. In the fifth section various thermometric techniques are reviewed and the problems associated with thermal dosimetry are outlined. Finally, some comments on likely developments in the future are provided in the sixth section.

The topic of systemic hyperthermia in which the whole body is raised to 41°-42° C is not addressed here. The background to this approach and the relevant techniques are reviewed elsewhere (Bull 1983; Milligan 1984; Engelhardt 1985).

Radiofrequency and Microwave Applicators for Localised Hyperthermia

Radiofrequency Applicators

Radiofrequency applicators can be classified into two groups - electric or magnetic devices. In the frequency range commonly used for these methods (8-30 MHz), heating is predominantly due to conduction currents in the tissues. Dimensions of applicators and of the treated regions are considerably smaller than the wavelength and so insight into

Recent Results in Cancer Research. Vol 104

these techniques can be gained through quasi-stationary models in which electric and magnetic fields are assumed to be distributed in space as though they were static fields.

A common electric applicator consists of a small electrode, usually no more than 5–6 cm in diameter, placed above the lesion to be treated, with a second, larger electrode positioned beneath the patient so that the lesion is between the electrodes. A characteristic of this technique is the presence of large electric fields near the edge of the small electrode (Wiley and Webster 1982), and so a bolus is required between the electrode and the skin to prevent overheating of superficial tissue (Brezovich et al. 1981). When the thickness of the bolus is sufficient to contain the regions of high electric fields around the electrode, the impedance of the applicator is relatively insensitive to inevitable patient movements, thus simplifying treatment (Bini et al. 1983).

Calculations of the specific absorption rate (SAR) produced in inhomogeneous tissues by this technique have been reported, e.g. by Doss (1982) and Armitage et al. (1983). Tissue inhomogeneities disturb current flow and can produce undesirable hot or cold spots, as shown in Fig. 1. The tendency for the technique to produce excessive heating in superficial fat layers (Guy et al. 1974) can also be seen. This limits the clinical use of the tech-

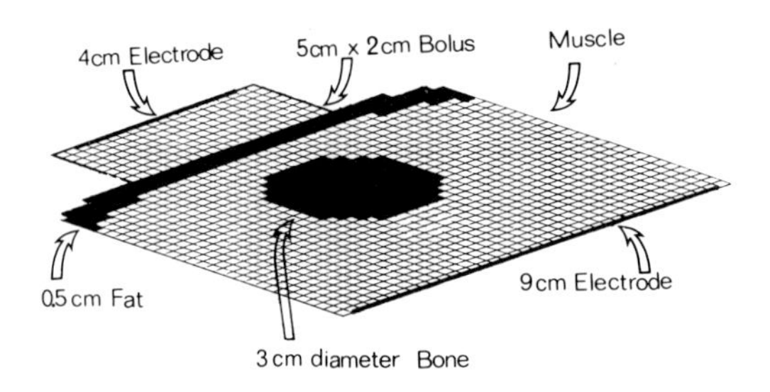

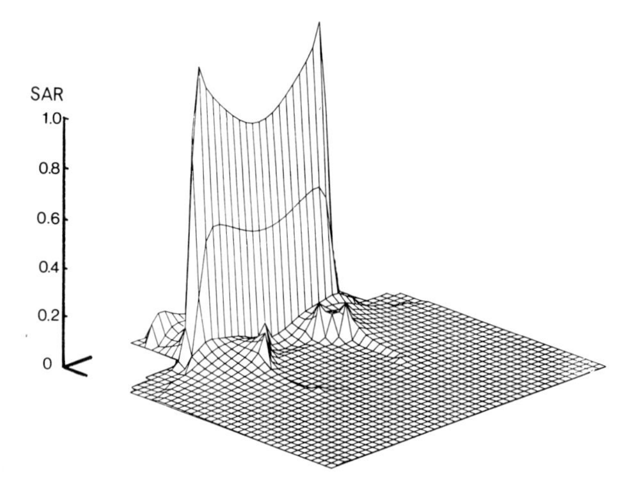

Fig. 1. Specific absorption rate (SAR) in a fat/muscle/bone model due to a 27.12-MHz electric applicator. The muscle ($\sigma = 0.6$ S m^{-1}) contains a 3-cm-diameter region of bone ($\sigma = 0.05$ S m^{-1}). A 4-cm electrode is spaced from the 0.5-cm-thick fat layer by a 2-cm-thick bolus ($\sigma = 0.7$ S cm^{-1}). The return electrode is 9 cm wide and is spaced 10 cm away from the first. SAR is normalised to the peak in the fat layer. The SAR in bolus within 0.5 cm of the electrode is not shown. The grid spacing is 0.25 cm

nique to patients with fat layers less than 1.5–2 cm although a possible solution in which the permittivity and conductivity of the adipose tissue is increased by injecting isotonic saline solution has been suggested by Bini et al. (1984).

Magnetic applicators for superficial treatments often consist of a planar coil with up to three or four turns (pancake coil) in which a 10- to 30-MHz current is impressed. The coil, positioned parallel to and about 3 cm above the skin, produces induced currents in the tissue which flow predominantly parallel to fat/muscle boundaries. The SAR peaks beneath the coil windings and is zero on the axis of the coil (Guy et al. 1974). Coil applicators do not require bolus and so the site being treated is easily observed. The size and shape may also be chosen to suit particular treatments (Lerch and Kohn 1983). Coupling efficiency of coil applicators is low, since the coil must be spaced from the tissues to avoid excessive heating of skin and fat (Hand et al. 1982). Recently, Bach Andersen et al. (1984) described a 150-MHz inductive applicator with linear current-carrying conductors. This applicator produces a more uniform SAR distribution than planar coils but both types have similar penetration (3–4 cm) in tumour and muscle tissues. A larger inductive applicator based on similar principles to Bach Andersen's device and consisting of a single, square current-carrying loop is described by Kato and Ishida (1983).

Microwave Applicators

At frequencies above about 200 MHz, microwave applicators based on cylindrical waveguides with various cross sections or on microstrip or similar techniques have been developed for use in direct contact with the skin or with a bolus in contact with the skin. Since the dimensions of these applicators can be made comparable with the wavelength, energy is transmitted into the tissue in a wave propagating from the aperture of the applicator.

For simplicity, applicators have often been designed with rectangular cross section and operated at a frequency 10%–30% greater than the lowest frequency at which propagation can occur in the applicator. For a given operating frequency, the linear dimensions of an applicator may be reduced from those of an air-filled device if it is loaded with a suitable dielectric material. A drawback of these simple applicators is that only approximately 60% of the aperture provides effective heating. Applicators using different excitation and/or geometry which may achieve more uniform heating have been developed (e.g., Stuchly and Stuchly 1978; Kantor and Witters 1983; Vaguine et al. 1982; Lin et al. 1982). In particular, waveguide applicators with a ridged cross section offer a number of advantages (Turner 1983). Recently, Hand and Hind (1986) reviewed the designs and performances of several types of waveguide applicators.

Microwave applicators based on microstrip or related techniques can offer a number of advantages including small size, low weight and the ability to conform to tissue surfaces. Mendecki et al. (1979) described a 2450-MHz printed circuit antenna loaded with dielectric powder which could conform closely to tissue contours. A microstrip ring applicator was reported by Bahl et al. (1980). The operating frequency, bandwidth and heating characteristics of microstrip applicators can depend critically on the load presented to them (Bahl and Stuchly 1980; Sandhu and Kolozsvary 1983) although predictable performance can be achieved if suitably thick bolus is used. Johnson et al. (1984) developed a series of low-profile applicators which are relatively insensitive to tissue load. They consist of resonant patches sandwiched between dielectric slabs and operate at frequencies between 200 MHz and 915 MHz. By incorporating ferrite material into this type of applicator, an operating frequency as low as 27 MHz can be achieved (Johnson et al. 1985). Tanabe et al. (1983) described a microstrip spiral antenna which offers good coupling, broad bandwidths and a circularly symmetrical treatment field. These authors have developed a 915-MHz array of such applicators designed to heat large areas of the chest wall. An array of rectangular patch radiators designed for the same purpose was reported by Sandhu and Kalozsvary (1984).

The dependence of heating characteristics on the size and operating frequency of applicators has been discussed by several authors (e.g., Guy 1971; Turner and Kumar 1982; Hand and Johnson 1986); in general, applicators small in relation to the wavelength exhibit relatively poor penetration. Figure 2 shows the results of a calculation of the absorbed power density beneath the centre of an applicator with a 10 × 10 cm aperture. Figure 3 shows the results of similar calculations for various sized applicators operating at 434 MHz.

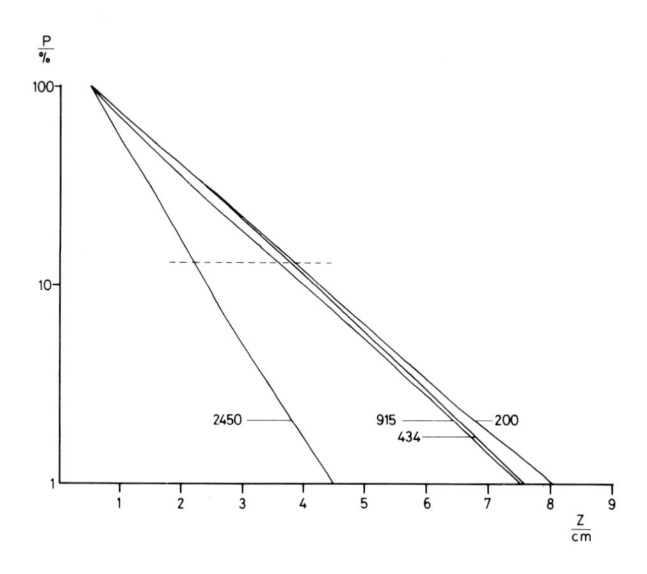

Fig. 2. Frequency dependency of penetration into uniform muscle from applicators with 10 cm × 10 cm aperture. *P* is the absorbed power density normalised to the peak value at depth *Z* = 0.5 cm

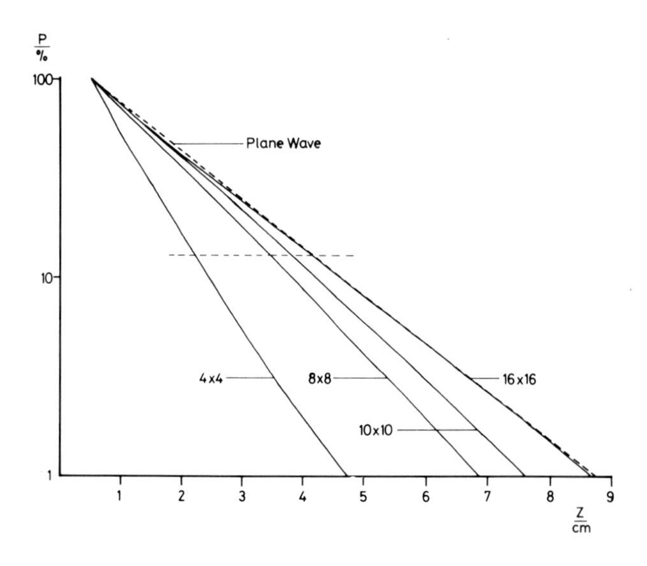

Fig. 3. Penetration into uniform muscle from 434-MHz applicators for various aperture sizes. *P* is the absorbed power density normalised to the peak value at depth value at depth *Z* = 0.5 cm

Moderate focussing of energy deposition and the ability to treat larger areas and to produce more uniform temperature distributions are amongst possible advantages of systems employing more than one applicator. The choice of operating frequency for a multiapplicator system is determined essentially by a trade-off between the degree of localisation and the penetration required. Those systems operating at frequencies below 100 MHz and which induce regional hyperthermia are discussed in the next section. At high frequencies good localisation is possible (e.g. Gee et al. 1984) but penetration in tissues with high water content is limited.

Arcangeli et al. (1984) modelled the SAR distribution within the thorax due to 50 dielectrically loaded applicators spaced equidistantly around the chest cross section and showed that SAR could be maximised within the lung when phases and amplitudes of the applicators were optimised. Takahashi et al. (1985) describe a 433-MHz waveguide applicator in which some convergence of the fields from four elements is achieved. Hand et al. (1985) compared SAR distributions due to 3×3 and 4×4 arrays of small aperture sources with those of single apertures with the same overall dimensions. Frequencies of 300, 434 and 915 MHz were considered. When the elements were driven with equal phases and amplitudes, the useful area under the aperture was increased, although penetration into the muscle-like medium remained the same as for the single-aperture applicator. When suitable adjustments to the relative phases (and amplitudes, in the 915-MHz case) were made, some ‘focussing’ and a moderate improvement in effective penetration depth was achieved. The dimensions of the focal volumes were approximately 60% of the wavelength in the medium. Burdette et al. (1985) described an hexagonal packed, 7-element, 915-MHz array which produced greater power deposition at depth than a single-element applicator. Planar arrays such as those described by Burdette et al. (1985) and Hand et al. (1985) achieve only moderate improvements in penetration, and power densities near the surface may still limit clinical applications. Different excitations of aperture applicators may prove more useful (Bach Andersen 1985a; Taylor 1984). Nevertheless, the ability to adjust the amplitudes of individual elements in a multielement applicator should improve the uniformity of temperature distributions over those usually observed during the clinical use of single applicators.

Regional Hyperthermia Devices

Radiofrequency (RF) heating techniques employing capacitive electrodes have often been used in attempts to induce deep body hyperthermia. As mentioned earlier, the large electric fields close to the edges of the electrodes can lead to tissue burns unless a sufficiently thick bolus is included between electrode and tissue. Attempts to spare superficial tissues by sequentially switching the RF power between several pairs of electrodes have had limited success (LeVeen et al. 1980; Scott 1957). Unless properly isolated, de-energised electrodes can provide a low-impedance return path leading to significant current densities in tissues beneath them whilst the depth of heating beneath the energised electrodes is limited because they are often separated by a distance greater than their diameter (Armitage et al. 1983).

Interest in RF capacitive techniques in Japan has resulted in the development of commercially available machines operating at 13.56 MHz and 8 MHz which employ large electrodes (e.g. 20 cm or 25 cm diameter). Since such electrodes are separated in clinical use by a distance comparable with, or smaller than, their diameter, significant absorbed power densities can be expected deep in the body (Hiraoka et al. 1984; Song et al. 1986).

However, an inherent characteristic of the capacitive technique is that the electric fields are predominantly perpendicular to the boundaries of superficial tissue layers, leading to high absorbed power density in the low-permittivity adipose tissue. Cooling the electrodes can minimise the problem in some cases (Matsuda et al. 1984) but this becomes ineffective when fat layers are greater than approximately 1.5 cm thick (Hiraoka et al. 1985). This may be a severe limitation in the treatment of many European and North American patients. A further limitation is that inhomogeneities in electrical conductivity of tissues within the body affect the RF current flow and current density (Armitage et al. 1983; Oleson 1982a) and lead to undesirable hot spots, making the qualitative prediction of heating patterns difficult.

There has also been interest in inductive applicators intended for heating deep-seated tumours. One type, the Magnetrode (Storm et al. 1981), consists of a single cylindrical electrode formed from copper sheet and placed around the patient, concentric with his or her longitudinal axis (Fig. 4). A 13.56-MHz current is driven through the electrode and the alternating magnetic field produced creates eddy currents within the patient. Theoretical calculations and measurements in homogeneous circular cylindrical phantoms show the SAR produced by a concentric electrode of this type is zero on the axis of the phantom and varies as the square of the distance from the axis. Oleson (1982a), Armitage et al. (1983) and Hill et al. (1983) have shown that for inhomogeneous, realistic anatomical phantoms (not necessarily concentric with the applicator), significant variations in SAR can occur over relatively short distances where tissues of low and high conductivity are adjacent, e.g. around the sternum and sacrum. Clinical experience has shown that patients often experience pain in these regions.

It is emphasised that the resulting temperature distribution associated with any heating technique depends not only on SAR but also on geometry, thermal conductivity and specific heat of the tissues, the degree of skin cooling and blood flow. Several assessments of the temperature distributions associated with the concentric coil based on the bioheat transfer equation suggest that, in general, tissues within about 6 cm of the skin can be expected to reach unacceptably high temperatures before the perfused regions of deep-lying tumours reach therapeutic temperatures (Halac et al. 1983; Paulsen et al. 1984b). These findings have been borne out by general clinical experience (Oleson et al. 1983b; Sapozink et al. 1984). Proponents of the technique (Storm et al. 1982) argue that effective heating at depth may be achieved in vivo when the rate of heating is sufficiently low to allow thermal redistribution to occur throughout the heated volume. However, Paulsen et al. (1984b) have modelled transient temperature distributions and suggest that this approach is unlikely to achieve effective heating in deep-seated tumours.

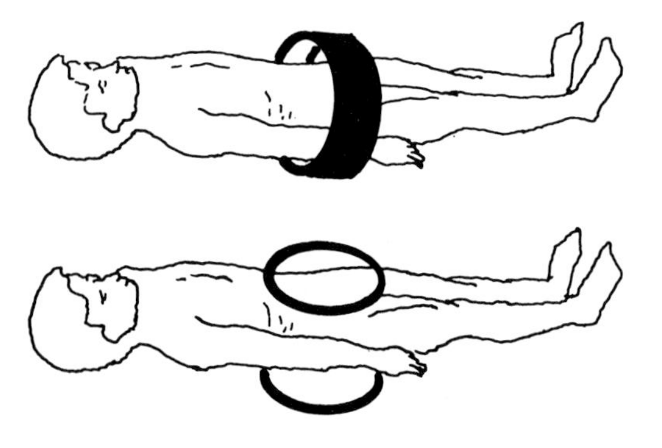

Fig. 4. Inductive applicator - concentric coil

Fig. 5. Inductive applicator - coaxial coils

Another inductive technique uses a coaxial pair of coils placed on either side of the body (Fig. 5). Unlike the concentric coil described above, this arrangement can produce absorbed power density at positions off the common axis of the coils in coronal and sagittal planes in the patient. Oleson et al. (1983a) suggest that the coil diameters should be within the range 18–28 cm and that the separation of the coils should be approximately 30 cm. Smaller coils lead to excessive heating of superficial tissues whilst larger coils can lead to problems of excessive heating in the patient's flanks. Some improvement in power deposition may be achieved by using solenoidal coils rather than single current loops. The method has been used to treat a range of tumours in the thorax and abdomen (Oleson 1982b; Corry et al. 1982a). In practice, the power deposition is non-uniform and can be strongly dependent upon geometry and the electrical properties of the patient. Oleson et al. (1983a) point out that such dependencies of the power deposition on the load presented to the coils may compromise the performance of systems in which coils are scanned over the patient (e.g. von Ardenne et al. 1977). Because of the somewhat complementary nature of the heating characteristics of the coaxial pair of coils and the concentric coil, some patients may benefit from alternate treatments with these magnetic induction systems.

Another device under investigation for regional hyperthermia is the helical coil applicator in which the axial component of the current in the winding is associated with an axial electric field which is relatively uniform across planes perpendicular to the coil axis. This type of coil applicator produces an electric field distribution which is conducive to power deposition at depth (Morita and Bach Andersen 1982; Brezovich et al. 1982; Bach Andersen 1985b) and should not be confused with the inductive applicators described above which produce an axial magnetic field. Ruggera and Kantor (1984) used self-resonant single-layer helical coils in which the length of the coil winding was adjusted to be either $\lambda/2$ or λ (half-wave or full-wave operation) and showed that the maximum of the axial electric field could be located in the central parts of these coils. Operating frequencies of 13.56, 27.12 or 40.68 MHz were chosen. It was suggested that the uniformity of heating across transverse planes could be optimised when the coil length was twice the coil diameter for half-wave operation (or four times the coil diameter for full-wave operation). Hagmann (1984) and Hagmann and Levin (1984) analysed the helical coil applicator and suggested that a modest improvement in the ratio of power deposition in deep muscle to that in superficial fat could be achieved by reducing the operating frequency and increasing the pitch angle and/or the radius of the helical winding. Hagmann and Levin (1984) also pointed out that these design parameters were interdependent for self-resonant coils and greater optimisation may be achieved if the coils were externally tuned to a lower frequency or if a travelling wave mode of operation was adopted.

The development of aperture-type applicators for regional hyperthermia is promising since a linearly polarised electric field parallel to the patient's longitudinal axis can be readily achieved. However, a frequency below 100 MHz must be used to achieve effective heating at depths and so a major consideration for this type of applicator is the design of an aperture source with suitable dimensions.

Paglione et al. (1981) reported single-ridged, water-filled applicators designed to operate at 27.12 MHz. These applicators had effective apertures ranging from 25.4 cm × 7.6 cm to 29.6 cm × 13.7 cm with a bolus incorporated to provide skin cooling and to avoid excessive heating in superficial tissues due to high electric fields close to the aperture. According to van Rhoon et al. (1984), the effective penetration depth in muscle phantoms of these applicators is approximately 6 cm. Using an iterative technique to solve the electromagnetic problem, Van den Berg et al. (1983) developed a two-dimensional model to predict heating patterns within the pelvic region due to a pair of ridged waveguide applicators. The clinical use of this type of applicator has been described by Marchal et al. (1985).

A multiple-aperture applicator has been developed by Turner (1983, 1984). The annular phased array (APA) consists of 16 radiating apertures arranged in pairs within an octagonal array, with each aperture directing its energy towards the centre of the array where the patient is positioned. In versions of the device currently used in clinical investigations, all apertures are driven in the same phase. Some control of amplitude is provided in that the 16 apertures are divided into quadrant groups of 4, with each quadrant selected to be 'on' or 'off'. The width of the applicator (i.e. the dimension along the patient's longitudinal axis) is 46 cm and is formed by a pair of 20 cm × 23 cm aperture sources. The internal diameter of the array is 51 cm. The space between the aperture sources and the patient is filled with a distilled water bolus to improve energy transfer to the patient, to reduce stray field levels and to control skin temperature. An operating frequency in the range 50–100 MHz is used. Measurements in phantoms and large animals indicate that fairly uniform power deposition is achieved at depth at these frequencies.

Iskander et al. (1982) used a moment method to calculate SAR in a realistic cross section of the thorax heating by an APA operating at 70 MHz. In a later paper, Iskander and Khoshdel-Milani (1984) calculated steady state temperature distributions for this case and stressed the importance of blood perfusion rates and thermal parameters used in the model. Paulsen et al. (1984a) selected CT scans from six patients with tumours in pelvic, abdominal or thoracic regions and, using a finite element technique, determined electric fields and SAR within these cross sections due to an APA operating at 100 MHz. The maximum electric fields in the tissue were usually in the central areas of the cross sections but only a twofold variation between maximum and minimum fields was found. SAR in tumours ranged from 20% to 60% of maximum SAR. Steady state temperature distributions were also predicted for a range of blood perfusion rates in the different tissue regions. In an extension of this study, Strohbehn et al. (1986) calculated SAR and temperature distributions for two further pelvic tumours and for a tumour located in the thigh. In a pelvic case, flushing the bladder with deionised water resulted in more effective tumour heating. Considerable improvement was also achieved if the amplitude and phase of the aperture sources of the APA were adjusted so that tumour SAR was increased. In all the cases considered by Paulsen et al. (1984a) and Strohbehn et al. (1986), tumours with little or no blood flow located in regions with moderate blood flow appeared to be the easiest to heat but even in these cases significant fractions of the tumour failed to reach target temperature (42° or 43° C).

Current two-dimensional computer models probably present an optimistic estimate of the clinical performance of hyperthermia devices and some of their limitations have been discussed by Lagendijk and de Leeuw (1986). Clinical experiences suggest that the greatest potential of an APA device is in treatment of tumours in the pelvic region; systemic heating is often a limiting factor in the treatment of abdominal tumours whilst the large water bolus presents a difficulty to patients with tumours in the thorax (Gibbs 1984; Emani et al. 1984). Sapozink et al. (1984, 1985b) reported the difficulty of maintaining a high fraction of tumour temperature measurements above 43° C for extended periods, a finding consistent with the predictions of the computer models.

Other regional hyperthermia applicators under development include a coaxial transverse electromagnetic (TEM) wave applicator (Lagendijk 1983) and a coaxial segmented cylindrical applicator (Raskmark and Bach Andersen 1984). These devices each have advantages and disadvantages with respect to the APA but it is likely that all three devices will encounter similar problems in clinical use.

Interstitial Techniques

The use of invasive techniques for heating local volumes of tissue has increased rapidly over the past 4 or 5 years. From the physical point of view these techniques offer features unobtainable with other methods of heating. For example, localised heating is possible not only in superficial regions but also in deep sites since energy deposition in the tissues is limited almost entirely within the implanted region. In addition the uniformity of temperature distributions produced within relatively large volumes of tissues can compare very favourably with that associated with various non-invasive methods. The interstitial methods adopted are also easily combined with existing techniques of brachytherapy. In general, invasive heating methods fall into one of three categories – RF electrodes, microwave antennas and heating of implanted ferromagnetic material.

Radiofrequency Electrodes

Doss and McCabe (1976) described a method of inducing localised hyperthermia in which electrical current of low frequency (200 kHz–1 MHz) was passed through the treatment volume defined by electrodes which were either interstitial or a combination of interstitial and superficial in direct contact with the skin. At these frequencies the displacement currents are small and the resistive currents are dominant. This technique was referred to as localised current fields (LCFs). It has subsequently been developed and used in the treatment of patients with tumours in a wide range of anatomical sites (Sternhagen et al. 1978; Joseph et al. 1981; Manning et al. 1982; Vora et al. 1982; Aristazabal and Oleson 1984; Brezovich et al. 1984b; Cosset et al. 1984; Frazier and Corry 1984).

Strohbehn (1983) reported the results of a two-dimensional calculation of the absorbed power density and temperature distribution produced in homogeneous tissue for linear arrays of needle electrodes. The absorbed power density falls off rapidly (r^{-2}) away from each electrode and, if all electrodes are driven with the same voltage, is greater around electrodes near the end of the array than those near the centre. The temperature distributions are dependent upon the spacings between electrodes and between arrays and the blood flow rate in the tissue. For blood flow rates similar to that of resting muscle (~3 ml/100 g min) relatively uniform temperatures can be achieved with the electrodes 1–2 cm apart. If the blood flow rate is an order of magnitude greater, however, electrode spacing should be 1 cm or less. In this model, improved temperature distributions were obtained if equal absorbed power density was produced at each electrode. A practical means of achieving this in which the RF currents are switched between electrodes in a suitable temporal and spatial sequence is described by Astrahan and Norman (1982).

It is important that the implanted electrodes are parallel to avoid gross non-uniformities in the temperature distribution. To avoid excessive heating of normal tissue around the entry and exit sites of the electrodes, it is necessary to insulate the electrodes from the tissue in these regions. For example, Joseph et al. (1981) sheathed the electrodes with either heat-shrink plastic or polytetrafluorethylene catheters. Cosset et al. (1984) describe the use of tubes with flexible plastic ends attached to metallic central regions in which the RF power is brought to the central region by a coaxial cable. Provision is also made for a thermistor probe to be inserted into the assembly. The use of conventional thermistors or thermocouples in the presence of the relatively low frequency currents used in these methods of interstitial hyperthermia is free from severe artefacts which are associated with other electromagnetic heating techniques.

Microwave Antennas

Taylor (1978, 1980) reported the design of an implantable microwave antenna formed from subminiature semirigid coaxial cable terminated by extending the central conductor beyond the outer conductor by a distance of about one-fourth of a wavelength in the tissue. The heating pattern associated with this type of directly implanted antenna is elliptical - typically 1-2 cm in diameter and 3-4 cm long. The length of the antenna can be varied by choosing the operating frequency but there is little frequency dependence in the radial dimension of the heated volume since the fall off in electric field for these antennas is determined primarily by near-field structure rather than attenuation in the tissue (Strohbehn et al. 1979; Swicord and Davis 1981; King et al. 1983). Several variations in the design of implantable microwave antennas, aimed at improved performance, have been reported. For example, di Sieyes et al. (1981) investigated the performance of a non-insulated antenna, a fully insulated antenna and antennas in which the central conductor was either fully insulated or exposed to the tissue only at its tip and found that the fully insulated antenna was optimum. Details of practical antennas, typically 1 mm in diameter and designed for insertion into nylon catheters, are given by Trembly et al. (1982), Strohbehn et al. (1982) and Lyons et al. (1984). A coaxial antenna with part of the outer conductor removed to form a slot is described by Samaras (1984).

The clinical applications of a single implanted antenna would appear to be limited because of the small extent of the heating pattern in the radial direction. To overcome this, the use of arrays of these antennas have been studied by several groups of workers. Strohbehn et al. (1982) have reported the results of a two-dimensional calculation of temperature distributions produced in a homogeneous medium by circular arrays of antennas. The effects of blood flow, number of antennas, spacing between antennas and operating frequency on the distribution were considered. Four antennas spaced at 2 cm could be expected to produce therapeutic temperatures over an area of about 2×2 cm for blood flow rates similar to that of resting muscle [3 ml/(100 gmin)]. If the blood flow rate was an order of magnitude greater then 6 antennas spaced at about 1.4 cm were required to heat approximately the same area whilst ten antennas at approximately 1.5-cm spacing were required to produce an acceptable temperature distribution over an area 5 cm in diameter. In the case of this larger area the mean density of antennas was 1/2 cm^2, compared with 1 or more/cm^2 for the smaller areas. This study showed that acceptable frequencies for such an array were within the range 0.5-1 GHz but that the temperature distribution was controlled primarily by the geometry of the array. An interstitial microwave array system is described by Wong et al. (1984) whilst the improvement to temperature distributions achieved by air cooling the antenna/catheter implantation is discussed by Trembly et al. (1984) and Bicher et al. (1984a).

An advantage of interstitial microwave antennas over implanted RF electrodes is that fewer implants are required to heat a given volume of tissue. Whilst this is a marked advantage when administering hyperthermia alone it is less important when the technique is combined with brachytherapy since the number of implants is then determined by radiation dosimetry requirements. A disadvantage of the microwave techniques compared with RF electrodes is that it is more difficult to adjust the heating pattern along the antenna since for a given frequency there are preferred lengths for the antenna and, in general, heating is greatest near the drive point of the antenna. In addition, conventional thermometry faces the usual problems encountered in the presence of microwave fields.

Reports of the clinical use of interstitial microwave antennas have been given by Coughlin et al. (1983), Salcman and Samaras (1983) and Bicher et al. (1984b).

Heating of Ferromagnetic Implants

Localised heating of tissues may be achieved by implanting ferromagnetic materials into the region of interest and subjecting them to an RF magnetic field. An early report of the use of this method in experimental animals was given by Medal et al. (1959). Later reports include those of Moidel et al. (1976), who investigated the use of several implant materials to produce controlled, localised hyperthermia in the brain, and Burton et al. (1971), who used a related method to coagulate brain tissue. The study of Burton et al. was one of the first to take advantage of the fact that under certain conditions the heating rate of an implant is particularly dependent upon the magnetic permeability of the implanted material. Self-regulating 'thermo-seeds' of nickel-palladium with Curie points consistent with tissue coagulation were developed to avoid the elaborate procedures necessary to achieve predictable and repeatable temperatures.

Recently there has been renewed interest in developing these techniques for localised hyperthermal treatments. Stauffer et al. (1984a, 1984b) and Atkinson et al. (1984) have discussed induction heating of tissues and implanted ferromagnetic material. In tissues, the absorbed power density ($W\,m^{-3}$) at a distance r from the patient's 'axis' is proportional to $(H\,r\,f)^2$ where H and f are the amplitude and frequency of the magnetic field. On the other hand, long cylindrical ferromagnetic implants arranged parallel to H produce a heating power ($W\,m^{-3}$) per unit length proportional to $(H^2 f^{1/2} \mu^{1/2} a)$ where μ and a are the permeability and radius of the implant. Heat production by the implants is considerably reduced when they are perpendicular to the direction of H. Because of the different frequency dependencies of power absorption, considerably greater heating of the implants than induction heating of the tissues can be achieved if the frequency of the magnetic induction field is sufficiently low. Stauffer et al. (1984a) suggest that frequencies below about 2 MHz should be acceptable for heating implants within a dielectric load up to 11 cm diameter. Atkinson et al. (1984) argue that a versatile system could be designed using an induction frequency of 200 kHz or lower and 1 mm diameter implants with $\mu > 8$. The density of implants should be about 1 cm^{-2}. This method of heating tissue by conduction from the hot implants has the advantage that it is independent of dielectric inhomogeneities present in the target volume, a characteristic which is particularly useful in treating tumours in the oral cavity. A greater advantage of the technique is the potential to use 'constant temperature seeds' in which the permeability, and hence the heating power, of the implants decreases rapidly above the maximum temperature desired in the hyperthermal treatment.

Brezovich et al. (1984a) describe the use of a nickel-copper alloy (70.4% nickel: 29.6% copper) and show that the heating power of 0.9-mm-diameter implants of this material decreases rapidly with increasing temperature above about 45° C. Workers at the University of Arizona, Tucson (L. Demer, M. Daneto, and T. C. Cetas, unpublished work) are investigating nickel-silicon alloys which pass through a broad Curie point transition between 40° and 50° C. Numerical simulations (Brezovich et al. 1984a; Matloubieh et al. 1984) suggest that the use of such 'constant temperature' seeds should result in improved temperature distribution compared with 'constant power seeds' in the presence of tissue and blood flow inhomogeneities and irregular implant geometries.

Ultrasound Applicators

Detailed dicussions of the acoustic properties of tissue are found in Dunn and O'Brien (1978) and Wells (1977). For the purposes of this chapter, three points should be noted. Firstly, at the frequencies of interest in hyperthermia (300 kHz to 3 MHz) the penetration of ultrasound into soft tissue is greater than that associated with microwave techniques. Secondly, diffraction spreading of the beam from conveniently sized transducers (4–10 cm diameter) is small because of the small wavelength of ultrasound (of the order of 1 mm in soft tissue). This means that well-defined beams and good localisation are possible. Final-

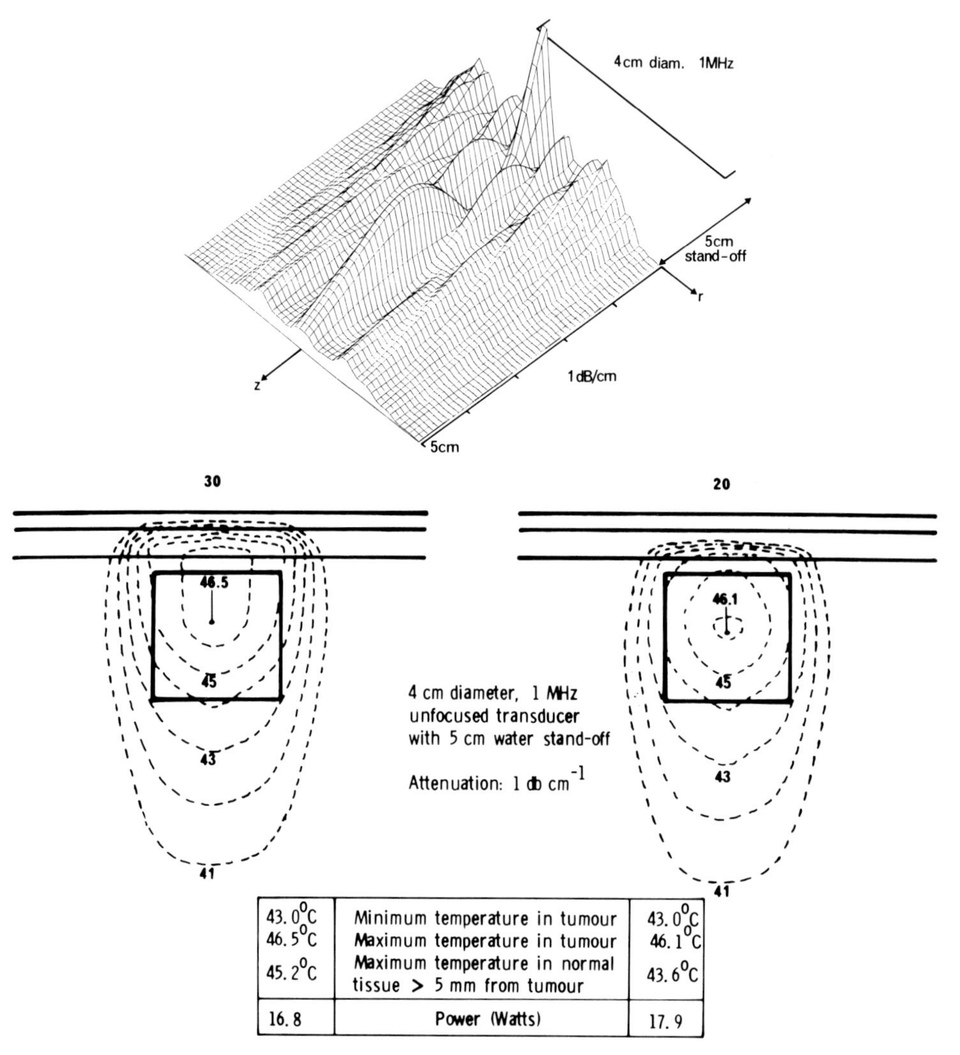

43.0°C	Minimum temperature in tumour	43.0°C
46.5°C	Maximum temperature in tumour	46.1°C
45.2°C	Maximum temperature in normal tissue > 5 mm from tumour	43.6°C
16.8	Power (Watts)	17.9

Fig. 6. *Above:* Distribution of absorbed power density due to unfocussed 1-MHz ultrasound transducer 4 cm in diameter. A 5-cm-thick water stand-off is assumed to be between the transducer and the tissue having an attentuation of 1 dB/cm. *Below:* Two-dimensional temperature distributions due to absorbed power density distribution shown *above.* Thermal properties of skin, fat and muscle layers are those assumed by Hand et al. (1982). In particular, the blood flow in the 'tumour' region is assumed equal to that in muscle. The temperature of the water bolus is 30° C *(on the left)* and 20° C *(on the right)*

ly, the significant differences in acoustic impedance between soft tissue and either gas or bone lead to considerable reflection with little transmission of the beam across such interfaces.

Ultrasound transducers suitable for hyperthermia are usually cut from a piezoelectric material such as lead zircinate titanate 4 (PZT4). Several groups of workers have used transducers in the form of circular discs with plane faces driven at frequencies between 1 MHz and 3 MHz to treat patients with tumours in superficial sites (Marmor et al. 1982; Marchal et al. 1982; Corry et al. 1982b). Such applicators produce adequate temperatures at depths of 3–4 cm in soft tissue (Fig. 6). Recently, a thin flexible ultrasound applicator has been developed by Dickinson (1984a). A prototype consists of a 3 × 3 matrix of 2-MHz piezoelectric elements, each 2 cm square, bonded to a copper foil envelope. The nine elements can be driven in parallel, or in groups of three or individually giving control of the heating pattern. When the elements are driven individually and sequentially, near field effects can be contained in a thin water stand off because of the small size of the elements.

The ability to focus ultrasound permits selective deposition of energy at depth, with the possibility of producing heating in deep tissues. The methods commonly used to focus ultrasound for hyperthermia are based on transducers with a spherical concave face or plane transducers in combination with an acoustic lens. Penttinen and Luukkala (1976) describe a calculation of the field distribution from a concave transducer. As shown in Fig. 7 focal volumes are characteristically ellipsoidal, typically a few millimetres in diameter and a few centimetres long (i.e. length/diameter ~10), and smaller than tumour volumes usually presented at hyperthermia clinics. One method of overcoming this problem is to move the transducer mechanically over the tissue. For example, Lele (1983) described a system which can selectively heat tumours at a depth of 3 cm. In this example the transducer is scanned in a conical manner so that the ultrasound is introduced into the tissue through a 'window' of greater area than the area presented to the beam by the tumour target. A technique such as this is necessary since the high gain (intensity at focus/intensity at skin) of a stationary focused transducer is a result of the ultrasound energy (accounting for attenuation where appropriate) passing through an area of skin much larger than the cross-sectional area of the focus. If the focussed transducer is simply scanned uniformly over distances comparable with or greater than its diameter, the areas at the skin and at

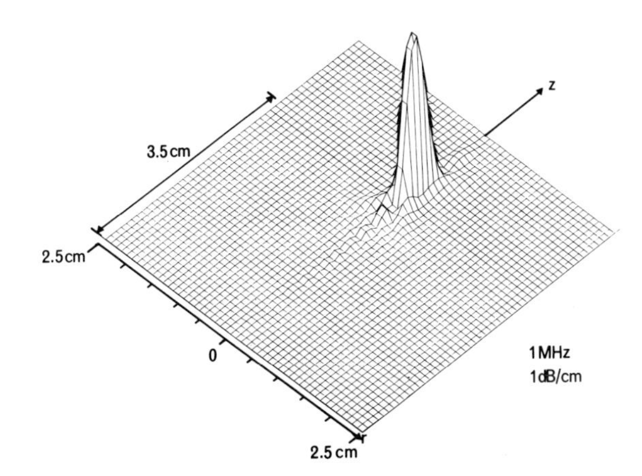

Fig. 7. Absorbed power density distribution due to a 5-cm-diameter 1-MHz focussed ultrasound transducer in direct contact with tissue having an attenuation of 1 dB/cm. The focal length is 3.5 cm

depth through which the ultrasound passes become comparable, causing reduction in gain and a displacement of the maximum absorbed power density to a more superficial position (Dickinson 1984b).

An alternative ultrasonic technique for heating tissue at depth has been reported by workers at Stanford University (Fessenden et al. 1982; Fessenden 1984). The system is based on six transducers mounted on a concave spherical surface and immersed in degassed water contained by a flexible membrane. Each transducer is operated near to its resonant frequency of 350 kHz. The water provides not only the medium for coupling the ultrasound into the tissues but also a means of varying the geometric focus through changes in its thickness. The technique has produced selective heating of perfused volumes of tissue of 200 cm^3 at depths of 7–10 cm.

It has been reported that some patients have experienced pain during hyperthermal treatments involving ultrasound, especially when tumours overlay bone or nerve tissue (Mayer 1984). One reason for this, particularly when lower frequencies are used, is that significant heating can be produced beyond the tumour target. Systems in which focussed transducers are scanned exhibit minimal overshoot and may prove useful in reducing this clinical problem. The principal drawback to the use of ultrasound in hyperthermia, however, is reflection at interfaces between soft tissues and either gas or bone. For this reason the treatment of intrathoracic and deep abdominal tumours where bowel gas could be expected to be within the field are excluded. In some sites surgical itervention can overcome this fundamental problem. For example, Britt et al. (1983) have used a focussed 2-MHz ultrasound transducer to heat brain tissue following craniotomy. Potential advantages of this ultrasonic technique over those methods employing invasive microwave antennas are that irregular tumour volumes may be heated more precisely and since the heating source is non-invasive, treatments could be carried out over a period of weeks or months.

Thermometry and Thermal Dosimetry for Clinical Hyperthermia

For a thermometer to be useful in hyperthermia studies its characteristics should be in general agreement with the following guidelines. The resolution and accuracy of the temperature measurement should be 0.1°–0.2° C because of the strong temperature dependence of biological responses. Such a resolution is easily achieved by many of the thermometers discussed in this section but to ensure this accuracy, the thermometer usually requires frequent calibration against a suitable standard. The spatial resolution required is 5–10 mm in view of the compromise which must be made between the need to make temperature measurements in large temperature gradients and the degree of invasion which can be tolerated by the patient when invasive thermometry is used or the technical and/or physical limitations inherent in various non-invasive thermometric techniques. A temporal resolution of the order of 1 s is adequate to follow the time-varying temperatures which are observed during hyperthermal treatments. Invasive thermometers should have the smallest diameter possible and certainly be no larger than 16 gauge to minimise trauma to the patient. It should be noted that some relaxation of these guidelines may be necessary if a particular thermometric method has other useful characteristics. For example, various tomographic techniques offer non-invasive measurement but take longer than 1 s to make the measurement.

The goal of non-invasive thermometry has attracted much attention. Various ultrasound techniques have been investigated for this purpose (Nasoni et al. 1982; Robert et al. 1982; Christensen 1983) but progress has been limited by major problems which lead to non-unique temperature reconstructions. Fallone et al. (1982) and Bentzen and Overgaard (1984) have used X-ray tomography in phantom studies and have shown that adequate spatial and temperature resolutions can be achieved. Several groups have investigated the potential of magnetic resonance imaging techniques in which the spin-lattice relaxation time (T_1) has been the temperature-sensitive parameter (Parker et al. 1983; Kato et al. 1985). Although these early studies of tomographic techniques in thermometry are encouraging, many questions about their future role in hyperthermia remain to be answered. For example, the effects of physiological changes such as oedema and perfusion on the temperature-sensitive parameters need to be investigated. Cetas (1984) has reviewed the considerable problems which must be overcome before tomographic thermometry becomes an accepted technique in the hyperthermia clinic.

The use of microwave radiometers for non-invasive temperature sensing in tissues located within 3 cm from the skin is more promising. Several improvements to these devices have been reported (Mamouni et al. 1977; Carr et al. 1982; Ludecke and Kohler 1983) and the development of multireceiver systems (Semet et al. 1984), correlation techniques (Mamouni et al. 1983; Haslam et al. 1984) and multifrequency radiometers together with multispectral analysis (Miyakawa 1981; Schaller 1984; Plancot et al. 1984) should lead to systems in which spatial resolution (both laterally and discrimination in depth), temperature resolution and response times satisfy the requirements previously listed.

It is clear that routine temperature measurements during hyperthermal treatments will continue to use invasive methods for the foreseeable future. Invasive thermometry is based on conventional thermistors and thermocouples or on devices developed especially for use in electromagnetic fields (Vaguine et al. 1984; Wickersheim et al. 1985). The advantages and disadvantages of each type of sensor have been discussed in detail by several authors (e.g. Cetas 1982, 1985; Christensen 1983; Hand 1984, 1985). Particular attention has been drawn to sources of artefact. Chakraborty and Brezovich (1982) discussed the use of thermocouples in RF fields and outlined methods to minimise errors whilst Hynynen et al. (1983) and Hand and Dickinson (1984) discussed problems associated with thermocouple thermometry in ultrasound fields. Other sources of error due to thermal conduction along probes have been discussed by Harmark (1983), Fessenden et al. (1984), Bach Andersen et al. (1984), Dickinson (1985), Lyons et al. (1985) and Samulski et al. (1985). Waterman (1985) has drawn attention to factors which can modify the time constants of thermometers.

Temperature measurements made at a single point are meaningless when attempting to assess hyperthermal treatments because of the considerable temperature gradients present. Multisensor probes (thermocouples, thermistors or non-perturbing sensors) when used with computer-controlled data acquisition equipment can provide detailed information on the variations in temperature with time at, say 20–30 points. Alternatively, the thermometer probes may be moved in a stepwise manner through a catheter to measure the temperature profile along that track (Gibbs 1983; Sapozink et al. 1985a). The concept of thermal dose which accounts for the temporal variations in temperature has been introduced in an attempt to quantify treatment although a common definition of this parameter has yet to be agreed (Gerner 1985).

It is clear, regardless of the technique adopted, that invasive thermometry can provide temperatures from only a very limited sample of the treated volume. A means of improving knowledge of the complete temperature distribution may be to use the information available from invasive measurements in a thermal model of the hyperthermal treatment. Detailed discussions of the use of the bioheat transfer equation in simulations are given by Roemer and Cetas (1984) and Cetas (1985). A major complication in such models is the need to account for some vascular structure, and alternative models to the bioheat transfer equation have been suggested by Chen and Holmes (1980), Lagendijk et al. (1984) and Weinbaum and Jiji (1985).

Summary

This chapter has dicussed some recent technical developments and trends in clinical hyperthermia. Several techniques for the treatment of tumours within 3–4 cm of the body surfaces were described. Each technique has its minor advantages and disadvantages; all techniques employing a single applicator produce temperature distributions with considerable gradients. The introduction of microwave and ultrasound techniques using multiple applicators in which there is some control of the pattern of the energy deposition within the treatment area should improve superficial treatments in this respect.

A number of electromagnetic devices for regional hyperthermia are being developed and evaluated. The theoretical predictions of their performances are beginning to suggest restrictions to their use; the limited clinical experience is in general agreement with these predictions. Scanned and focussed ultrasound beams may offer the unique possibility of non-invasive, deep, yet localised hyperthermia in some locations. Such systems are at an early stage of their development; if they prove successful, their controlled and safe use will require detailed information of the temperature distributions produced.

Invasive methods for inducing hyperthermia can produce relatively good temperature distributions. The development of 'constant temperature seeds' is promising. Both RF and microwave interstitial systems offering individual control of power to several channels should lead to improved temperature distributions.

In general, non-invasive thermometry in clinical hyperthermia remains a distant goal, although developments in microwave radiometry may lead to systems with suitable spatial, temporal and temperature resolutions for use in superficial treatments. Invasive thermometry techniques can provide temperature measurements from several points or from along tracks within the treatment volume. The development of computer models to infer temperature distributions from the limited information available will be a major step in quantifying hyperthermal treatments.

References

Arcangeli G, Lombardini PP, Lovisolo GA, Marsiglia G, Piatelli M (1984) Focussing of 915 MHz electromagnetic power on deep human tumours: a mathematical model study. IEEE Trans Biomed Eng BME-31: 47–52

Aristazabal SA, Oleson JR (1984) Combined interstitial radiation and localized current field hyperthermia: results and conclusions from clinical studies. Cancer Res (Suppl) 44: 4757s–4760s

Armitage DW, LeVeen HH, Pethig R (1983) Radiofrequency induced hyperthermia: computer simulation of specific absorption rate distributions using realistic anatomical models. Phys Med Biol 28: 31–42

Astrahan MA, Norman A (1982) A localized current field hyperthermia system for use with 192-iridium interstitial implants. Med Phys 9: 419–424

Atkinson WJ, Brezovich IA, Chakraborty DP (1984) Usable frequencies in hyperthermia with thermal seeds. IEEE Trans Biomed Eng, BME-31: 70–75

Bach Andersen J (1985a) Theoretical limitations on radiation into muscle tissue. Int J Hyperthermia 1: 45–55

Bach Anderson J (1985b) Electromagnetic heating. In: Overgaard J (ed) Hyperthermic oncology 1984, vol 2. Taylor and Francis, London, pp 113–128
Bach Andersen J, Baun A, Harmark K, Heinzl L, Raskmark P, Overgaard J (1984) A hyperthermia system using a new inductive applicator. IEEE Trans Biomed Eng, BME-31: 21–27
Bahl IJ, Stuchly SS (1980) Analysis of a microstrip covered with a lossy dielectric. IEEE Trans Microwave Theory Tech, MTT-28: 104–109
Bahl IJ, Stuchly SS, Stuchly MA (1980) A new microstrip radiator for medical applications. IEEE Trans Microwave Theory Tech, MTT-28: 1464–1468
Bentzen SM, Overgaard J (1984) Non invasive thermometry by subtraction X-ray computed tomography. In: Overgaard J (ed) Hyperthermic oncology 1984, vol 1. Taylor and Francis, London, pp 557–560
Bicher HI, Moore DW, Wolfstein RW (1984a) A method of interstitial thermoradiotherapy. In: Overgaard J (ed) Hyperthermic oncology 1984, vol 1. Taylor and Francis, London, pp 595–598
Bicher HI, Wolfstein RS, Fingerhut AG, Frey HS, Lewinsky BS (1984b) An effective fractionation regime for interstitial thermoradiotherapy – preliminary clinical results. In: Overgaard J (ed) Hyperthermic oncology 1984, vol 1. Taylor and Francis, London, pp 575–578
Bini M, Ignesti A, Millanta L, Rubino N, Vanni R, Lauri G, Lenzi A, Pace M, Pirillo P (1983) Apparatus for RF and microwave hyperthermia. Alta Frequenza, LII: 176–178
Bini M, Distante V, Ignesti A, Lauri G, Millanta L, Pace M, Pirillo M, Rubino N, Vanni R (1984) A new procedure prevents fat-tissue overheating in radiofrequency hyperthermia. In: Overgaard J (ed) Hyperthermic oncology 1984, vol 1. Taylor and Francis, London, pp 845–848
Brezovich IA, Lilly MB, Durant JR, Richards DB (1981) A practical system for clinical radiofrequency hyperthermia. Int J Radiat Oncol Biol Phys, 7: 423–430
Brezovich IA, Young JH, Atkinson JW, Wang MT (1982) Hyperthermia consideration for a conducting cylinder heated by an oscillating electric field parallel to the cylinder axis. Med Phys 9: 746–748
Brezovich IA, Atkinson WA, Chakraborty DP (1984a) Temperature distributions in tumor models heated by self-regulating nickel-copper alloy thermoseeds. Medical Physics 11: 145–152
Brezovich IA, Atkinson WJ, Lilly MB (1984b) Local hyperthermia with interstitial techniques. Cancer Research (Suppl) 44: 4572s–4756s
Britt RH, Lyons BE, Pounds DW, Prionas DS (1983) Feasibility of ultrasound hyperthermia in the treatment of malignant brain tumours. Med Instrum 17: 172–177
Bull JM (1983) Systemic hyperthermia: background and principles. In: Storm FK (ed) Hyperthermia in cancer therapy. Hall, Boston, pp 401–405
Burdette EC, Benson T, Magin RL, Loane J, Lee SW (1985) Patch antenna phased array microwave hyperthermia applicator. In: Abstracts for 33rd annual meeting radiation research society, Los Angeles, May 1985. Abstract Ad-16, p 12
Burton C, Hill M, Walker AE (1971) The RF thermoseed – a thermally self regulating implant for the production of brain lesions. IEEE Trans Biomed Eng BME-18: 104–109
Carr KL, El-Mahdi AM, Shaeffer J (1982) Passive microwave thermography coupled with microwave heating to enhance early detection of cancer. Microwave J 25 (5): 125–136
Cetas TC (1982) Invasive thermometry. In: Nussbaum GH (ed) Physical aspects of hyperthermia. American Institute of Physics, New York, pp 231–265
Cetas TC (1984) Will thermometric tomography become practical for hyperthermia treatment monitoring? Cancer Res (Suppl) 44: 4805s–4808s
Cetas TC (1985) Thermometry and thermal dosimetry. In: Overgaard J (ed) Hyperthermic oncology 1984, vol 2. Taylor and Francis, London, pp 91–112
Chakraborty DP, Brezovich IA (1982) Error sources affecting thermocouple thermometry in RF electromagnetic fields. J Microwave Power 17: 17–28
Chen MM, Holmes KR (1980) Microvascular contributions in tissue heat transfer. Ann NY Acad Sci 335: 137–150
Christensen DA (1983) Thermometry and thermography. In: Storm FK (ed) Hyperthermia in cancer therapy. Hall, Boston, pp 223–232
Corry PM, Barlogie B, Frazier OH, Choksi J, Headley D (1982a) Treatment of bulky human neoplasms with a magnetic induction system. Rad Res 91: 422
Corry PM, Barlogie B, Tilchen EJ, Armour EP (1982b) Ultrasound-induced hyperthermia for the treatment of human superficial tumours. Int J Radiat Oncol Biol Phys 8: 1225–1229

Cosset J-M, Dutreix J, Dufour J, Janoray P, Damia E, Haie C, Clarke D (1984) Combined interstitial hyperthermia and brachytherapy: Institute Gustave Roussy technique and preliminary results. Int J Rad Oncol Biol Phys 10: 307-312
Coughlin CT, Douple EB, Strohbehn JW, Eaton WL, Trembly BS, Wong TZ (1983) Interstitial hyperthermia in combination with brachytherapy. Radiology 148: 285-288
Dickinson RJ (1984a) A non-rigid mosaic applicator for local hyperthermia. In: Overgaard J (ed) Hyperthermic oncology 1984, vol 1. Taylor and Francis, London, pp 671-674
Dickinson RJ (1984b) An ultrasound system for local hyperthermia using scanned focused transducers. IEEE Trans Biomed Eng BME-31: 120-125
Dickinson RJ (1985) Thermal conduction errors of manganin-constantan thermocouple arrays. Phys Med Biol 30: 445-453
di Sieyes DC, Douple EB, Strohbehn JW, Trembly BS (1981) Some aspects of optimisation of an invasive microwave antenna for local hyperthermia treatment of cancer. Med Phys 8: 174-183
Doss JD (1982) Calculation of electric fields in conductive media. Med Phys 9: 566-573
Doss JD, McCabe CW (1976) A technique for localized heating in tissue: an adjunct to tumor therapy. Med Instrum 10: 16-21
Dunn F, O'Brien WD (1978) Ultrasonic absorption and dispersion. In: Fry FJ (ed) Ultrasound: its applications to medicine and biology, vol 1. Elsevier, Amsterdam, pp 393-439
Emani B, Perez C, Nussbaum G, Leybovich L (1984) Regional hyperthermia in treatment of recurrent deep-seated tumors: preliminary report. In: Overgaard J (ed) Hyperthermic oncology 1984, vol 1. Taylor and Francis, London, pp 605-608
Engelhardt R (1985) Whole body hyperthermia. Methods and results. In: Overgaard J (ed) Hyperthermic oncology 1984, vol 1. Taylor and Francis, London, pp 263-276
Fallone BG, Moran PR, Podgorsak EB (1982) Non-invasive thermometry with a clinical X-ray CT scanner. Med Phys 9: 715-721
Fessenden P, Anderson TL, Marmor JB, Pounds D, Sagerman R, Strohbehn JW (1982) Experience with a deep heating ultrasound system. Rad Res 91: 415
Fessenden P (1984) Ultrasound methods for inducing hyperthermia. Front Radiat Ther Onc 18: 62-69
Fessenden P, Lee ER, Samulski TV (1984) Direct temperature measurement. Cancer Res 44: 4799s-4804s
Frazier OH, Corry PM (1984) Induction of hyperthermia using implanted electrodes. Cancer Res (Suppl) 44: 4864s-4866s
Gee W, Lee SS, Bong NK, Cain CA, Mittra R, Magin RL (1984) Focussed array hyperthermia applicator. Theory and experiment. IEEE Trans Biomed Eng BME-31: 38-46
Gerner EW (1985) Definition of thermal dose. Biological isoeffect relationships and dose for temperature-induced cytotoxity. In: Overgaard J (ed) Hyperthermic oncology, vol 2. Taylor and Francis, London, pp 245-251
Gibbs FA (1983) 'Thermal mapping' in experimental cancer treatment with hyperthermia: description and use of a semiautomatic system. Int J Radiat Oncol Biol Phys 9: 1057-1063
Gibbs FA (1984) Regional hyperthermia: a clinical appraisal of non-invasive deep-heating methods. Cancer Res (Suppl) 44: 4765s-4770s
Guy AW (1971) Electromagnetic fields and relative heating patterns due to a rectangular aperture source in direct contact with bilayered biological tissue. IEEE Trans Microwave Theory Techn, MTT-19: 214-223
Guy AW, Lehmann JF, Stonebridge JB (1974) Therapeutic applications of electromagnetic power. Proc IEEE 62: 55-75
Hagmann MJ (1984) Propagation on a sheath helix in a coaxially layered lossy dielectric medium. IEEE Trans Microwave Theory Tech MTT-32: 122-126
Hagmann MJ, Levin RL (1984) Analysis of the helix as an RF applicator for hyperthermia. Electron Lett 20: 337-338
Halac S, Roemer RB, Oleson JR, Cetas TC (1983) Magnetic induction heating of tissue: numerical evaluation of tumor temperature distributions. Int J Radiat Oncol Biol Phys 9: 881-891
Hand JW (1984) Microwaves and ultrasound in clinical hyperthermia: some physical aspects of heating and thermometry. In: Bajzer Z, Baxa P, Franconi C (eds) Applications of physics to medicine and biology. World Scientific Publishing, Singapore, pp 309-336

Hand JW (1985) Thermometry in hyperthermia. In: Overgaard J (ed) Hyperthermic oncology 1984, vol 2. Taylor and Francis, London, pp 299-308

Hand JW, Dickinson RH (1984) Linear thermocouple arrays for in vivo observation of ultrasonic hyperthermia fields. Br J Radiol 57: 656

Hand JW, Hind AJ (1986) A review of microwave and RF applicators. In: Hand JW, James JR (eds) Physical techniques in clinical hyperthermia. Research Studies Press, Letchworth, pp 98-148

Hand JW, Johnson RH (1986) Field penetration from electromagnetic applicators for localised hyperthermia. In: Bruggmoser G, Hinkelbein W, Engelhardt R, Wannemacher M (eds) Locoregional high-frequency hyperthermia and temperature measurement. Springer, Berlin Heidelberg New York Tokyo, pp 7-17 (Recent results in cancer research, vol 101)

Hand JW, Ledda JL, Evans NTS (1982) Considerations of radiofrequency induction heating for localised hyperthermia. Phys Med Biol 27: 1-16

Hand JW, Hind AJ, Cheetham JL (1985) Multi-element microwave applicators for localised hyperthermia. Strahlentherapie 161: 535

Harmark K (1983) Design of multipoint thermometer. Strahlentherapie 159: 373

Haslam NC, Gillespie AR, Haslam CGT (1984) Aperture synthesis thermograph - a new approach to passive microwave temperature measurements in the body. IEEE Trans Microwave Theory Tech MTT-32: 829-835

Hill SC, Christensen DA, Durney CH (1983) Power deposition patterns in magnetically-induced hyperthermia: a two dimensional low frequency numerical analysis. Int J Rad Oncol Biol Phys 9: 893-904

Hiraoka M, Jo S, Takahashi M, Nishida H, Abe M (1984) Clinical application of RF capacitive heating for deep seated tumors. In: Matsuda T, Kikuchi M (eds) Hyperthermic oncology. Jap Soc Hyperthermic Oncol, Tokyo, pp 190-191

Hiraoka M, Jo S, Akita T, Takahashi M, Abe M (1985) Effectiveness of RF capacitive hyperthermia in the heating of human deep-seated tumors. In: Abe M, Takahashi M, Singahara T (eds) Hyperthermia in cancer therapy. Mag Bros, Tokyo, pp 98-99

Hynynen K, Watmough DJ, Mallard JR, Fuller M (1983) Local hyperthermia induced by focussed and overlapping ultrasonic fields - an in vivo demonstration. Ultrasound Med Biol 9: 621-627

Iskander MF, Khoshdel-Milani O (1984) Numerical calculations of the temperature distribution in realistic cross-sections of the human body. Int J Radiat Oncol Biol Phys 10: 1907-1912

Iskander MF, Turner PF, Dubow JB, Kao J (1982) Two dimensional technique to calculate the EM power deposition pattern in the human body. J Microwave Power 17: 175-185

Johnson RH, James JR, Hand JW, Hopewell JW, Dunlop PRC, Dickinson RJ (1984) New low profile applicators for local heating of tissue. IEEE Trans Biomed Eng, BME-31: 28-37

Johnson RH, James JR, Hand JW, Dickinson RJ (1985) Compact 27 MHz applicators. Strahlentherapie 161: 537-538

Joseph CP, Astrahan M, Lipsett J, Archambeau J, Forell B, George FW (1981) Interstitial hyperthermia and interstitial iridium 192 implantation: a technique and preliminary results. Int J Radiation Oncology Biol Phys 7: 827-833

Kantor G, Witters DM (1983) The performance of a new 915 MHz direct contact applicator with reduced leakage. J Microwave Power 18: 133-142

Kato H, Ishida T (1983) A new inductive applicator for hyperthermia. J Microwave Power 18: 331-336

Kato H, Ishida T, Kano E, Oikawa S (1985) Non invasive thermometry with NMR-CT: temperature dependency of T, of the implanted tumor. In: Abe M, Takahashi M, Sugahara T (eds) Hyperthermia in cancer therapy. Mag Bros, Tokyo, pp 104-105

King RWP, Trembly BS, Strohbehm JW (1983) The electromagnetic field of an insulated antenna in a conducting or dielectric medium. IEEE Trans Microwave Theory Techn, MTT-31: 574-583

Lagendijk JJW (1983) A new coaxial TEM radiofrequency/microwave applicator for non-invasive deep-body hyperthermia. J Microwave Power 18: 367-376

Lagendijk JJW, de Leeuw AAC (1986) The development of applicators for deep body hyperthermia. In: Bruggmoser G, Hinkelbein W, Engelhardt R, Wannemacher M (eds) Locoregional high-frequency hyperthermia and temperature measurement. Springer, Berlin Heidelberg New York (Recent results in cancer research, vol 101), pp 18-35

Lagendijk JJW, Schellekens M, Schipper J, van der Linden PM (1984) A three-dimensional description of heating patterns in vascularised tissues during hyperthermia treatment. Phys Med Biol 29: 495–507

Lele PP (1983) Physical aspects and clinical studies with ultrasonic hyperthermia. In: Storm FK (ed) Hyperthermia in cancer therapy. Hall, Boston, pp 333–367

Lerch IA, Kohn S (1983) Radiofrequency hyperthermia: the design of coil transducers for local heating. Int J Radiat Oncol Biol Phys 9: 939–948

LeVeen HH, Ahmed N, Piccone VA, Shugaar S, Falk G (1980) Radiofrequency therapy: clinical experience. Ann NY Acad Sci 335: 362–371

Lin JC, Kantor G, Ghods A (1982) A class of new microwave therapeutic applicators. Rad Sci 17: 119s–123s

Ludeke KM, Kohler J (1983) Microwave radiometric system for biomedical tissue temperature and emissivity measurements. J Microwave Power 18: 277–283

Lyons BE, Britt R, Strohbehn JW (1984) Localized hyperthermia in the treatment of malignant brain tumors using an interstitial microwave antenna array. IEEE Trans Biomed Eng BME-31: 53–62

Lyons BE, Samulski TD, Britt RH (1985) Temperature measurements in high thermal gradients: 1. The effects of conduction. Int J Radiat Oncol Biol Phys 11: 951–962

Mamouni A, Bliot F, Leroy Y, Moschetto Y (1977) Radiometer for temperature and microwave properties measurements of biological substances. In: Proceedings 7th European Microwave Conference. Microwave Exhibitions and Publishers, Sevenoaks, Kent, pp 703–717

Mamouni A, Leroy Y, Van de Velde JC, Bellarbi L (1983) Introduction to correlation microwave thermography. J Microwave Power 18: 285–293

Manning MR, Cetas TC, Miller RC, Oleson JR, Connor WG, Gerner EW (1982) Clinical hyperthermia: results of a phase I trial employing hyperthermia alone or in combination with external beam or interstitial radiotherapy. Cancer 49: 205–216

Marchal C, Bey P, Metz R, Gaulard ML, Robert J (1982) Treatment of superficial human cancerous nodules by local ultrasound hyperthermia. Br J Cancer 45 (Suppl V) 243–245

Marchal C, Bey P, Jacomino J, Hoffstetter S, Gaulard ML, Robert J (1985) Preliminary technical, experimental and clinical results on the use of HPRL27 system for the treatment of deep seated tumors by hyperthermia. Int J Hyperthermia 1 (2): 105–116

Marmor JB, Pounds D, Hahn GM (1982) Clinical studies with ultrasound-induced hyperthermia. Nat Cancer Inst Mon 61: 333–337

Matloubieh AY, Roemer RB, Cetas TC (1984) Numerical simulation of magnetic induction heating of tumors with ferromagnetic seed implants. IEEE Trans Biomed Eng, BME-31: 227–234

Matsuda T, Sugiyama A, Nakata Y (1984) Fundamental and clinical studies of radiofrequency hyperthermia and radiation therapy. In: Overgaard J (ed) Hyperthermic oncology 1984, vol 1. Taylor and Francis, London, pp 349–352

Mayer JL (1984) Ultrasound hyperthermia – the Stanford experience. Front Radiat Ther Oncol 18: 126–135

Medal R, Shorey W, Gilchrist RK, Barker W, Hanselman R (1959) Controlled radiofrequency generator for production of localized heat in intact animal. Am Med Assoc Arch Surgery 79: 427–431

Mendecki J, Friedenthal E, Botstein C, Sterzer F, Paglione R (1979) Therapeutic potential of conformal applicators for induction of hyperthermia. J Microwave Power 14: 139–144

Milligan AJ (1984) Whole-body hyperthermia induction techniques. Cancer Res (Suppl) 44: 4869s–4872s

Miyakawa M (1981) Study of microwave thermography application to the estimation of subcutaneous temperature profiles. Trans IECE Japan, E64: 786–792

Moidel RA, Wolfson SK, Selker RG, Weine SB (1976) Materials for selective heating in a radiofrequency electromagnetic field for the combined chemothermal treatment of brain tumours. J Biomed Mater Res 10: 327–334

Morita N, Bach Andersen J (1982) Near-field absorption in a circular cylinder from electric and magnetic line sources. Biolectromagnetics 3: 253–274

Nasoni RL, Bowen T, Dewhirst MW, Roth H (1982) In vivo temperature dependence of the speed of sound in mammalian tissue and its possible use in hyperthermia. Nat Cancer Inst Monogr 61: 501–504

Oleson JR (1982a) Hyperthermia by magnetic induction: I. Physical characteristics of the technique. Int J Radiat Oncol Biol Phys 8: 1747–1756

Oleson JR (1982b) A clinical comparison of heating patterns produced by magnetic induction vs paired coaxial electrodes. Strahlentherapie 158: 388
Oleson JR, Cetas TC, Corry PM (1983a) Hyperthermia by magnetic induction: experimental and theoretical results for coaxial coil pairs. Rad Res 95: 175-186
Oleson JR, Heusinkveld RS, Manning MR (1983b) Hyperthermia by magnetic induction: clinical experience with concentric electrodes. Int J Radiat Oncol Biol Phys 9: 549-556
Paglione R, Sterzer F, Mendecki J, Friedenthal E, Botstein C (1981) 27 MHz ridged waveguide applicators for localised hyperthermia treatment of deep-seated malignatnt tumours. Microwave Journal 24: 71-80
Parker DL, Smith V, Shelden P, Crooks LE, Fussell L (1983) Temperature distribution measurements in two-dimensional NMR imaging. Med Phys 10: 321-325
Paulsen KD, Strohbehn JW, Lynch DR (1984a) Theoretical temperature distributions produced by an annular phased array type system in CT-based patient models. Rad Res 100: 536-552
Paulsen KD, Strohbehn JW, Hill SC, Lynch DR, Kennedy FE (1984b) Theoretical temperature profiles for concentric coil induction heating devices in a two dimensional axi-asymmetric, inhomogeneous patient model. Int J Rad Oncol Biol Phys 10: 1095-1107
Penttinen A, Luukkala M (1976) Sound pressure near the focal area of an ultrasonic lens. J Phys D App Phys 9: 1927-1936
Plancot M, Chive M, Gioaux G, Prevost B (1984) Thermal dosimetry in microwave hyperthermia process based on radiometric temperature measurements: principles and feasibility. In: Overgaard J (ed) Hyperthermic oncology 1984, vol 1. Taylor and Francis, London, pp 863-866
Raskmark P, Bach Andersen J (1984) Focussed electromagnetic heating of muscle tissue. IEEE Trans Microwave Theory Tech MTT-32: 887-888
Robert J, Marchal C, Drocourt M, Escayne JM, Thouvenot P, Gaulard ML, Tosser A (1982) Ultrasound velocimetry for hyperthermia control. In: Gautherie M, Albert E (eds) Biomedical thermology. Liss, New York, pp 555-560
Roemer RB, Cetas TC (1984) Applications of bioheat transfer simulations in hyperthermia. Cancer Res (Suppl) 44: 4788s-4798s
Ruggera PS, Kantor G (1984) Development of a family of RF helical coil applicators which produce transversely uniform axially distributed heating in cylindrical fat-muscle phantoms. IEEE Trans Biomed Eng, BME-31: 98-106
Salcman M, Samaras GM (1983) Interstitial microwave hyperthermia for brain tumors: results of a phase I clinical trial. J Neuro Oncol 1: 225-236
Samaras GM (1984) Intracranial microwave hyperthermia: heat induction and temperature control. IEEE Trans Biomed Eng BME-31: 63-69
Samulski TY, Lyons BE, Britt RH (1985) Temperature measurements in high thermal gradients: II analysis of conduction effects. Int J Radiat Oncol Biol Phys 11: 963-971
Sandhu TS, Kolozvary AJ (1983) Effect of bolus/tissue inhomogeneities on the resonance frequency of MW microstrip applicators. Radiation Research 94: 594
Sandhu TS, Kolozvary AJ (1984) Conformal hyperthermia applicators. In: Overgaard J (ed) Hyperthermic oncology 1984, vol 1. Taylor and Francis, London, pp 675-678
Sapozink MD, Gibbs FA, Gates KS, Stewart JR (1984) Regional hyperthermia in the treatment of clinically advanced deep seated malignancy: results of a pilot study employing an annular array applicator. Int J Radiat Oncol Biol Phys 10: 775-786
Sapozink MD, Gibbs FA, Sandhu TS (1985a) Practical thermal dosimetry. Int J Radiat Oncol Biol Phys 11: 555-560
Sapozink MD, Gibbs FA, Thomson JW, Eltringham JR, Stewart JR (1985b) A comparison of deep regional hyperthermia from an annular array and a concentric coil in the same patients. Int J Radiat Oncol Biol Phys 11: 179-180
Schaller G (1984) Inversion of radiometric data from biological tissue by an optimisation method. Electron Lett 20: 380-382
Scott BO (1957) The principles and practice of diathermy. Heinemann, London, pp 100-101
Semet C, Mamouni A, van de Velde JC, Hochedez-Robillard M, Leroy Y (1984) Système de thermographie microonde multisonde à balayage electronique. Innov Technol Biol 5: 200-209
Song CW, Rhee JG, Lee CKK, Levitt SH (1986) Capacitive heating of phantom and human tumors with an 8 MHz radiofrequency applicator (Thermotion RF-8). Int J Radiat Oncol Biol Phys 12: 365-372

Stauffer PR, Cetas TC, Fletcher AM, DeYoung DW, Dewhirst MW, Oleson JR, Roemer RB (1984a) Observations on the use of ferromagnetic implants for inducing hyperthermia. IEEE Trans Biomed Eng, BME-31: 76–90

Stauffer PR, Cetas TC, Jones RC (1984b) Magnetic induction heating of ferromagnetic implants for inducing localized hyperthermia in deep seated tumours. IEEE Trans Biomed Eng, BME-31: 235–251

Sternhagen CJ, Doss JD, Day PE, Edwards WS, Doberneck RC, Herzon FS, Powell TD, O'Brien GF, Larkin JM (1978) Clinical use of radio-frequency current in oral cavity carcinomas and metastatic malignancies with continuous temperature control and monitoring. In: Streffer C (ed) Cancer therapy by hyperthermia and radiation. Urban und Schwarzenberg, Munich, pp331–334

Storm FK, Harrison WH, Elliott RS, Kaiser LR, Silberman AW, Morton DL (1981) Clinical radiofrequency hyperthermia by magnetic-loop induction. J Microwave Power 16 (2): 179–184

Storm FK, Harrison WH, Elliott RS, Silberman AW, Morton DL (1982) Thermal distribution of magnetic loop induction hyperthermia in phantoms and animals: effect of the living state and velocity of heating. Int J Radiat Oncol Biol Phys 8: 865–871

Strohbehn JW (1983) Temperature distributions from interstitial RF electrode hyperthermia systems: theoretical predictions. Int J Radiation Oncology Biol Phys 9: 1655–1667

Strohbehn JW, Bowers ED, Walsh JE, Douple EB (1979) An invasive microwave antenna for locally-induced hyperthermia for cancer therapy. J Microwave Power 14: 339–350

Strohbehn JW, Trembly BS, Douple EB (1982) Blood flow effects on the temperature distributions from an invasive microwave antenna array used in cancer therapy. IEEE Trans Biomed Eng BME-29: 649–661

Strohbehn JW, Paulsen KD, Lynch DR (1986) Use of the finite element method in computerised thermal dosimetry. In: Hand JW, James JR (eds) Physical techniques in clinical hyperthermia. Research Studies Press, Letchworth, pp383–451

Stuchly SS, Stuchly MA (1978) Multimode square waveguide applicators for medical applications of microwave power. In: Proceedings 8th European Microwave Conference, Microwave Exhibitions Publishers, Sevenoaks, Kent, pp553–557

Swicord ML, Davis CC (1981) Energy absorption from small radiating coaxial probes in lossy media. IEEE Trans Microwave Theory Tech MTT-29: 1202–1209

Takahashi Y, Nikawa Y, Mori S, Nakagawa M, Kikuchi M (1985) Electromagnetic field convergent applicator for microwave hyperthermia at 433 MHz. In: Abe M, Takahashi M, Sugahara T (eds) Hyperthermia in cancer therapy. Mag Bros, Tokyo, pp132–133

Tanabe E, McEwen A, Norris CS, Fessenden P, Samulski TU (1983) A multi-element microstrip antenna for local hyperthermia. In IEEE MTT-S international microwave symposium digest (IEEE 83 CH 1871-3). IEEE, New York, pp183–185

Taylor LS (1978) Electromagnetic syringe. IEEE Trans Biomed Eng BME-25: 303–305

Taylor LS (1980) Implantable radiators for cancer therapy by microwave hyperthermia. Proc IEEE 68: 142–149

Taylor LS (1984) Penetrating electromagnetic wave applicators. IEEE Trans Antennas and Propogation, AP-32: 1138–1142

Trembly BS, Strohbehn JW, King RWP (1982) Practical embedded insulated antenna for hyperthermia. In: Hansen EW (ed) Proceedings 10th annual northeast bioengineering conference, 15–16 Mar 1982. Dartmouth College, Hanover. IEEE, New York, pp105–108

Trembly BS, Richter HJ, Mechling JA (1984) The effect of antenna surface cooling on the temperature distribution of an interstitial microwave antenna array. Presented at 4th international symposium on hyperthermic oncology, 2–6 July 1984, Aarhus, Denmark

Turner PF (1983) Electromagnetic hyperthermia devices and methods. MSc thesis, Department Electrical Engineering, University of Utah, June 1983

Turner PF (1984) Regional hyperthermia with an annular phased array. IEEE Trans Biomed Eng BME-31: 106–114

Turner PF, Kumar L (1982) Computer solution for applicator heating patterns. Nat Cancer Inst Monogr 61: 521–532

Vaguine VA, Tanabe E, Giebeler RH, McEwen AH, Halin GM (1982) Microwave direct-contact applicator system for hyperthermia therapy research. Nat Cancer Inst Monogr 61: 461–464

Vaguine VA, Christensen DA, Lindley JH, Watson TE (1984) Multiple sensor optical thermometry system for application in clinical hyperthermia. IEEE Trans Biomed Eng BME-31: 168–172 van denBerg PM, de Hoop AT, Segal A, Praagman N (1983) The computational model of the electromagnetic heating of biological tissue with application to hyperthermic cancer therapy. IEEE Trans Biomed Eng BME-30: 797-805

van Rhoon GC, Visser AG, van den Berg PM, Reinhold HS (1984) Temperature depth profiles obtained in muscle-equivalent phantoms using the RCA 27 MHz ridged waveguide. In: Overgaard J (ed) Hyperthermic oncology, vol 1. Taylor and Francis, London, pp 499-502

von Ardenne M, von Ardenne Th, Böhme G, Reitnauer PG (1977) Selektive Lokalhyperthermie der Krebsgewebe. Homogenisierte Energiezufuhr auch in tief liegende Gewebe der Hochleistungs-Dekawellen-Spulenfeld + Rasterbewegung des Doppelsystems. Arch Geschwulstforsch 47: 487-523

Vora N, Forell B, Joseph C, Lipsett J, Archambeau JO (1982) Interstitial implant with interstitial hyperthermia. Cancer 50: 2518-2523

Waterman FM (1985) The response of thermometer probes inserted into catheters. Med Phys 12: 368-372

Weinbaum S, Jiji LM (1985) A new simplified bioheat equation for the effect of blood flow on local average tissue temperature. Trans ASME J Biomech Eng 107: 131-139

Wells PNT (1977) Biomedical ultrasonics. Academic Press London

Wickersheim KA, Sun MH, Heinemann SO (1985) 16-channel fiberoptic thermometry system with multisensor arrays for thermal mapping. In: Abstracts for 33rd annual meeting radiation research society, Los Angeles, May 1985. Abstract Fb-7, p 62

Wiley JD, Webster JG (1982) Analysis and control of the current distributions under circular dispersive electrodes. IEEE Trans Biomed Eng BME-29: 381-385

Wong TZ, Strohbehn JW, Smith KF, Trembly BS, Douple EB, Coughlin CT (1984) An interstitial microwave antenna array system (IMAAH) for local hyperthermia. Presented at 4th international symposium on hyperthermic oncology, 2-6 July 1984, Aarhus, Denmark

The Biological Basis for Tumour Therapy by Hyperthermia and Radiation

C. Streffer and D. van Beuningen

Institut für Medizinische Strahlenbiologie, Universitätsklinikum Essen, Hufelandstrasse 55, 4300 Essen 1, FRG

Introduction

The use of hyperthermia in treatment of cancers has a long tradition. One of the oldest medical texts describes the treatment of a breast tumour with hyperthermia. The description is found in the *Edwin Smith Surgical Papyrus;* an Egyptian papyrus roll which can be dated back to about 3000 B.C. (Breasted 1930; Overgaard 1985). Treatment with hyperthermia is also mentioned in medical reports of greek physicians. Parmenides believed that he could cure all illnesses including tumours if he had the ability to induce fever. Hippocrates described the favourable role of fever: *"Quae medicamenta non sanant, ferum sanat. Quae ferum non sanat, ignis sanat. Quae vero ignis non sanat, insanobilia repotari oportet".*

The more recent interest in the use of hyperthermia for cancer treatment started with the observation of the German physician W. Busch that a sarcoma disappeared after a prolonged infection with erysipelas, which results in high fever (Busch 1866). These and similar findings have led to studies on the use of bacterial toxins which were extracted from bacteria causing erysipelas. Among these studies, that by the surgeon W. B. Coley from New York should especially be mentioned. He extracted a toxin with which he treated a number of tumour patients (Coley 1893).

Although it is difficult to evaluate the direct effect of hyperthermia in these treatments with whole body hyperthermia and unspecific immunotherapy, these studies have stimulated further investigations into the use hyperthermia, especially local hyperthermia, for the treatment of tumours. The names F. Westermark (1898), N. Westermark (1927), K. Overgard (1934) and M. v. Ardenne (1971, 1975) should be mentioned in this connection.

In the late 1960s, a new interest in hyperthermia was initiated in several laboratories with the application of new techniques which came from studies of cell biology and radiobiology. The observation that hyperthermic treatment enormously increased the effects of radiation and chemotherapeutic drugs was very stimulating. Investigations into the mechanisms of the action of heat increased tremendously. This has led to several International Symposia on Cancer Therapy by Hyperthermia, Drugs and Radiation, which took place in Washington DC in 1975 (Wizenberg and Robinson 1975), in Essen in 1977 (Streffer et al. 1978), in Fort Collins in 1980 (Dethlefsen and Dewey 1982) and in Aarhus in 1984 (Overgaard 1985). The number of participating scientists and clinicians has increased from symposium to symposium. Especially the number of clinical contributions grew steadily. It has been demonstrated that hyperthermia can improve tumour therapy if it is used as a palliative treatment modality. The curative value still has to be proven. A number of studies have been started for this purpose.

Recent Results in Cancer Research. Vol 104

Biological Basis for the Action of Hyperthermia and of Combination with Ionizing Radiation

Mammalian Cell-Survival After Exposure to Hyperthermia

It is the current understanding that cancer therapy is achieved by removal or killing of neoplastic cells. By surgery an attempt is made to remove the tumour as completely as possible. Using radiotherapy and chemotherapy the tumour cells have to be killed in situ. This means that the reproductive ability of these cells has to be inhibited. These goals must be reached by keeping the irreversible damage in the normal tissues to a minimum so that the reduction of functional integrity is tolerable. For many normal tissues functional integrity is dependent on the number of stem cells which have survived the treatment. Hyperthermia is usually used as a modality which modifies radiosensitivity or chemosensitivity of cells or which adds its cell-killing potential to the effects of ionizing radiation or drugs. A very important question has been asked and studied in this connection: Are the malignant cells in a tumour more thermosensitive than the surrounding normal cells from which the malignant cells have probably developed?

There are two phenomena which might support such an assumption: It is possible that the process of malignant transformation might involve a step which induces a higher thermosensitivity by itself. Malignant cells are usually mutants of normal cells. Quite a number of thermosensitive mutants have been isolated. The second phenomenon which has to be studied is whether the physiological conditions and their microenvironment have been altered in tumours in such a way that the thermosensitivity of the cells is enhanced (Hahn 1982). Both phenomena require extensive discussion.

The assay which is normally used for studying the thermosensitivity of cells is the test of colony-forming units (CFUs). If mammalian cells are incubated at temperatures above 37° C (up to about 47° C), survival, defined as cells with reproductive integrity, decreases usually with increasing incubation time. If survival is plotted on a logarithmic scale against incubation at a constant temperature, different types of dose effect curves are observed.

The first type, which appears comparatively simple, was found after heating HeLa cells for periods up to 5 h at temperatures between 41° and 45° C (Gerner et al. 1975). The dose effect curves have a linear shape. It follows that survival is expotential. The curves can be described by an equation of the type:

$$S = S_o e^{-kt}$$

where S is the survival of clonogenic cells at any time t, S_o is the number of clonogenic cells at the start of the experiment ($t=0$), k is a constant representing the inactivation rate at a given temperature, and t is the duration of incubation at a given temperature. In analogy to dose effect curves which have been obtained for survival after exposure to ionizing radiation, a value for D_{37} or D_o can be calculated. This value represents the time of incubation at a given temperature which results in the reduction of cell survival to $1/e$ of the initial cell number (equal to 37% of S_o).

A different type of survival curve was obtained for CHO cells by Dewey et al. (1971). For these experiments the cells were incubated with elevated temperatures in the range of 43.5°-46.5° C. The survival curves bend and apparently reach a linear shape at later incubation times if the survival of the cells is again plotted on a logarithmic scale against incubation time at a constant temperature (Fig. 1). In analogy to radiobiology it is said that these dose effect curves are characterized by a shoulder. For the exponential part of the dose effect curve an analogous D_o can be calculated. It is interesting that the constant of the inactivation rate which determines the steepness of the exponential part of the dose effect curves approximately doubles when temperature is increased by 1° C. By definition this means that D_0 approximately halves under these conditions. This correlation is observed although the constants for the inactivation rates vary from cell line to cell line over a wide range.

A third type of dose effect curve is seen when cells are incubated at elevated temperatures which are comparatively low (usually below 43° C). Cell survival decreases after short incubation times and in their first part the dose effect curves look very similar to those described previously. However, after a longer duration of incubation survival decreases less than expected and the survival curves bend and show a much shallower slope. Cell inactivation becomes smaller for a certain incubation time than during the initial part of the experiment (Dewey et al. 1977; Sapareto et al. 1978) (Fig. 1). Survival curves with such a shape have been found for many cell lines after heating with mild hyperthermia. Similar dose effect curves have been observed in some experiments in radiobiology. It has been demonstrated that they are found after irradiation of a heterogeneous cell population with different radiosensitivities (Streffer and van Beuningen 1985). After small radiation doses the dose effect curve is determined by the more radiosensitive cell and the dose effect curve is steep. After higher radiation doses the dose effect curve is determined by the less radiosensitive cell population; the survival curve becomes shallower. Several experiments have been undertaken to prove an analogous situation for cell survival after treatment with mild hyperthermia. The results of such studies were negative in almost all cases. From these experiments it can be concluded that protection of cell subpopulations by external environmental conditions, either physical or chemical, is very unlikely (Hahn 1980). Furthermore a different thermosensitivity of cells in the various phases of the cell generation cycle can be excluded for these observations, as similar dose effect curves have also been found with synchronous cells in the G_1-phase (Sapareto et al. 1978). It has been observed that plateau phase cells also show this type of survival curve (Li and Hahn 1980b). Plateau phase cells have stopped or decreased cell proliferation. Thus, most cells are found in the G_0-phase. It is generally agreed today that the cells become more thermoresistant ("thermotolerant") during such a treatment. This phenomenon will be described and discussed later.

Several authors have reported that malignant cells are more thermosensitive than the normal cells from which the malignant cells have deveolped by transformation. Such ob-

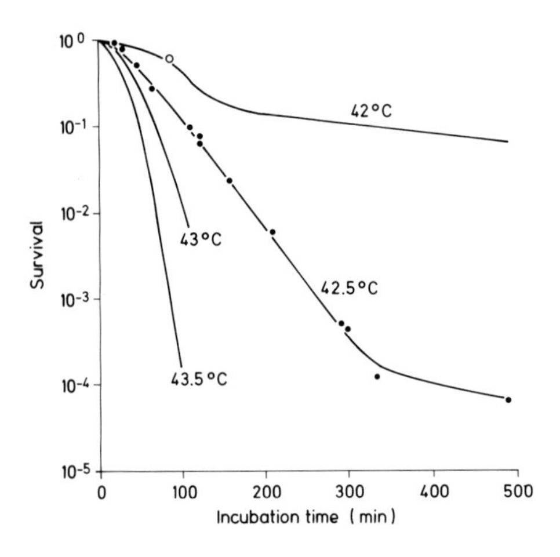

Fig. 1. Survival curves of asynchronous CHO cells heated at different temperatures for varying periods. (Redrawn from Dewey et al. 1977)

servations have underlined the possible potentiality of hyperthermia as a treatment modality in tumour therapy. This subject has frequently been reviewed (Cavaliere et al. 1967; Suit and Shwayder 1974; Strom et al. 1977; Hahn 1982). However, if one looks into the literature more carefully, one finds that the reported data are far from uniform. Quite often the comparing studies tested the thermosensitivity by determining the cell number as a measure of cell survival (Giovanella et al. 1973, 1976). Such data can be misleading if colony-forming ability is not investigated.

Chen and Heidelberger (1969) found that transformed mouse prostate cells were more thermosensitive than the original normal cells. The transformation was performed in vitro by carcinogenic hydrocarbons. Kase and Hahn (1975) studied the thermosensitivity of a human fibroblast cell line and compared it with the thermosensitivity of a cell line which was obtained from the fibroblasts by transformation with the virus SV 40. During exponential growth the malignant cells were somewhat less thermoresistant than the normal cells. However, the difference between the two cell lines disappeared when the heating was performed at high cell densities. Hahn (1980) further studied several transformed cell lines which were all obtained by transformation from 10T½ cells (mouse embryo fibroblasts) and compared cell survival after heat treatment. In no case was a significantly higher thermosentivity found with the transformed cells than with the parental cells, when the cells were heated during the plateau phase. Harisiades et al. (1975) even found that the cell survival of hepatoma cells was higher than of normal liver cells after the same heat treatment.

Several authors have studied the cell survival of cell lines from human tumours where the cells came from different individual tumours of the same tumour type. In extensive studies the thermosensitivity of 11 human melanoma cell lines was determined. Dose effect curves were observed which varied tremendously. Exponential survival curves and survival curves with broad shoulders were found for human melanoma cells (Rofstad et al. 1985). Thus, it can be concluded from these data that in some cases the transformation of cells may lead to a higher thermosensitivity for the malignant cell but in other cases it is just the opposite and the thermoresistance of the malignant cells may be even higher (perhaps higher than that of the normal cells) (Fig. 2). Especially cells in plateau phase apparently do not show differences in thermosensitivity between transformed cells and the par-

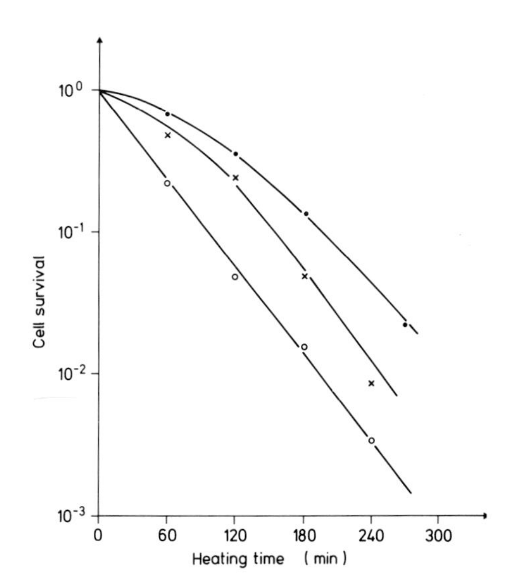

Fig. 2. Heat survival curves for cells from human melanomas treated at 43.5° C. (Redrawn from Rofstad et al. 1985)

ental normal cells. Also tumour cells which differ in thermosensitivity during exponential growth lose this difference in the plateau phase (van Beuningen and Streffer 1985). The statement that a “selective heat sensitivity of cancer cells” exists is very optimistic and cannot be proven by the experimental data if the cell survival is tested after heat treatment in vitro. Cells which are grown and heated in vitro show apparently no characteristic difference in survival which is dependent on the malignant or normal state of the cells.

Cell Survival During the Cell Cycle

In radiobiology, studies of cellular radiosensitivity during the different phases of the cell generation cycle have found great interest (Alper 1979; Streffer and van Beuningen 1985). Thermosensitivity of cells also changes during the cell cycle, but the highest sensitivity is usually observed during other cell cycle phases than those which have been found most sensitive after exposure to ionizing radiation. The general behaviour with respect to heat sensitivity has been described by Westra and Dewey (1971). Synchronized CHO cells were used for these experiments; the cells were synchronized by physical treatment (shaking), which allows the selective isolation of mitotic cells (Terasima and Tolmach 1963). The harvested cells are collected at 4° C; this treatment arrests the mitotic cells in the cell cycle. Cells from several harvests can be pooled and after raising the temperature to 37° C the cells progress through the cell generation cycle in a synchronous way.

Now the synchronous cell population can be treated either with heat or with ionizing radiation during the different phases of the cell cycle. After exposure to 6.0 Gy X-rays, the lowest cell survival was observed during mitosis. It increased during the progression of the cells through the generation cycle and reached the highest value after exposure during the late S-phase (Westra and Dewey 1971) (Fig. 3). A different picture was observed for the

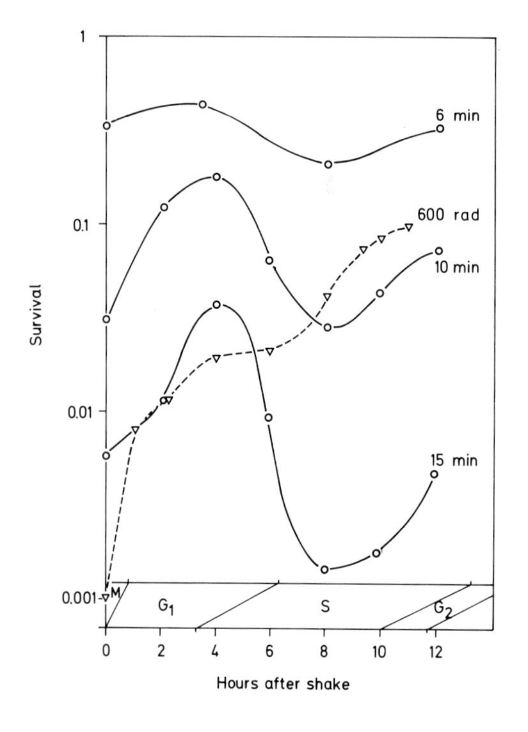

Fig. 3. Suvival of synchronous CHO cells heated or irradiated during various phases of the cell cycle. (Redrawn from Westra and Dewey 1971)

thermosensitivity of the synchronous CHO cells. The cells were heated at 45° C for 6, 10, or 15 min. The lowest cell survival also occurred after heating of mitotic cells. With the progression through the cell cycle the thermoresistance of the CHO cells increased and reached a maximum during the late G_1-phase. During the S-phase the thermosensitivity increased. The highest sensitivity was observed during the second half of the S-phase. Similar data were obtained by other authors. Especially the high thermosensitivity of the S-phase has been observed by a number of authors (Schlag and Lücke-Huhle 1976; Kim et al. 1976; Lücke-Huhle and Dertinger 1977; Bhuyan et al. 1977). Kim et al. (1976) also found an increased thermosensitivity of cells in G_2-phase.

After the hyperthermic treatment a division delay also occurs. The reasons for this effect apparently are an inhibition of cell migration through S-phase and an arrest in the G_2-phase (Kal and Hahn 1976; Schlag and Lücke-Huhle 1976). The arrested S-phase cells die immediately after the heat treatment (Lücke-Huhle and Dertinger 1977). It has been suggested that the newly synthesized DNA pieces and the DNA-synthesizing complex disaggregate and this effect, which will be discussed later, leads to cell death (Streffer 1986). The enormous thermosentivity of mitotic cells can apparently be explained by a disaggregation of the microtubules which form the spindle apparatus. The heated mitotic cells cannot complete mitosis. As a consequence many tetraploid cells appear (Coss et al. 1982). However, in further studies little or no differences were found for the various cell cycle phases in HeLa cells (Palzer and Heidelberger 1973) as well as in kidney cells (Reeves 1972).

In general it appears that cells in late S-phase and in mitosis have a high thermosensitivity. This finding is of special interest, as cells in late S-phase are usually radioresistant. Thus, ionizing radiation and hyperthermia act complementarily. These observations support the suggestion that the combination of ionizing radiation and hyperthermia is a very promising modality in tumour therapy.

Modification of Cell Survival by the Microenvironment

The micromilieu by which the cells are surrounded is very important for their thermosensitivity (Vaupel and Kallinowski, this volume; Gerweck and Epstein 1985; Reinhold 1985). This micromilieu is determined by two important factors: (1) physiological factors; in the case of hyperthermic treatment the blood flow plays a very important role. These phenomena will be reviewed and discussed by Vaupel and Kallinowski in the next chapter. (2) Metabolic factors, which will be described in the next part of this chapter.

It has been well established that the absence of oxygen or low oxygen pressure increases the radioresistance of cells (Alper 1979). Also it has been shown frequently that due to a lesser density of blood vessels in tumours than in normal tissues the oxygen pressure is frequently lower in tumours than in normal tissues (Vaupel et al. 1983). Therefore it has often been discussed that hypoxic cells in tumours increase the radioresistance of these tumours and that the occurrence of such cells is one important reason for the failure of radiotherapy in many cases (Hall 1978; Streffer 1985). Therefore information about the thermosensitivity of hypoxic cells is of great interest. In a number of studies a higher heat-induced cell killing was observed in hypoxic cells than in euoxic cells (Hahn 1974; Harisiadis et al. 1975; Kim et al. 1975; Gerweck 1977; Power and Harris 1977; Schulman and Hall 1974; Gerweck et al. 1979).

Hahn (1974) studied the cell survival of Chinese hamster cells after incubation at 43° C in the presence or absence of oxygen during heating and found that the cell killing was independent of these conditions. Bass et al. (1978) observed a slightly higher survival rate of HeLa cells when the cells were heated at 43° C under hypoxic conditions than under euoxic conditions. In these studies acute hypoxia was induced in the cells and medium by flushing the culture flasks with nitrogen instead of air. Durand (1978) investigated the survival of V-79 cells grown as spheroids under hypoxic and euoxic conditions. Large spheroids in which the proliferating cells decreased became more and more thermoresistant under euoxic conditions. If hypoxia was induced, the thermosensitivity increased, however. The thermosensitivity of small spheroids was not modified by hypoxia. Durand (1978) suggests that hypoxia per se is not responsible for the modified thermosensitivity but accompanying metabolic changes.

In further investigations cell respiration was used in order to reduce the oxygen in the culture medium. This was achieved by using high cell densities (Hahn 1974) or large numbers of feeder cells which were irradiated with high radiation doses (Kim et al. 1975). Under these conditions it has been found that the hypoxic cells are more thermosensitive than euoxic cells. However, the experimental conditions are such that not only the oxygen but also other nutrients are heavily consumed. Also the pH may have changed so that interference of these parameters with hypoxia cannot be excluded. In a very careful investigation Gerweck et al. (1979) compared the modifying action of acute and chronic hypoxia on the thermosensitivity of Chinese hamster cells. It was observed that acute hypoxia did not change the cell killing, which was induced by 3 h heating at 42° C. However, an increased cell killing took place when the cells were kept under chronic hypoxic conditions for 18 h and longer. Under these conditions the hypoxic cells still had a higher survival after irradiation with 7.5 Gy than euoxic cells (Fig. 4).

As hypoxic cells in a tumour in situ will probably live longer under hypoxic conditions the investigations of Gerweck et al. (1979) certainly represent a realistic situation for tumour therapy. In general it appears that in contrast to ionizing radiation hypoxic cells are certainly not more resistant to hyperthermia than euoxic cells. There may even exist a higher thermosensitivity for hypoxic cells in tumours.

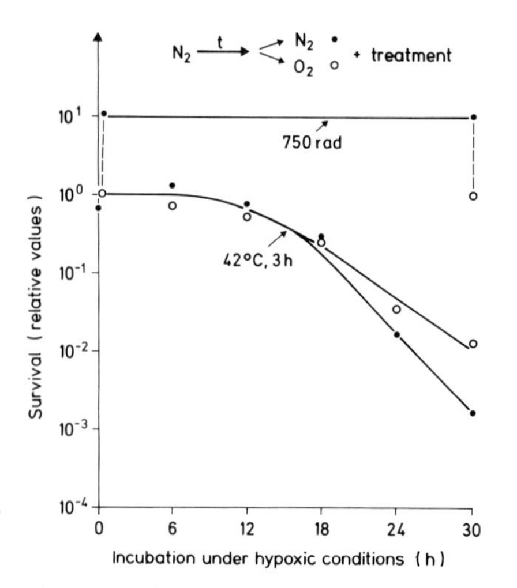

Fig. 4. Survival curves of CHO cells after heat treatment and irradiation under chronic hypoxic conditions. Cells were heated or irradiated under euoxic or hypoxic conditions following 0–30 h culturing under hypoxic conditions at 37° C. The data were normalized to the survival level obtained after heating or irradiation under oxygenated conditions without prior culturing under hypoxia. (Redrawn from Gerweck et al. 1979)

A wide discussion has taken place as to whether the pH of the microenvironment has an influence on cellular thermosensitivity. Von Ardenne et al. (1969; 1976) have shown that glucose infusions lead to a decrease of pH in tumours and this effect may be responsible for an increased cell killing by hyperthermia. Several studies have demonstrated that reduction in medium pH to approximately 7.0 and below induces an increased cell death after heating the cells in vitro to 42°-45° C (Overgaard 1976; Gerweck 1977; Freeman et al. 1977; Meyer et al. 1979; Gerweck and Richards 1981). In most cases the modifying effect on cell survival was comparatively small in the pH range 7.0-7.4 and the effect became much larger with the pH below 7.0, especially in the range 6.5-6.7. If a dose effect curve was observed in which the thermosensitivity apparently decreased during heating at pH 7.4 (type III survival curve, see above), this effect was dramatically reduced at a low pH (Fig. 5). Gerweck (1977) also studied the time dependence of the pH changes in order to obtain thermosensitization. It turned out that a decreased pH was necessary either shortly before or during the incubation at elevated temperatures. The effect was highest under the latter conditions. If the pH of the medium was decreased after the heat treatment, a significant reduction of cell survival was not observed.

These data are especially interesting as it has been frequently found that the pH is lower in tumours than in normal tissues (Gerweck 1978). Such observations have also been made in human tumours, although this phenomenon is not found in all tumours and a considerable heterogeneity of pH was observed within individual human tumours (Wike-Hooley et al. 1985). It has been suggested that high lactate production and low blood flow in tumours are mainly responsible for the decrease of pH. Quite often the microenvironment of tumour cells is characterized by hypoxia, nutrient deprivation and decreased pH. It has been postulated that this situation is even enhanced during a hyperthermic treatment so that a further sensitization of tumours to heat takes place. Therefore these parameters need further discussion.

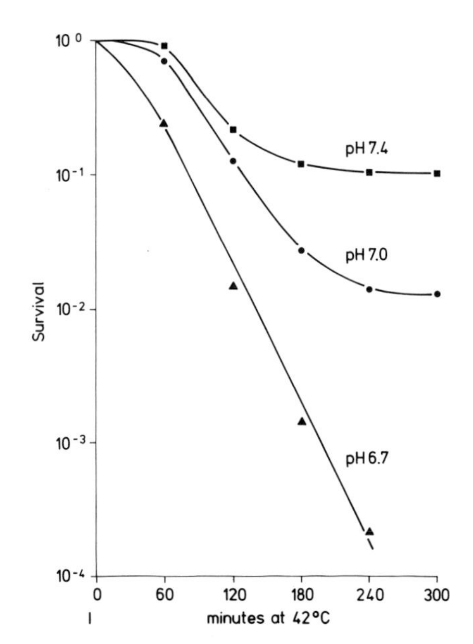

Fig. 5. pH modification of cell survival. The cells were heated during the midportion of a 5-h exposure at the indicated pH. (Redrawn from Gerweck 1977)

Molecular and Metabolic Changes by Hyperthermia

The foregoing discussion has demonstrated that pH and oxygen or hypoxia respectively can modify the cellular response to hyperthermia. These parameters are closely connected to cellular metabolism and cannot be seen in isolation. Immediately during a hyperthermic treatment considerable molecular and metabolic changes occur. While after exposure to ionizing radiation DNA damage is the most important effect leading to reproductive cell death (Alper 1979), the mechanism of cell killing by hyperthermia alone is less clear. It appears evident that events in the cytoplasm as distinct from the cell nucleus are important (Hahn 1982; Streffer 1982). At the molecular and metabolic level heat predominantly induces two principal effects (Streffer 1985):

1. Conformational changes as well as destabilization of macromolecules and of multimolecular structures
2. Increased rates of metabolic reactions during the heat treatment followed by disregulation of metabolism mainly after hyperthermia

Conformational Changes of Multimolecular Structures

The conformation of biological macromolecules is mainly stabilized by covalent bonds between subunits of the macromolecules, by hydrogen bridges, by interactions of ionic groups within the macromolecules and with their environment as well as by hydrophilic/hydrophobic interactions of groups within the macromolecules and their environment. The last three classes of bonds and interactions are comparatively weak bonds: they can be altered easily, for instance by an increase in temperature. Thus, especially in proteins such changes can be induced which lead to disturbances of the native protein conformation (Privalow 1979; Lepock 1982; Lepock et al. 1983; Leeper 1985; Streffer 1985; Streffer 1986).

The range of temperatures in which structural transitions of proteins occur depends very much on the specific proteins. For a number of proteins such transitions have been demonstrated in the range of 40°-45° C, in which cell killing also takes place and which is used for hyperthermic treatment in tumour therapy. Furthermore, heat-induced structural changes in proteins are extremely dependent on the pH (Privalow 1979). The pH determines whether the various ionic groups of amino acid residues exist in the protonated or deprotonated form (Streffer 1964). A number of these groups, which stabilize protein conformation by ionic interactions, have pK values near the physiological pH. These pK values are dependent on temperature; therefore, a change in temperature will also alter the protonation and by this effect the ionic state of the amino acid residues is altered. Such alterations contribute to conformational changes of proteins during hyperthermia (Streffer 1964; Wallenfels and Streffer 1964, 1966).

Lepock et al. (1983) have reported that the fluorescence of proteins bound to membranes and the quenching of fluorescence by paranaric acid is altered by heating isolated cytoplasmic and mitochondrial membranes. It is concluded that these conformational changes of membrane proteins are responsible for observed effects on membranes after heating. It has frequently been suggested that membranes are the main cellular targets for hyperthermia in bacteria as well as in mammalian cells (Wallach 1978; Hahn 1982).

Much attention has been paid to the lipids of membranes and their influence on membrane fluidity in relation to cell killing (Yatvin et al. 1982). Such correlations have been observed for bacterial systems. Bowler et al. (1973) observed an increase in membrane permeability after hyperthermia and a loss of membrane-bound ATPase. These effects correlated with cell killing. Yatvin (1977) studied a mutant of *E. coli* K12 which required unsaturated fatty acids. With an increasing incorporation of these fatty acids into the membrane the fluidity of the membrane as well as cell killing by heating was enhanced. The thermosensitivity of the bacteria increased in a proportional manner with microviscosity when cells were grown in unsaturated fatty acid at different temperatures (Fig. 6) (Dennis and Yatvin 1981). Also anesthetics, like procaine, increase bacterial thermosensitivity as well as membrane fluidity (Yatvin et al. 1982). Furthermore the membrane-active drug chlorpromazine was able to increase bacterial thermosensitivity (Shenoy and Singh 1985).

However, the situation appears to be somewhat different in mammalian cells. The incubation of V-79 Chinese hamster cells with cholesterol resulted in a higher microviscosity of membranes but the thermosensitivity of the cells was not changed (Yatvin et al. 1983). On the other hand, Cress et al. (1982) observed a positive correlation between cholesterol content of plasma membranes and cell killing in several cell lines. Li et al. (1980) found remarkable similarities between the action of hyperthermia or ethanol on cell killing and interpreted these findings as a modification of membrane fluidity. However, in a further study, cell killing by hyperthermia and its modification by ethanol apparently correlated closer with protein denaturation in membranes than with lipid fluidity (Massicotte-Nolan et al. 1981).

Studies with electron spin resonance have demonstrated that lipid transitions occur in the temperature ranges around 7°-8° C and 23°-26° C in mitochondria and whole cell homogenates, while conformational transitions in proteins were also observed between 40° and 47° C (Lepock et al. 1982, 1983). Investigations on membrane-bound receptors have shown in several studies that they are inactivated or lost from the membranes. Concanavalin A induced capping and cell survival responded in a similar manner to hyperthermia (Stevenson et al. 1981). Calderwood and Hahn (1983) observed a heat-induced inhibition of insulin binding to the plasma membrane of CHO cells, which was apparently caused by a decrease in the number of available insulin receptors and this effect correlated well with cell killing. Similar effects were found for the binding of monoclonal antibodies to murine lymphoma cells (Mehdi et al. 1984). By studies with scanning electron microscopy it has been shown that proteins, which go through both phospholipid layers

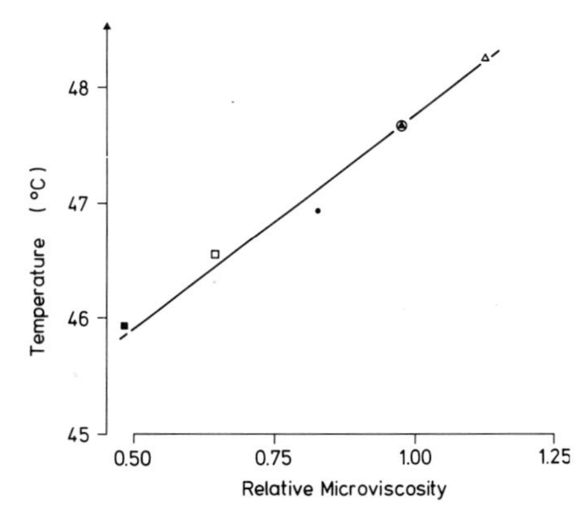

Fig. 6. Temperature required to kill 90% bacterial cells in 3 h as a function of relative membrane viscosity. *Open symbols,* growth in oleic acid; *closed symbols,* growth in linoleic acid. Growth at 25° C *(squares),* 37° C *(circles)* and 41° C *(triangles).* (Redrawn from Dennis and Yatvin 1981)

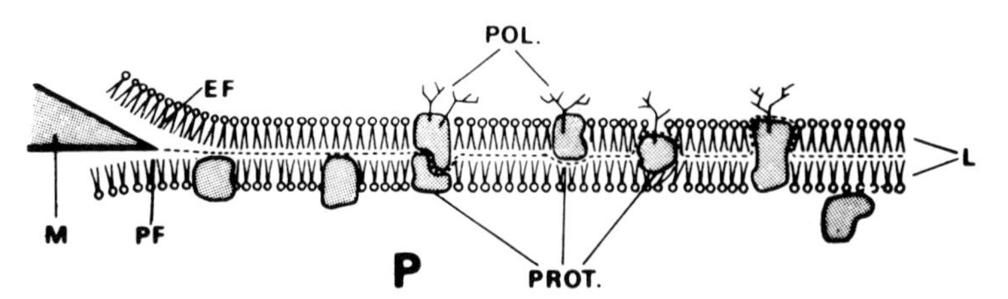

Fig. 7. Structure of membranes, phospholipid-bilayer with proteins

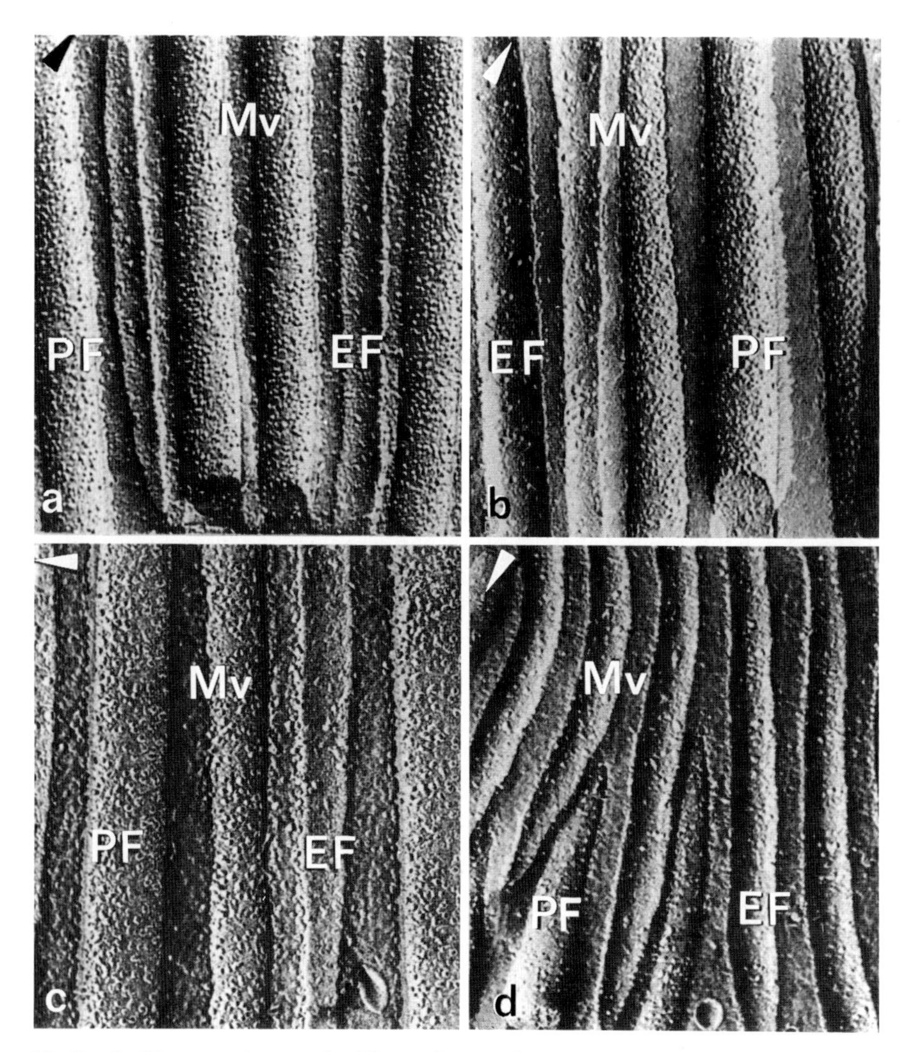

Fig. 8 a–d. Electron micrograph of freeze-fractured small intestine mouse microvilli *(Mv)*. *PF*, inner membrane; *EF*, external membrane. **a** Control, ×77000; **b** non-exteriorized intestine after heating at 41° C for 30 min, ×98000; **c** exteriorized intestine immediately after heating at 41° C for 30 min ×77000; **d** 3 h after heating at 41° C for 30 min, ×77000. Note the reduction of IMP particles especially in **d.** (Issa 1985)

of the membrane (intermembrane protein particles, IPP) (Fig. 7) and which stabilize the membrane, are removed by hyperthermia (Fig. 8) (Streffer 1985). However, not only the cytoplasmic membrane has been damaged by hyperthermia but also the intracellular membranes, which, for instance, show dramatic disturbances in the small intestine after heating (Fig. 9) (Breipohl et al. 1983). The loss of the receptor activities and other membrane proteins might be caused by conformational changes of the corresponding membrane proteins. However, a clear decision with respect to the role of lipids or proteins may be difficult. The cooperative state between both molecular species in membranes is dependent on both membrane fluidity and protein conformation (Wallach 1978). Studies

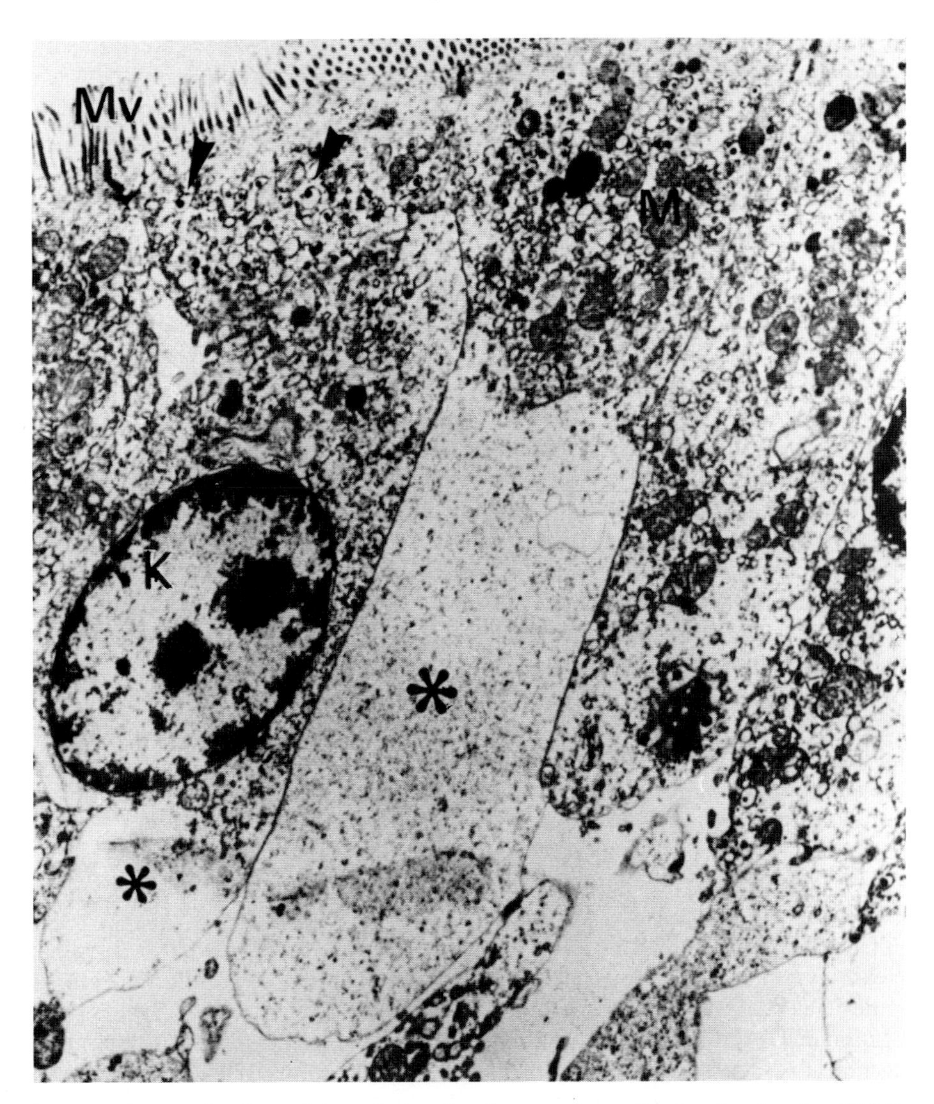

Fig. 9. Electron micrograph of small intestinal cells of mouse immediately after heating (41° C, 30 min. *K*, cell nucleus; *M*, mitochondrium; *Mv*, microvilli; *arrows*, membrane-bound vesicles. The basal part of the enterocyte *(asterisk)* shows an amorphous ground substance. Cell organelles are concentrated in the apical part of the cells, × 7600. (Issa 1985)

with Raman spectroscopy on erythrocyte membranes show that temperature-dependent transitions involve concerted processes in which hydrophobic amino acid residues and lipids participate (Verma and Wallach 1976). These membrane changes also induce alterations in the ion permeabilities. An increased influx of Ca^{2+} and efflux of K^+ has frequently been observed (Leeper 1985). These effects will be discussed in connection with the mechanisms of cell killing.

Mitotic cells have been found to be very heat sensitive. Hyperthermic treatment apparently prevents the aggregation of the globular proteins to the spindle apparatus or cause the disaggregation of spindles. In consequence mitotic cells are unable to complete the mitotic division, and cells with a tetraploid genome enter into the G_1-phase (Coss et al. 1982). In fast proliferating cell systems many tetraploid cells are seen (van Beuningen et al. 1978). In a similar way microtubules of the cytoskeleton disaggregate during hyperthermic treatment and reaggregate during incubation at 37° C (Lin et al. 1982). A correlation between cell killing and disturbances of the cytoskeleton has been observed; in dead cells no reassembly of the cytoskeleton occurs (Cress et al. 1985). In heavily heat damaged cells these structures have been completely lost. Again it appears reasonable to assume that conformational changes of the proteins take place.

Inhibition of DNA, RNA and Protein Synthesis

Heating of cells to 42°–45° C leads to a very sudden inhibition of DNA, RNA and protein synthesis as measured by the incorporation of labelled precursors into these macromolecules (Hahn 1982; Streffer 1982). In some cellular systems the synthesis of all three molecular species has been measured (Mondovi et al. 1969; Henle and Leeper 1979; Reeves 1982). Generally it has been observed that the degree of inhibition is correlated to the temperature by which the cells are heated. The duration of heating influences the period during which the macromolecular synthesis is affected (Streffer 1982). Protein and RNA synthesis recover comparatively rapidly while DNA synthesis is reduced for a longer period after an identical heat treatment.

Not only the initiation of DNA synthesis in new replicons but also the elongation of the nascent DNA in these replicons is inhibited during and after a hyperthermic treatment (Gerner et al. 1979; Henle and Leeper 1979; Wong and Dewey 1982). Heat causes an increased amount of single-stranded DNA. DNA elongation recovers faster than replicon initiation after a heat treatment (Warters and Stone 1983). Therefore the prolonged depression of heat-induced DNA synthesis is apparently connected to the inhibition of the initiation processes. Warters and Stone (1983) have also reported that heat treatment caused a long-term inhibition of ligation of replicative DNA fragments into chromosome-sized DNA. Nascent DNA was observed in replicated fragments with a size as small as the length of a replicon after hyperthermia.

In agreement with these data the activity of the poly (ADP-ribose)-synthetase decreases in human melanoma cells after heating to 42° and 44° C (Streffer et al. 1983; Tamulevicius et al. 1984). This enzyme is firmly bound to the chromatin and is involved in ligation of DNA fragments as well as in DNA repair processes (Shall 1984). In this connection it is interesting that a correlation has been observed between inhibitors of poly (ADP-ribose)-synthetase and cell killing by heat in Chinese hamster cells (Nagle and Moss 1983). The observation that elevated temperatures cause chromosome aberrations in S-phase cells but not in G_1-phase cells (Dewey et al. 1980) may be due to the inhibition of DNA ligation.

Enzymatic studies have shown that the DNA polymerase β, which is involved in unscheduled DNA synthesis for DNA repair, is more thermosensitive than the DNA polymerase α (Dube et al. 1977). A positive correlation was observed between the hyperthermic cell killing as well as the heat-sensitizing effect on cell killing by ionizing radiation and the inhibition of DNA polymerase β; the correlation between cell killing and depression of DNA polymerase α was much poorer (Spiro et al. 1982). From an Arrhenius plot an inactivation energy of 152 kcal/mol and a breaking point of the Arrhenius plot at 42.5° C were obtained which were similar to those for cell killing. Such data were not found for DNA polymerase α. The positive correlation with DNA polymerase β has also been reported for thermotolerant CHO cells (Dewey and Esch 1982) but not for thermotolerant HeLa cells (Jorritsma et al. 1984). Also the correlation between DNA polymerase β and cell killing does not agree with the finding that the enzyme activity is inhibited to the same degree in G_1- or S-phase (Spiro et al. 1982). The suggestion that the inactivation of DNA polymerase β is an important part of the mechanism by which heat-induced cell killing occurs is of great interest. However, the described inconsistencies should not be overlooked.

Simard and Bernhard (1967) from electron microscopic observations have reported that a heat treatment at 42° C destroys the structure of nucleoli, which are the sites for the synthesis of rRNA in the cell nucleus. Also the processing of the 45 S RNA, a precursor of rRNA, to the functional 18 S rRNA is apparently blocked by heating (Warocquier and Scherrer 1969; Ashburner and Bonner 1979). After recovery from a heat shock, synthesis of rRNA becomes quite heat resistant (Burdon 1985). The RNA synthesis for certain proteins (heat shock proteins) is even enhanced.

For the translational processes of protein synthesis a complex has to be formed between mRNA and ribosomes to polysomes. In mammalian cells elevated temperatures lead to a breakdown of the active polypeptide-synthesizing polysomes. McCormick and Penman (1969) have reported a rapid disaggregation of polysomes when HeLa cells have been heated to 42° C. Panniers and Henshaw (1984) found markedly decreased levels of the initiation complex for polypeptide synthesis, which is formed between the 40 S ribosome subunit and methionine-tRNA in Ehrlich cells after heating at 43° for 20 min. After a short hyperthermic treatment apparently all components of the synthesizing machinery are still present, so that a reaggregation can occur and protein synthesis recovers. Phosphorylation and dephosphorylation of the initiation factor eIF-2 may play a role in this connection (Burdon 1985).

Glucose Metabolism and Hyperthermia

Glucose metabolism is closely linked to the metabolism of lipids and a number of amino acids as well as to energy metabolism in general. These pathways are very dependent on the extra- and intracellular milieu. A number of differences exist between tumours and normal tissues with respect to glycolysis (Warburg et al. 1926), pH (Gerweck 1982) and nutrients (Hahn 1982), which may be very important for hyperthermic treatment. However, it must be stressed that the variability between individual tumours even of the same entity and localization is extremely high. Nevertheless some metabolic patterns may be characteristic in some tumours. Thus in hepatomas the glycolytic pathway is increased and gluconeogenesis is reduced in comparison with normal liver tissue (Weber 1983).

The rate of glycolysis is dependent on various factors, such as pH and oxygen tension. Glycolysis influences the intracellular milieu. Glucose loading of rats decreases the pH in normal tissues and in tumours. This effect has been measured in the extracellular space (von Ardenne 1982; Jähde and Rajewsky 1982). It is generally assumed that these processes lead to an accumulation of lactic acid in the heated tissues. In this connection it is of considerable importance that cell killing by hyperthermia is enhanced if cells are heated at low pH (Gerweck 1982).

Dickson and Calderwood (1979) have observed a reduction of glycolysis under anaerobic incubation conditions (in vitro). The effects are considerably enhanced, when the rats obtain a high glucose load of 6 mg/g body wt. before the hyperthermic treatment. In these studies the rat tumours were heated in situ to 42° C and the biochemical measurements have been performed with tissue slices in vitro. This experimental design apparently does not reflect the metabolic situation during hyperthermia in vivo. In contrast Strom et al. (1977) have reported for several tumours that glycolysis has been unaffected by temperatures up to 44° C. Extensive studies with a mouse adenocarcinoma transplanted intramuscularly into the hindleg have demonstrated that the glucose level (3.2 µmol/g tissue) decreases to about one-third after 1 h local hyperthermia at 43° C and remains decreased for the next 24 h (Streffer et al. 1981; Hengstebeck 1983). The lactate levels increase slightly during the first 40 min of hyperthermia and decrease at later times. Similar patterns are found for a number of glycolytic intermediates, like glucose-6-phosphate, fructose-1.6-bisphosphate, dihydroxyacetonphosphate and pyruvate. On the other hand a large increase has been observed in the levels of β-hydroxybutyrate and acetoacetate several hours after hyperthermia. This latter effect will be discussed later. These data demonstrate that glycolysis is apparently not inhibited during the hyperthermic treatment, as long as enough glucose is available.

Burdon et al. (1984) have observed an increase in lactate formation in HeLa cells, glioma cells and some other cells directly after heating for 2–4 h at 42° C (increase 1%–40%). The effect varied greatly from cell line to cell line, however. No change was seen in normal human glial cells.

Even if very localized heating is performed, the metabolism of the tumour cannot be considered in isolation. A steady exchange of metabolites, like glucose and lactate, occurs between the various tissues. In these processes liver metabolism is especially involved and plays a central role, e.g. glucose is transported from the liver to peripheral organs including the tumour, and lactate flows in the reverse direction. The hyperthermic treatment of the tumour therefore also induces a decrease of the glucose level and a temporary increase in lactate in the liver (Hengstebeck 1983). The increase in hepatic lactate levels during hyperthermia may be caused by an impaired blood flow and oxygenation through glycolysis in the liver itself, as the temperature increases also in the liver to about 40° C under the experimental conditions. A decrease in blood flow has been seen in patients during whole body hyperthermia (Wike-Hooley et al. 1983).

Another possibility could be that an enhanced flow of lactate from peripheral organs and tissues to the liver occurs, as the tumour and muscle degrade more glucose to lactate during the hyperthermic treatment. This is supported by the findings that the levels of lactate increase in the blood during hyperthermia in rats (42.5° C) (Burger and Engelbrecht 1967) as well as in dogs (42° C) (Frankel and Ferrante 1966). Also a hyperthermia-induced increase of leakage of lactate from the tumour may be possible through membrane changes as the lactate level is much higher in the tumour than in the blood and in other tissues. These effects lead to an increased hepatic lactate level, as has been observed by Frascella and Frankel (1969) in rats during hyperthermia.

The following decrease in the hepatic lactate level, which can occur during a long-lasting hyperthermic treatment or after the heating, can be caused by a depletion of glucose and other energy sources during hyperthermia. In perfused rat liver the reduction of glucose is counteracted when exogenous palmitate is added to the perfusate as a further energy source (Collins et al. 1980).

Hyperthermic treatment speeds up metabolic reaction rates including glycolysis (Schubert et al. 1982; Streffer 1982). More energy is needed by the heated cells in order to maintain ion gradients through membranes, structural characteristics in the various cell compartments, etc. Thus, investigations with [^{14}C]glucose (uniformly labelled) have demonstrated an enhanced turnover of glucose during whole body hyperthermia of mice at 40° and 41° C, as measured by expired labelled CO_2 (Fig. 10) (Streffer 1982; Schubert et al. 1982). Under the same conditions a strong decrease of liver glycogen has been observed. The hepatic glycogen value remains low during the following 24 h and only a slow rise is induced by a glucose load after the hyperthermic treatment. This latter result supports the finding of Skibba and Collins (1978) that gluconeogenesis is reduced in perfused rat liver if the temperature is raised to 42° C. Under normal conditions the lactate, which flows from peripheral tissues to the liver, can be used for glucose formation. This pathway of gluconeogenesis is apparently impaired by hyperthermia. From the consideration of lactate generation it can generally be concluded that the glycolytic rate is usually not impaired in tumours and in the liver of mammals during a hyperthermic treatment; it may even be increased.

If a tumour cell obtains its energy supply predominantly from glycolysis, inhibition of this pathway may enhance the effect of hyperthermic cell killing considerably. This has been demonstrated by Song et al. (1977) as well as by Kim et al. (1978), with 5-thio-D-glucose and some other substances (Kim et al. 1984). On this basis it is understandable that the sensitizing effects on cell killing are especially expressed under hypoxic conditions, which induce an extreme increase of the lactate levels. Kim et al. (1984) have observed that the flavone derivative quercetin increases the thermosensitivity of HeLa cells. This drug inhibits the lactate transport across the cytoplasma membrane and produces an intracellular acidification and an inhibition of glycolysis (Belt et al. 1979). The decreased pH (Gerweck 1982) and the inhibition of glycolysis apparently enhances the thermosensitivity of the cells (Fig. 11).

As has been pointed out cells and tissues evidently need more energy at elevated temperatures and the metabolic turnover rates are enhanced (Streffer 1982). Glucose and liver glycogen are the primary sources for energy production through metabolism. At least in small rodents these reservoirs are much utilized. Also an extensive degradation of lipids

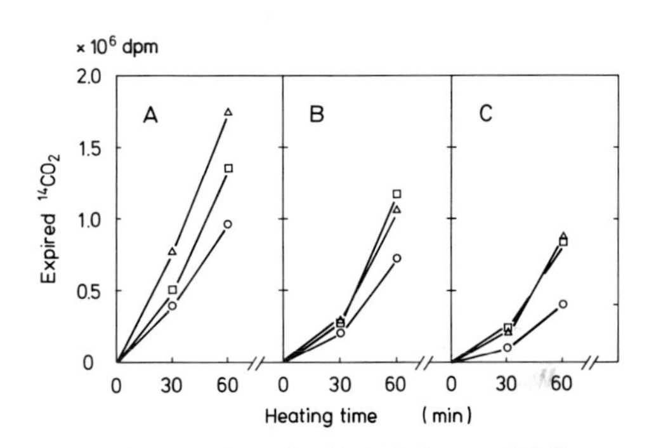

Fig. 10 A–C. Expiration of $^{14}CO_2$ from mice during 1 h of hyperthermia (*0*, 37° C; □, 40° C, △, 41° C) after injection of 3 µCi [^{14}C]-U-glucose without (**A**), or with a glucose load of 1.5 mg (**B**) and 6.0 mg/g body wt. (**C**). *dpm*, disintegrations/min. (Redrawn from Schubert et al. 1983)

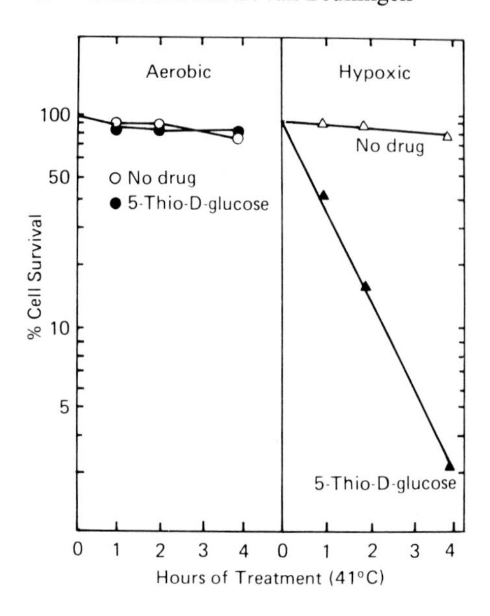

Fig. 11. Cell survival under oxic and hypoxic conditions with and without the addition of 5-thio-glucose. (Redrawn from Kim et al. 1978)

occurs (Tamulevicius et al. 1984). In particular the free fatty acids and esterified fatty acids are decreased in an adenocarcinoma and in liver of mice after hyperthermia at 43° C. On this basis it can be understood that cells which are deprived of glucose show a high thermosensitivity under hypoxia (Kim et al. 1980; Hahn 1982).

If the hyperthermic stress is very strong, the liver glycogen and glucose levels are depressed and the glucose content cannot be returned to normal values in liver but also in other tissues including the tumour. The acidic metabolites β-hydroxybutyrate and acetoacetate are formed and increase severalfold over the normal value in tumours and liver under these conditions (Streffer et al. 1981; Streffer 1982). An increase in these acidic metabolites has also been observed in the blood of patients after whole body hyperthermia (Adam et al. 1983). Apparently liver metabolism is changed to conditions which are usually seen in metabolic acidosis. A sufficient utilization of acetyl-CoA in the citrate cycle apparently does not take place after hyperthermia and the acetyl-CoA is used for the formation of acetocetate and β-hydroxybutyrate. These alterations may be at least partially responsible for liver damage which has been frequently found after whole body hyperthermia (Wike-Hooley et al. 1983).

In this connection it is of further interest that the glucose utilization is reduced *after* the hyperthermic treatment. This effect has been observed with whole body hyperthermia (Schubert et al. 1982; Streffer 1982) and in a mouse tumour (Hengstebeck 1983). When the animals obtain glucose before the hyperthermic treatment, the metabolic alterations are less severe. Under these conditions the complete depletion of carbohydrate sources is prevented, which apparently has a protective effect for the intermediary metabolism. Thus, a glucose load, which is given to mice with an adenocarcinoma before hyperthermia, results in an increase in lactate level in the tumour until 12 h after hyperthermia although the glucose has returned to normal values several hours earlier (Hengstebeck 1983). These data and the reduced CO_2 expiration from glucose, which is observed after hyperthermia (Schubert et al. 1982; Streffer 1982), show that apparently the citrate cycle is more severely damaged than glycolysis by hyperthermia. (After the hyperthermia treatment the glucose can be metabolized to lactate through glycolysis but the degradation to CO_2 through the citrate cycle is reduced). The oxidized metabolite of lactate, pyruvate, does not increase in the same way as the lactate level (Streffer et al. 1981). This effect leads to an increased lactate/pyruvate ratio (Fig. 12), which is a good indicator for the state of oxygenation of the cytosol in cells and shows the decrease of intracellular oxygen tension under these conditions. The redox ratios are coupled to the levels of free $NAD^+/NADH$ by the following equation, in which K is the equilibrium constant:

$$K = \frac{[\text{lact.}] \times [\text{NAD}^+]}{[\text{pyr}] \times [\text{NADH}] \times [\text{H}^+]}$$

The lactate/pyruvate ratio is representative of the redox status in the cytosol and the β-hydroxybutyrate/acetoacetate ratio of that in the mitochondria. The ratios react very quickly to the intracellular oxidative state, as the activity of the corresponding enzyme is high (Williamson et al. 1967).

Glutathione in its reduced and oxidized form is also coupled to the intracellular redox equilibria through glutathione reductase. It has been found that the reduced glutathione increases in cultured cells after hyperthermia (Leeper 1985). Usually the oxidized form has not been determined. If under the experimental conditions, which lead to an increase in the intracellular level of the reduced glutathione, the redox ratios are also shifted to the reduced state, the observed effect could be merely caused by a biochemical reduction of oxidized glutathione. Therefore it is not possible to know whether the observed effects are caused by new glutathione synthesis or by a metabolic shift of the redox ratio from the oxidized to the reduced form. From the other redox ratios such as lactate/pyruvate and β-hydroxybutyrate/acetoacetate it can be concluded that this latter phenomenon contributes at least to a certain degree to the observed increase in the reduced glutathione. Leeper (1985) has reviewed the data and discussed the possible biological relevance.

These redox ratios reflect the oxidative state and through this connection the oxygen tension as well as the pH within the cellular compartments. As both parameters are very important for cell killing by hyperthermia (Hahn 1982), more emphasis should be put on determining them. Data such as the determination of redox ratios have the great advantage that they reflect the intracellular situation in contrast to direct measurements of pH and of oxygen tension which usually only reflect the extracellular situation. During and after hyperthermia, especially with a glucose load, these lactate/pyruvate ratios are increased in the liver of rats (Frascella and Frankel 1969) and of mice as well as in a tumour (Hengstebeck 1983) (Fig. 12). This is consistent with heat-induced changes of vascularity and blood flow (Song et al. 1980; Vaupel 1983). A prognosis of to what degree the thermosensitivity is modified by the glucose load is difficult under these conditions. The glucose may decrease the pH and induce hypoxia, but the higher supply of nutrients could increase the thermoresistance (Hahn 1982). Such data have a high significance for the combination of ionizing radiation and hyperthermia in therapy. Irradiation should not be used at a time when tumour hypoxia is induced by hyperthermia and glucose.

Oxygen Consumption and ATP Levels and Hyperthermia

Mondovi et al. (1969) reported a decrease in respiration in Novikoff hepatoma cells incubated, after a heat treatment at 43° C, with glucose and succinate at 38° C for the measurement of the oxygen consumption. The reduction in respiration was not very great; however, a greater decrease in oxygen consumption has been observed by Durand (1978) in CHO cells after heating at 41° and 43° C, although the experimental design may have led to a depletion of metabolites and therefore less substrates are provided in the cells during

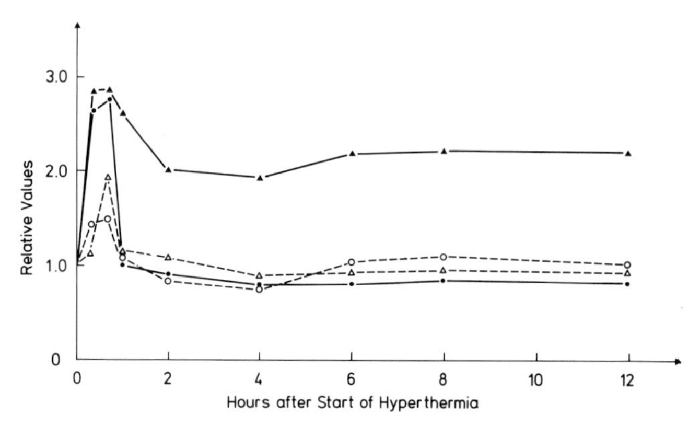

Fig. 12. Lactate/pyruvate ratio in mouse liver and adenocarcinoma after local heating of the tumour for 30 min at 43° C or injection of 6 mg/g body wt. glucose plus heating. ●, liver heating alone; ○, tumour heating alone; △, liver heating plus glucose; ▲, tumour heating plus glucose

hyperthermia. Similar data on oxygen consumption have been reported in a review by Strom et al. (1977). Thus, the mitochondrial respiratory chain seems to be more thermosensitive than glycolysis but less than the citrate cycle. The extent of change differs with the type of cells and tissues. More data are needed for clarification in respect of these metabolic processes.

Oxygen consumption and ATP synthesis in the mitochondria are coupled together. Several authors have described that ATP levels decrease in cells after a heat treatment (Francesconi and Mayer 1979; Ohyama and Yamada 1980; Lunec and Cresswell 1983; Mirtsch et al. 1984). In other experiments no change in the ATP content in cells has been observed (Henle et al. 1984; Nagle et al. 1982). Lunec and Cresswell (1983) found a decrease in ATP levels in murine lymphoma cells after heating at 44° C for 10–60 min, whereas no change occurred in Ehrlich ascites cells under the same conditions. Ohyama and Yamada (1980) observed that the radiation-induced decrease in ATP levels in thymocytes is partially reversed by a heat treatment at 43° C directly after the irradiation. Under these conditions it appears that enhanced ATP synthesis occurs, as demonstrated by increased incorporation of [^{32}P]phosphate into the ATP. Mirtsch et al. (1984) have also observed that [^{3}H]adenosine is incorporated into the ATP of heated melanoma cells at a higher rate than into unheated cells. If it is further taken into consideration that the ATP level is decreased in the heated cells, then this finding shows that the ATP turnover is possibly increased in the heated cells (Streffer 1985).

Unfortunately such experimental data on turnover have so far not been available. A clear correlation between the cellular ATP levels and cell killing by hyperthermia has not been shown in the reported studies. From the reported data it may even be expected that no direct effect of a disturbed energy metabolism through oxidative phosphorylation and cell killing exists (Streffer 1986).

Thermotolerance

It has already been mentioned that living cells including human as well as tumour cells can apparently increase their thermal resistance during or after an appropriate heat treatment. This phenomenon has been called thermotolerance. Whether this terminology is reasonable or not has been discussed (Hahn 1982). Thermotolerance can be induced in two ways:

1. It develops during continuous heating over several hours at comparatively low temperatures (42° C and below). This effect has been described with the explanation of the observed dose effect curves for cell killing (type 3). The phenomenon was observed first although not interpreted and recognized as thermotolerance by Palzer and Heidelberger (1973).
2. It develops by fractionation of heat treatments if the cells are incubated at 37° C during the interval between two heat treatments at 39°-47° C. Gerner and Schneider (1975) as well as Henle and Leeper (1976) demonstrated that the thermoresistance of the cells increased after the first heating. The slope of the exponential part of the survival curve became shallower (Fig. 13).

Part of the increased cell survival after fractionated heat treatments may be due to recovery from sublethal damage (Hahn 1982), which is defined in analogy to recovery from damage after exposure to ionizing radiation (Elkind et al. 1967). However, it is extremely difficult to separate recovery from development of thermotolerance; therefore, both phenomena are discussed together here. Thermotolerance develops not only with respect to cell survival but also with respect to synthesis of biological macromolecules (Dewey et al. 1979), energy metabolism (Lunec and Cresswell 1983) and other phenomena.

The experimental design of fractionated heat treatments has profound effects on the development of cellular thermotolerance. If the incubation during the interval between acute heat fractions was 40° C instead of 37° C a decrease rather than an increase of thermoresistance occurred (Henle and Leeper 1976). This "step-down" heating was studied very carefully by Jung (1982). A pretreatment at a high temperature (43° C) followed by a mild hyperthermic treatment (40° C) led to a steady increase in the slope in the survival curve (decrease in D_o). The cells become more thermosensitive (Joshi and Jung 1979). In contrast a pretreatment at a low temperature (40° C) followed directly by a treatment at

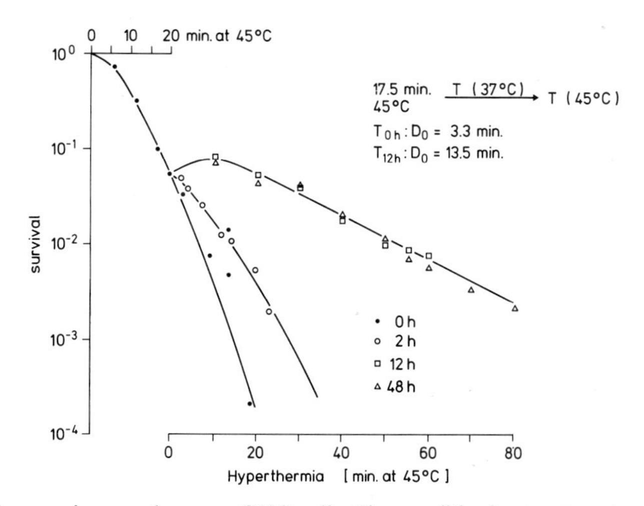

Fig. 13. The induction of thermotolerance in asynchronous CHO cells. The conditioning treatment was 45° C over 17.5 min. *Upper abscissa,* duration of heat treatment for the single treatment; *lower abscissa,* duration of the second heat treatment. The interval between both treatments is given in the figure. (Redrawn from Henle and Leeper 1976)

43° C induces thermotolerance. Especially the slope of the survival curve becomes shallower and D_0 increases. This phenomenon is called "step-up" heating (Jung 1982).

It has been also shown that the cellular microenvironment is important for thermotolerance development. Thus, less thermotolerance is induced when the pH is lowered (Goldin and Leeper 1981; Gerweck et al. 1982; Eickhoff and Dikomey 1984). This pH dependence of thermotolerance development was demonstrated for acute conditioning heating (Goldin and Leeper 1981) as well as for continuous heating (Gerweck et al. 1982). Furthermore it was observed that the decay of thermotolerance was slower at low pH. As under these conditions cell division was delayed, the influence of cell division on the decay was suggested (Gerweck et al. 1982). Presence or absence of oxygen has apparently no influence either on the extent or on the kinetics of thermotolerance (Gerweck and Bascomb 1982).

Several authors have investigated the kinetics for the development of thermotolerance. The degree of thermotolerance as well as the rate for the development depend on the severity of the conditioning heat treatment. The more severe the pretreatment the longer was the development until maximal thermotolerance was reached but the degree of thermotolerance also increased (Henle et al. 1979; Nielsen and Overgaard 1982). Li et al. (1982) suggested from their observations on Chinese hamster cells (HA-1), which were treated in the plateau phase, that a reduction of the treatment temperature by 1° C requires a doubling of the treatment duration in order to obtain a similar level of thermotolerance. However, at low temperatures such as 41° C and 42° C (conditioning treatment) the kinetics of thermotolerance development were apparently different.

Thermotolerance does not become an inherent property of the cell; after some time it decays again and the cells reach their original thermosensitivity. In HeLa cells after cell division thermotolerance immediately disappeared. However, division was not required for the decay, as this also occurred in plateau phase cells (Gerner et al. 1979). Nielsen and Overgaard (1982) observed that the decay of thermotolerance was dependent on its induced level. The authors suggested that the rate of decay was constant. Although it has been found that thermotolerance is not always abolished by cell divison completely (Hahn 1982), the comparison of proliferating cells with plateau phase cells has established a division-dependent decay of thermotolerance (Gerweck and Delaney, 1984). The mechanism through which thermotolerance is induced and develops is not yet clear. Li and Hahn (1980) have proposed an operational model. The authors suggest three phases:

1. The induction of thermotolerance ("trigger") can occur at all temperatures above normal growth temperatures.
2. The expression of thermotolerance ("development") generally occurs only at temperatures below 42° C.
3. Decay of thermotolerance.

Many studies have been undertaken in order to evaluate the molecular events which are involved in these processes. Intracellular levels of reduced glutathione (GSH) have been discussed in this connection (Mitchell et al. 1983; Russo et al. 1984; Konings and Penninga 1984). Continuous heating at 42.5° C (2 h) or acute heating at 43° or 45.5° C resulted in a rapid increase in GSH. Intracellular GSH reduction by either diethylmaleate (DEM) or buthionine sulfoximine (BSO) leads to thermosensitization of the cells. Konings and Penninga (1983, 1984) found that cell survival was only changed by such DEM concentrations, which also reduced the sulphhydryl groups of proteins. The involvement of radical and redox reaction in the development of thermotolerance was also proposed from the observation of a positive correlation between the tolerance development and the induction of superoxide dismutase (Loven et al. 1985). Also the role of catalase has been discussed in this connection (Issels et al. 1984 cited by Leeper 1985). In all these studies only the reduced but not the oxidized form of glutathione has been determined. The situation is far from clear.

Henle et al. (1983) assumed that naturally occurring intracellular polyols were necessary for the development of thermotolerance. Biochemical investigations could not confirm this hypothesis (Monson et al. 1984). Several authors have suggested that the modification of membranes especially with respect to fluidity are involved in the development of thermotolerance (Yatvin 1977; Dennis and Yatvin 1981; Anderson et al. 1981; Cress et al. 1982). A positive correlation was observed between membrane fluidity and increase of thermosensitivity of bacterial cells *(E. coli)*. The membrane fluidity was enhanced by an increased incorporation of unsaturated fatty acids into the membranes. Lepock et al. (1981) did not observe such a correlation between membrane fluidity and cell killing of the mammalian cell line V-79. Cress et al. (1982) demonstrated that the stiffening of membranes by cholesterol was important. However, the correlation with cell survival was only found when the cholesterol content was calculated on the basis of membrane proteins.

Much attention has been paid to the very widely observed phenomenon that a dramatic activation of heat shock protein genes occurs after heat exposure in cells which develop thermotolerance. These proteins have been found in many organisms (Schlesinger et al. 1982). Every investigated species synthesizes heat shock proteins (HSPs) in the molecular weight ranges of 60–74 kd and 80–90 kd (Burdon 1985). HSP gene transcription is increased during or directly after heating; later when non-HSP gene transcription recovers to normal levels HSP gene transcription is progressively switched off (Burdon 1985). A correlation between the synthesis of HSPs with the development of thermotolerance has been found in normal and malignant cells (Burdon et al. 1982; Subjek et al. 1982; Landry and Chretien 1983; Tomasovic et al. 1983; Tomasovic et al. 1984; Schamhart et al. 1984; Omar and Lanks 1984). The synthesis of these proteins is largely reduced by actinomycin D and cycloheximide.

After continuous heating as well as after acute heat shocks the kinetics of thermotolerance development followed very well the time course of HSP synthesis in CHO cells. If the severity of the heat treatment was increased both effects were delayed in the same way (Sciandra and Subjek 1984; Sciandra 1986). Li et al. (1982) observed a comparatively constant level of HSPs and thermotolerance over a period of 36 h in plateau phase HA-1 cells. Gerweck and Epstein (1985) observed that the degradation of HSPs was dependent on cell division and occurred simultaenously with the loss of thermotolerance. However, these correlations have not been established in all cases. After severe heat treatment some HSPs have been synthesized without an increase of thermoresistance (Sciandra 1986). Apparently the intracellular level of HSPs does not predict the degree of tolerance when the whole family of HSPs is considered.

Furthermore, actinomycin D cannot inhibit the synthesis of all HSPs, even if the substance is added to the medium before the heat treatment. These proteins were translated from preexisting mRNA; they are called "prompt" HSPs (Reiter and Penman 1983). Interestingly these proteins are exclusively associated with the nuclear matrix intermediate filament (NM-IF). These data also show the technical limitations of characterization of HSPs, which has been mostly performed with gel electrophoresis (Sciandra 1986). The intracellular localization certainly is very important with respect to their function and mechanism (Burdon 1985). HSPs have been found in nuclear structures and in association with elements of the cytoskeleton. Although their mechanism for achieving thermoresistance is not clear up to now, a tremendous amount of information has been produced to show close correlations. Thermotolerance is a phenomenon which can increase thermoresistance effectively and has to be considered seriously when hyperthermia is used in clinical tumour therapy.

Combination of Hyperthermia and Ionizing Radiations

Ionizing radiations are a very important treatment modality for cancer therapy. It was therefore logical to combine exposure to ionizing radiation with a hyperthermic treatment. Langendorff and Langendorff (1943) showed that an increase in temperature after X-irradiation results in a remarkable enhancement of the radiation effect. In the 1970s it was observed by several groups that cell survival strongly decreases when a combined exposure to ionizing radiation and heat was performed (Westra and Dewey 1971; Overgaard and Overgaard 1972, 1974; Ben-Hur et al. 1972, 1974; Robinson and Wizenberg 1974; Harisiades et al. 1975; Dietzel 1975; Leith et al. 1977; Streffer et al. 1979).

Heating during irradiation as well as with short intervals before or after irradiation increases cell killing in a supraadditive (synergistic) way. Both the shoulder (D_q) and the D_o of the dose effect curve, which is observed after irradiation at 37° C, are reduced when the temperature is elevated to 40°–45° C (Fig. 14). Frequently the shoulder of the dose effect curve disappears completely. The maximal effect is observed when the irradiation takes place during the heat treatment. With increasing intervals between irradiation and heating the radiosensitizing effect of hyperthermia decreases. An analysis of the dose effect curves shows that the combined effect is more than an additive effect which would result from the mere addition of the single effects. Such an effect is called supraadditive or synergistic. Frequently a heat exposure, which has no effect on cell killing, increases the effect of low linear energy transfer (LET) radiation, like X-rays. Thus, a radiosensitizing effect is obtained. For simplicity in the following for "supraadditivity", "synergism" and "radiosensitization" the technical term "radiosensitization" will be used.

In order to obtain a measure for the modifying dose of hyperthermia on radiation-induced cell killing the "thermal enhancement ratio" (TER) is frequently used. It has been defined as the ratio of the D_o which is obtained from the survival curve after irradiation

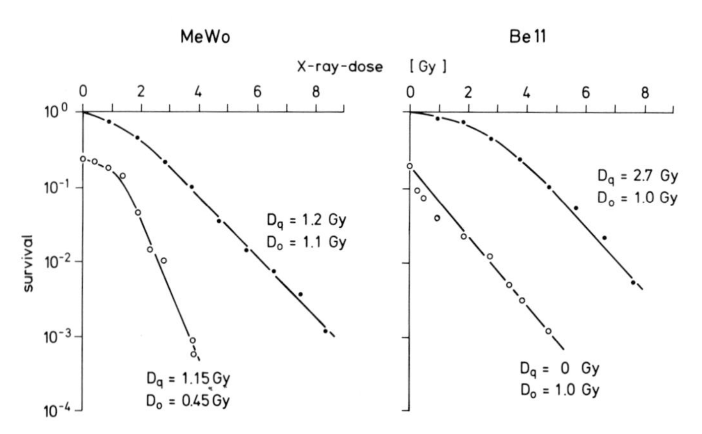

Fig. 14. Radiation survival curves of two melanoma cell lines (MeWo and Be 11). Immediately after the radiation dose, cells were heated at 42° C for 3 h. ●——●, X-rays alone; ○——○, X-rays plus heat

alone divided by the D_0 from the survival curve after irradiation plus heat (Hahn 1982). The "interaction coefficients" of ionizing radiation and heat have been calculated from such data dependent on the heat dose (Loshek et al. 1977a, b). Sapareto et al. (1978) performed experiments with increasing temperatures. Hahn (1982) concluded from these data that there exists an upper limit for radiosensitization.

The thermal enhancement ratio can also be calculated in analogy to relative biological effectiveness (RBE) for high-LET radiation. In this case the ratios of radiation doses with and without hyperthermia are calculated. Such radiation doses are compared which give the same biological effect. When the TERs are determined under these conditions for various radiation doses and the same heat treatment, the TER decreases with increasing radiation doses (Fig. 15). This effect becomes especially large when the shoulder of the survival curve after irradiation alone is large and is strongly reduced or disappears after irradiation plus heat (Streffer et al. 1983). In the case of a melanoma cell line the dose effect curves have been determined for exponentially growing cells and for plateau phase cells. Under these conditions the TERs for the plateau cells show in principle the same dose dependence as the exponential cells; however, they are somewhat smaller (Fig. 15). Robinson et al. (1974) measured the radiosensitizing effect for euoxic and hypoxic cells from mouse bone marrow in vitro and calculated the oxygen enhancement ratio (OER) for various temperatures. The OER decreased with increasing temperatures, meaning that radiosensitization of hypoxic cells increased with increasing heat exposure. Similar data were obtained by Kim et al. (1975) with HeLa cells. On the other hand Power and Harris (1977) found an increase in OER when V-79 cells were irradiated and heated in comparison with irradiation alone. In this connection investigations on the recovery from potential lethal damage (PLD) are very interesting (Rao and Hopwood 1985). The PLD repair was only reduced by a postradiation heat treatment but not by heating before irradiation. Severe hypoxia inhibited the PLD repair strongly. No change was observed by Kiefer et al. (1976) for the radiosensitization of yeast cells. A decrease in pH as well as in nutrient levels increases radiosensitization by heat (Holohan et al. 1984; Dewey 1984).

The radiosensitization by heat has mainly been studied in combination with low-LET radiation. Only few investigations have been performed with high-LET radiation such as neutrons (Ngo et al. 1977, Loshek et al. 1981; Streffer et al. 1983) and accelerated helium

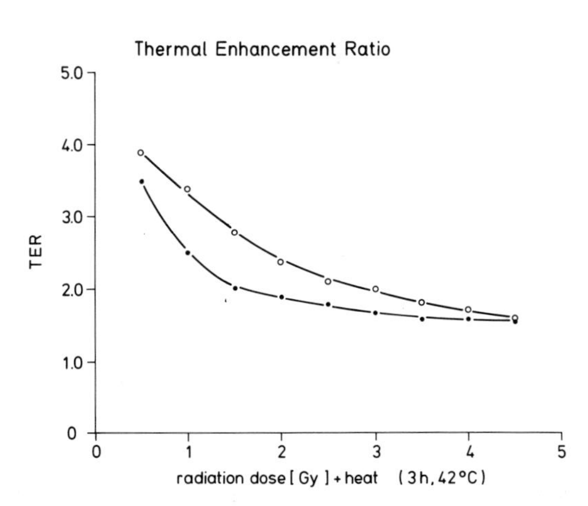

Fig. 15. Thermal enhancement ratio *(TER)* from human melanoma cells in exponential (○-----○) and plateau phase (●-----●). Survival of cells was measured

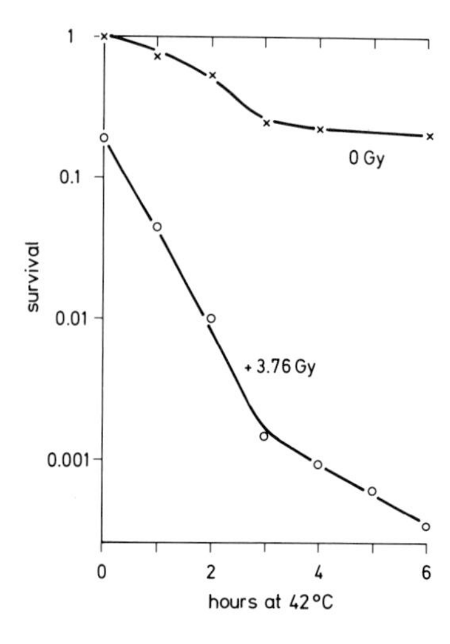

Fig. 16. Survival of human melanoma cells after continuous heating. × ----- ×, heat alone; ○ ----- ○, heat plus X-rays (3.76 Gy)

or carbon ions (Gerner et al. 1976; Gerner and Leith 1977). The interaction with high-LET radiation and heat is smaller than with low-LET radiation. In some studies only additive effects have been observed.

It has been mentioned above that an interaction between radiation and heat exposure takes place with the heat treatment before irradiation as well as after irradiation. Therefore it has been discussed which sequence has greater advantages for tumour therapy. It appears that physiological and metabolic changes, which have been observed with tumours after hyperthermia and which show increased hypoxia, favour the sequence irradiation followed by the heat treatment (Streffer 1986). Dewey (1984) has pointed out that the combination of hyperthermia with high-LET radiation would be difficult or even deleterious if the selective effect of radiation is reduced to the Bragg peak region. As the radiosensitization by heat is higher in the plateau region the localization of heating would be very important.

The further question is with which interval and fractionation schedules should both modalities be used. Does thermotolerance also develop after heating in combination with ionizing radiation? The data which have been reported in this field look somewhat conflicting on first sight. Streffer et al. (1980, 1983) and van Beuningen (1983) observed with human melanoma that thermotolerance which developed during continuous heating and after fractionated heating at 42° C was seen no longer or only to a small degree when an X-ray dose of 2 or 4 Gy directly preceded the heat treatment (Fig. 16). It was concluded that thermotolerance does not occur when heat is used as a radiosensitizing agent. Extensive studies with CHO cells demonstrated that thermotolerance decreased the radiosensitizing effect in the higher dose range but did not affect the shoulder region (Holohan et al. 1982; 1984). Cell survival after 2.0 Gy X-rays alone is 0.5, and 0.15 for one heat exposure (15 min at 45.5° C) 10 min or 10 h prior to the same X-ray dose. After two heat doses with an interval of 10 h followed by 2.0 Gy X-rays a cell survival of 0.05 has been observed. The second heat dose definitely has a radiosensitizing effect. This effect is not found with an X-ray dose of 6.0 Gy (Dewey 1984). It must also be pointed out that in these latter studies the first heat dose was not combined with irradiation whereas in the above-mentioned investigations the first heat dose was combined with irradiation.

On the other hand von Rijn et al. (1984) found with Reuber H5 hepatoma cells that thermotolerance decreased the extent of radiosensitization. Similar data were obtained by

Havemann (1983) with a murine mammary carcinoma cell line. The reduction of radiosensitization by thermotolerance was even predominantly observed on the shoulder in this case. Jorritsma et al. (1984) reported that radiosensitization was not modified by thermotolerance in HeLa cells.

Hahn (1982) has discussed the importance of cell proliferation kinetics in connection with fractionated studies. Unfortunately data about the distribution of cells in the generation cycle are available in few cases under the same conditions as those of the cell survival studies. EMT6 cells showed after 3.0 Gy X-rays the general accumulation of cells in G_2-phase with a delay of about 3 h, after a heat treatment at 43° C for 1 h a delay of cells in G_2-phase (about 9 h) as well as an accumulation in S-phase for about 6 h.

After the combined exposure the delay in G_2-phase lasted about 18 h (Kal et al. 1975; 1976). Hahn (1982) is certainly right when pointing out that such changes of cell distribution in the cell cycle have importance for fractionated studies. He further concludes that the modification of heat effects by radiation might be more remarkable than the reverse situation. Similar data have been obtained with V-79 cells grown as spheroids (Lücke-Huhle and Dertinger 1977).

The time course of these events is certainly dependent on the cell line. Thus, it has been demonstrated with two melanoma cell lines that after X-irradiation with doses of comparable cell killing effects the number of cells in G_2-phase increased immediately in one cell line for about 12 h (Be 11) while the so-called G_2-block appeared in the other cell line only about 12 h after irradiation. If the X-irradiation was combined with heating the G_2-block was shifted to later periods and it lasted much longer. In this latter case a strong increase in G_2-phase cells was seen even 48 h after the treatment (van Beuningen and Streffer, unpublished results).

In conclusion it can be stated that low-LET radiation strongly interacts with hyperthermia on cell killing. This effect is seen with lower temperatures, which have no cytotoxic effect by themselves. The survival curve after radiation exposure is modified in such a way that the shoulder (D_q) and the D_o are reduced. The reduction of recovery from sublethal radiation damage seems to be very important for the clinical use of hyperthermia. The TER for cell lines with a broad shoulder is especially high in the low radiation dose range. Thermotolerance generally does not interfere with radiosensitization by radiation doses in the shoulder region but reduces the sensitizing effect in the higher dose range. For clinical use of hyperthermia combined with low-LET radiation (2–4 Gy/dose fraction) thermotolerance plays a minor or even no role. In order to avoid interferences with heat-induced physiological and metabolic changes which increase radioresistance, e.g. by hypoxia, hyperthermic treatment should be performed after radiation exposure.

Mechanisms of Heat on Cell Killing

It has been shown and extensively discussed that hyperthermia can kill cells in the temperature range of around 42°–47° C. But cell killing will be remarkably enhanced if the heat exposure is combined with low-LET radiation with no or only a short interval between the two modalities. Under these conditions even lower temperatures than those mentioned above will have a strong effect (Fig. 16). These and other data suggest that the cytotoxic effect of heat alone on one hand and the interaction of heat with ionizing radiation, the radiosensitizing effect, on the other hand are based on different mechanisms which will be discussed separately. Heat can also interact with cytotoxic drugs. This subject will be discussed in another chapter of this volume (Engelhardt).

Before discussing these mechanisms possible definitions of a heat "dose" will be considered. In the field of ionizing radiation the radiation dose is very well defined on a purely physical basis: as absorbed energy per tissue mass (1 J absorbed per kilogram is equal to 1 Gray, Gy). In the field of hyperthermia we are far away from such a definition. From the data which have been presented it can easily be seen that apparently the temperature and the time during which the biological system has been heated to that temperature are important parameters. Heat dose depends on the two parameters. Pincus and Fischer (1931) found that above 44° C the exposure time could be reduced by a factor of 2 for 1° C temperature increase. Dewey et al. (1977) and Connor et al. (1977) observed the same correlation for cell killing above 42.5° C, but the time reduction was less below 42.5° C. Atkinson (1977) tried to define a dose unit from time-temperature relationships with the help of a standard survival curve of heat-treated mammalian cells. Several groups are using equivalent time at a reference temperature to describe biological responses to heat with isoeffects (Gerner 1985). It has been doubted whether such definitions can describe the situation properly for thermotolerance, radiosensitization and the various modifying factors. Gerner (1985) has discussed this subject in more detail. At the present time it seems reasonable to record temperature and time carefully when hyperthermia is used for tumour therapy.

Mechanisms of Exposure to Heat Alone

Several investigators have determined activation energies for cell-killing processes by means of an Arrhenius plot (Fig. 17). A value of 141 kcal/mol was obtained for the inactivation of CHO cells in the temperature range 43°–47° C. Below 43° C a change of the slope ("break") of the Arrhenius plot occurred. Such Arrhenius analyses are normally used in order to determine the activation energy of a chemical reaction. Activation ener-

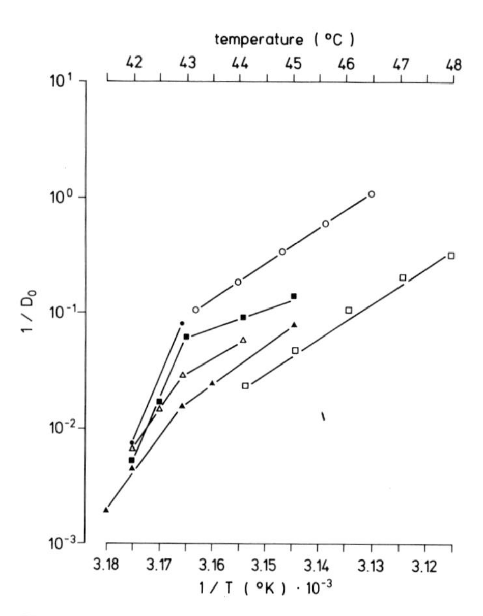

Fig. 17. Arrhenius plot for heat inactivation of various cell lines. *On the ordinate* the reciprocal of the D_0-values (inactivation rates) are plotted versus the reciprocal of the absolute temperature. (Redrawn from Leith et al. 1977)

gies of the observed values are in the same range as those which have been found for the denaturation of proteins (Privalow 1979). Therefore it has frequently been assumed that proteins are the molecules at risk for cell inactivation by heat and it can be assumed that the heat-sensitive target should be found among the cellular proteins. When the temperature is raised above 47° C the activation energy for the reaction which leads to cell inactivation changes again and becomes 20–30 kcal/mol. Below and above that temperature range of 42.5°–47° C, the cell-killing mechanisms by heat might be completely different from the mechanism which is active in the range 42.5°–47° C.

It has been discussed earlier that a transition of proteins from the native into the denatured state means a change of protein conformation. Until now it has not been possible to name a specific individual protein whose denaturation or conformational transition is responsible for cell inactivation. Perhaps it is even not necessary that a single protein is altered but a group of different proteins undergoes such structural changes during heat exposure, so that various cellular processes do not function properly. It has been explained that apparently membranes partly lose their proteins which are needed for stabilization of the membranes (Streffer 1985, 1986), for transport functions (Burdon et al. 1984) or as receptors (Calderwood and Hahn 1983; Calderwood et al. 1984; Stevenson et al. 1984). Such effects would explain the increased efflux of K^+ (Yi et al. 1983; Ruifrok et al. 1984). A good correlation exists between this effect and cell killing. Interestingly such a correlation has not been found for the combination of X-rays and heat (Ruifrok et al. 1984). Furthermore membrane changes lead to an increased Ca^{2+} influx into heated cells (Anghilari et al. 1984; Stevenson et al. 1984; Wiegant et al. 1984). In this connection it is also interesting that membrane-active phenothiazine drugs can enhance heat-induced cell killing (George and Singh 1982, 1985; Shenoy and Singh 1985).

It has already been pointed out that the fluidity of membranes can modify the thermosensitivity of cells. This must apparently been seen in a cooperative action between lipids and proteins. Besides the cytoplasmic membrane also changes of intracellular membranes certainly occur, as has been demonstrated by electron microscopic studies (Breipohl et al. 1983). Especially intracellular membranes of the endoplasmic reticulum are thermosensitive (Wallach 1977; Breipohl et al. 1983). The labilization of lysosomal membranes or increase of lysosome numbers have also been suggested as important in heat-induced cell killing (von Ardenne et al. 1969; Overgaard 1976). Hume et al. (1978) found an increase of lysosomal enzyme activity by histochemical methods after heating. On the other hand the lysosomal degradation of epidermal growth factor (EGF) is inhibited after heating of rat embryo fibroblasts (Magun and Fennie 1981). Also biochemical determination of lysosomal enzyme activities did not show any heat-induced changes (Tamulevicius and Streffer 1983). Lysosomes seem not to be the primary target for heat-induced cell killing (Hahn 1982).

Hahn (1982) has discussed the role of ATP production through oxidative phosphorylation and mitochondrial membrane for cell killing by heat. A decrease in ATP levels has been observed in heated cells (Francesconi and Mayer 1979; Ohyama and Yamada 1980; Lunec and Cresswell 1983; Mirtsch et al. 1984). However, at the same time the ATP turnover and hence also ATP synthesis are enhanced (Ohyama and Yamada 1980; Streffer 1985, 1986). Damage of ATP synthesis does not seem to be a primary event for heat-induced cell killing.

Conformational changes of proteins which form the structure of the cytoskeleton may be responsible for the observed disturbances of the cytoskeleton, which have been described earlier. In this connection it is interesting that some HSPs are apparently associated with the cytoskeleton (Burdon 1985) which might stabilize these structures. These are assumptions which certainly need more experimental evidence, to see whether they can be substantiated. However, in the case of mitotic cells good evidence exists that a disaggregation of the globular proteins of the spindle apparatus occurs which inhibits the regular mitosis and induces cell death (Coss et al. 1982). The mechanism is probably analogous to that discussed for the cytoskeleton above. Changes in intermediary metabolism, like glycolysis, fatty acid metabolism and the citrate cycle, can be remarkable during and after a hyperthermic treatment (Streffer 1986). The metabolic rates are increased during hyperthermia so that energy reservoirs are utilized and depleted. However, the data do not demonstrate that in these alterations the primary events for heat-induced cell killing can be found, but the processes can modify the microenvironment of cells, for instance the pH in such a way that thermosensitivity increases (Streffer 1986). Also a reduction of nutrients (Hahn 1982) and inhibition of glycolysis (Kim et al. 1978, 1980, 1984; Song et al. 1977) can sensitize the cells remarkably against heat.

Furthermore changes of RNA and DNA as well as the heat-induced inhibition of their biosynthesis have been discussed in connection with cell death (Hahn 1982; Streffer 1982). Heat-induced changes in DNA and RNA generally seem to occur only at higher temperatures than those which have been considered here. RNA synthesis recovers relatively rapidly; the same has been described for general protein synthesis earlier. It is not known whether the synthesis of some specific proteins or RNA species, for instance mRNA, is inhibited longer than it has been observed in general. The same situation is more or less found for DNA synthesis. However, the recovery of this process is much slower and apparently ligation of newly synthesized DNA pieces is delayed (Warters and Stone 1983). This long delay may induce disaggregation of the DNA structure and chromosomal aberrations which have been observed after heating cells in S-phase (Dewey et al. 1971). Thus, cell death of S-phase cells is probably caused through this mechanism. However, S-phase cells are a special case and even in a tumour most cells are not found in S-phase.

With respect to the cell nucleus it is interesting that an increase of nuclear protein content occurs after hyperthermic treatment of cells (Roti Roti 1982). This increase correlates with cell killing. The new proteins form a complex with the chromatin and may interfere with the formation of the DNA replication complex (Warters and Roti Roti 1982). The nature of these proteins as well as the mechanism, how these processes induce cell killing, remain unclear at the moment.

Thus, the proposed conformational changes of proteins in the various cellular structures remain a very intriguing possibility for a more general mechanism of heat-induced cell killing. These suggestions would also explain the most important modifications of cellular thermosensitivity. The micromilieu with its pH and ion concentration are very important for protein conformation as has been described earlier. In order to maintain the micromilieu and with it the conformation of proteins a steady supply of energy is needed; this fact would explain the modifications of sensitivity by metabolism. Also the development of thermotolerance with the synthesis of HSPs is in agreement with this proposal. Furthermore the trigger mechanism of thermotolerance shows an activation energy of about 120 kcal/mol (Li et al. 1982c) and would be in line with the suggestion. Such a mechanism would furthermore explain that heat effects including cell death develop rapidly, for instance in comparison with the effects of ionizing radiations, as the conformational changes will occur during heating. As a consequence changes in membrane function, cytoskeleton, etc. occur which lead to disturbances of vital cellular functions. If these alterations are severe and remain for a longer period they become irreversible and cells die. Besides these effects a number of metabolic reactions increase in their rates. The depletion of energy reservoirs or disturbances of energy metabolism will enhance thermosensitivity of cells.

Several authors have described models by which the cell-killing effect by heat and the corresponding heat survival curves can be discribed (Hahn 1982; Leeper 1985). Very recently Jung (1984, 1986) has introduced an interesting general concept for the action of heat on cell survival. The basic idea is that cell killing by heat takes place in two steps: (1) Heating produces non-lethal lesions and (2) the non-lethal lesions are converted into lethal lesions by further heating. Both processes occur randomly and depend only on temperature. On this basis Jung (1986) formulates an equation for cell survival (S) after heating at a certain temperature for time t:

$$S(t) = \exp\{(p/c)[1 - c \cdot t - \exp(-c \cdot t)]\}$$

where p is the rate constant for the production of non-lethal lesions per cell and c is the rate constant for the conversion of one non-lethal lesion into a lethal event. The values of the rate constants p and c are calculated for the temperature range 39° -45° C. In this temperature range the Arrhenius plot for p shows a breaking point at 42.5° C. The activation energy is 370 kcal/mol below this point and 185 kcal/mol above this point. The conversion process has an activation energy of 86 kcal/mol in this range without a breaking point. The formulated equations describe heat survival curves including the shoulder.

Mechanism of Radiosensitization

For cell killing by ionizing radiation it has been conclusively shown that the radiation dose to the cell nucleus determines the extent of the effect (Streffer et al. 1977). The question certainly remains: Is DNA the primary and only target? (Alper 1979). It is not astonishing therefore that many investigators have studied how heat interacts with radiation damage of the chromosomal level. Heat could increase the primary DNA damage or it could inhibit the repair of this damage.

Corry et al. (1977) have pointed out that the frequency of radiation-induced single-strand breaks increases when the irradiation is combined with heat; the number of double-strand breaks remains unchanged but several authors have observed no such effects by heat (Lunec et al. 1981; Radford 1983). The modification of the dose effect curve for radiation-induced cell killing in the shoulder region leads to the assumption that heat reduces the repair of DNA damage. This is substantiated by molecular biological studies. Rejoining of DNA strand breaks is reduced when heat is combined with X-rays either before or after irradiation (Clark and Lett 1978; Dikomey 1978, 1982).

From the temperature dependence an activation energy of about 140 kcal/mol was determined for the inhibition of DNA strand break rejoining in Ehrlich ascites cells (Jorritsma and Konings 1983). The authors found a relationship between repair of DNA damage and cell survival (Fig. 18). A similar correlation for residual DNA damage and cell killing was reported in CHO cells (Mills and Meyn 1983). Studies on the activity of DNA polymerase β underline the assumption that the repair of radiation-induced DNA damage is inhibited by a heat treatment. These studies have shown that the DNA polymerase β, which is involved in unscheduled DNA synthesis for DNA repair, is thermosensitive and its inactivation correlates with cell killing by irradiation plus heat. These data have been discussed earlier. These results are strongly in favour of the proposal that a heat-induced inhibition of DNA repair is closely involved in the process of radiosensitization by heat. Hahn (1982) has extensively discussed the interesting point that heat modifies the rate of misrepair.

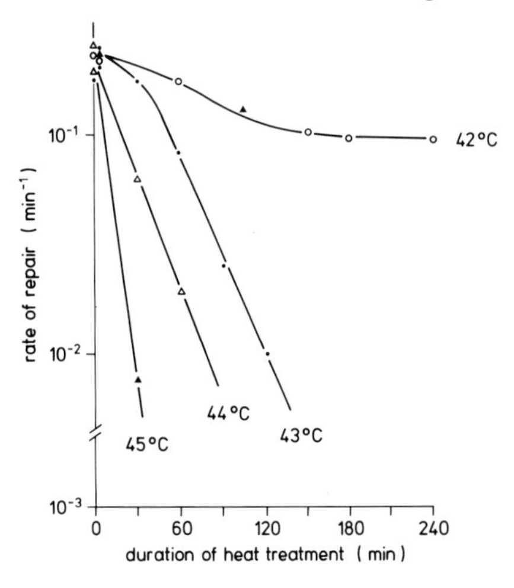

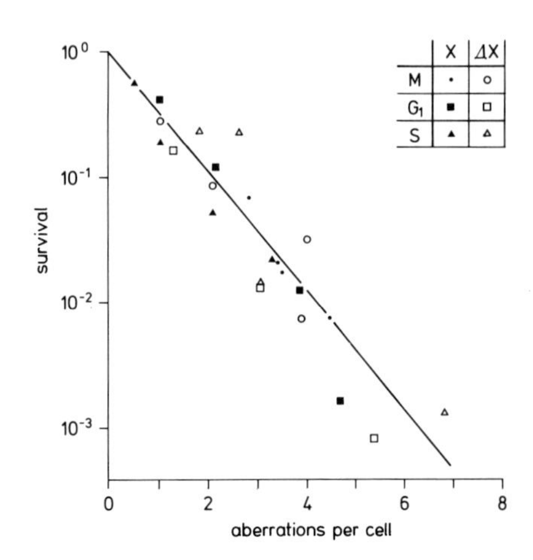

Fig. 18 *(left).* Kinetics of repair of radiation-induced DNA strand breaks after various time/temperature combinations. (Redrawn from Jorritsma and Konings 1983)

Fig. 19 *(right).* The relationship between cell survival and chromosome aberrations for CHO cells after irradiation alone or the combination with heat. (Redrawn from Dewey and Sapareto 1978)

Strong evidence that the heat-induced radiosensitization is due to an interaction with radiation damage on the chromosomal level comes from the studies of chromosomal damage. It has been demonstrated that a heat treatment after irradiation increases the number of chromosome aberrations (Dewey et al. 1971) or of micronuclei (Streffer et al. 1979) remarkably. Micronuclei are formed from acentric chromosome fragments or whole chromosomes which are not integrated into the daughter cell nuclei during mitosis (Heddle 1973). Dewey et al. (1978) demonstrated a positive correlation for the number of chromosome aberrations and cell killing. The correlation was obtained for G_1-phase cells as well as S-phase cells after the combined treatment (Fig. 19). On the other hand heat alone only induced chromosome aberrations in S-phase cells.

These data demonstrate that the mechanism of radiosensitization by heat must interfere with processes at the chromosomal level and results in enhanced irreparable chromosome damage in comparison to irradiation alone. The results conclusively show that cell killing by heat alone is caused by a mechanism different from that by irradiation plus heat. This conclusion also underlines the observation that heat radiosensitization is not or only slightly inhibited in thermotolerant cells. Radiosensitization takes place especially in the low-dose range, in the shoulder region of the survival curve where the expression of recovery processes is large.

Responses of Normal Tissues and Tumours

It has been extensively described that heating in the temperature range 40°–45° C can kill cells and that radiation-induced cell killing can be dramatically enhanced by combined exposure to radiation and heat. Furthermore it has been demonstrated that a number of factors, which determine the microenvironment of tissues and especially tumours, can modify the thermosensitivity of cells. For tumour therapy it is extremely important that more damage occurs in the tumour than in the normal tissue so that the tumour can be eradicated under conditions which induce only limited damage in the normal tissue.

For treatment of animal tumours a localization of the heat treatment to the tumours should be achieved as well as possible. Heating techniques are usually ultrasound, microwaves, and heating by waterbath. As the tumour mass is generally large in comparison to total body weight of rodents, these heating techniques quite frequently lead to a temperature increase in larger regions possibly in the total body. This is especially the case when a water-bath is used for the heating procedure. Even for localized heat treatments animals must be anaesthesized, while such anaesthesia is not necessary for the local heat treatment of human tumours. Anaesthesia might interfere with the hyperthermic action, as the drug modifies thermosensitivity or quite as often the temperature of the animal decreases below normal temperature after hyperthermia. Therefore it is necessary to consider these differences between animal studies and the clinical situation when data are extrapolated.

Responses of Normal Tissues to Heat

A number of studies have been performed on skin mainly of mice. In this connection the problem arises that the normal temperature of skin is below the body core temperature. It has been demonstrated that cells adapted to lower temperatures become more sensitive to hyperthermic treatments (Li and Hahn 1980). Furthermore the skin temperature might be even higher than the tumour temperature during the hyperthermic treatment when the heating is performed with a water-bath. This situation will induce comparatively greater damage in mouse skin than can be expected during the hyperthermic treatment for human skin.

A comprehensive study was performed on pig and human skin (Moritz and Henriques 1947). The authors demonstrated that only a small increase in heating is required to obtain a strong increase in necrosis if a critical level of heating has been reached. An activation energy of 150 kcal/mol was calculated for the induction of necrosis. As has previously been demonstrated for cell killing, again an increase of 1° C reduces the time of heating by a factor of 2 in order to obtain the same effect. Similar observations have been made for other normal tissues (Morris et al. 1977; Law et al. 1978). This steep temperature gradient of the hyperthermic effect underlines the necessity of a high uniformity of temperature in heated volumes.

It is characteristic for the hyperthermic effects that they occur very rapidly within a few hours to days, while radiation effects develop much more slowly (Fig. 20). If recovery

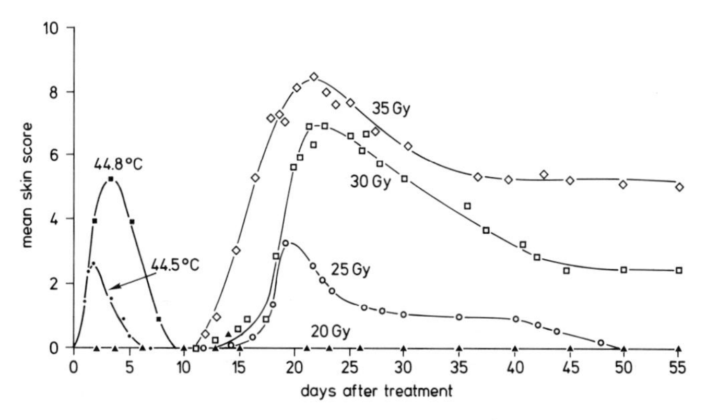

Fig. 20. Skin reaction in tails of adult mice following either hyperthermia (44.0° – 44.8° C for 30 min) or X-irradiation (20–35 Gy). (Redrawn from Hume and Myers 1984)

from the heat damage is possible then this recovery is also faster than the recovery processes from radiation effects. In mice, exposures of 1 h in the temperature range of 43°-44° C apparently do not cause irreversible damage to the skin. Above this range damage increases rapidly. The data obtained with mice seem to be conservative estimates in comparison to the effects observed with skin of pigs and humans (Hahn 1982). The data of Moritz and Henriques (1947) as well as of Henriques (1947) were confirmed by Martinez et al. (1980) after heating pig skin and adipose tissue with electromagnetic waves (1 MHz). No serious effects were observed at temperatures below 45° C (1 h).

Similar data were obtained for the thermosensitivity of cartilage, studied by measuring the stunting of the growth of mouse or rat tail. Heating at 43° C for 1 h induced stunting but no necrosis in about 10% of the animals. Increasing the temperature to 44° C produced necrosis in more than 50% of the animals (Field et al. 1976; 1978). A reduction of the heating time to 40 min avoided necrosis. A remarkable thermosensitization occurred when the blood supply was interrupted by clamping 20 min before heating. The baby rat tail became more thermosensitive, equivalent to an increase in heating time by a factor of about 3 (Morris et al. 1977). The clamping induces hypoxia but also pH and nutrients decrease and these latter factors may be very important.

Further investigations have been performed on the thermosensitivity of the testis (Hand et al. 1979) and especially the intestine (Hume et al. 1979; Henle 1982; Breipohl et al. 1983; van Beuningen et al. 1983; Milligan et al. 1984; Hume 1985). It has been found that the intestinal mucosa is very thermosensitive and also in this tissue the expression of heat damage is very rapid (Breipohl et al. 1983; Hume et al. 1983). This expression takes place especially in the villi where the non-proliferating epithelial cells (functional enterocytes) are heavily damaged and "fall off" from the villi (Fig. 21). The effects on the proliferating crypt cells are much less while after exposure to ionizing radiation the target cells are these proliferating crypt cells. Again this heat damage is already observed during the first hours after heating to 42°-43° C for 1 h (Hume et al. 1983; van Beuningen et al. 1983). The effects of heat on the villi may be caused by stromal oedema which may result from damaged capillaries (Hume 1985). The intracellular membranes in the enterocytes are also destroyed by the hyperthermic treatment.

The development of thermotolerance has been demonstrated in normal tissues after fractionated or prolonged heating at low temperatures (Hume 1985). This has been observed in skin (Law et al. 1979; Rice et al. 1982; Wondergem and Haveman 1983), in intestine (Hume and Marigold 1980) as well as in adipose tissue and muscle (Martinez et al. 1983). Development and decay of thermotolerance follow the same or similar rules which have been observed for cells in culture. The degree of thermotolerance increases the more severe the primary conditioning treatment is (Fig. 22); however, the development is delayed with increasing heat exposures. These phenomena have been discussed more extensively by Hume (1985).

Thermoradiosensitization in Normal Tissues

As it has been observed for cell killing the acute radiation effects in normal tissues are enhanced when the radiation exposure is combined with a heating treatment (Field 1978; Hahn 1982; Hume 1985). The response after the combined exposure to heat and X-rays is apparently qualitatively similar to that which is observed after irradiation alone (Law et al. 1978). If the thermal enhancement ratios (TERs) are calculated as has been described for cell killing, the values increase with increasing temperature in mouse ear, mouse foot,

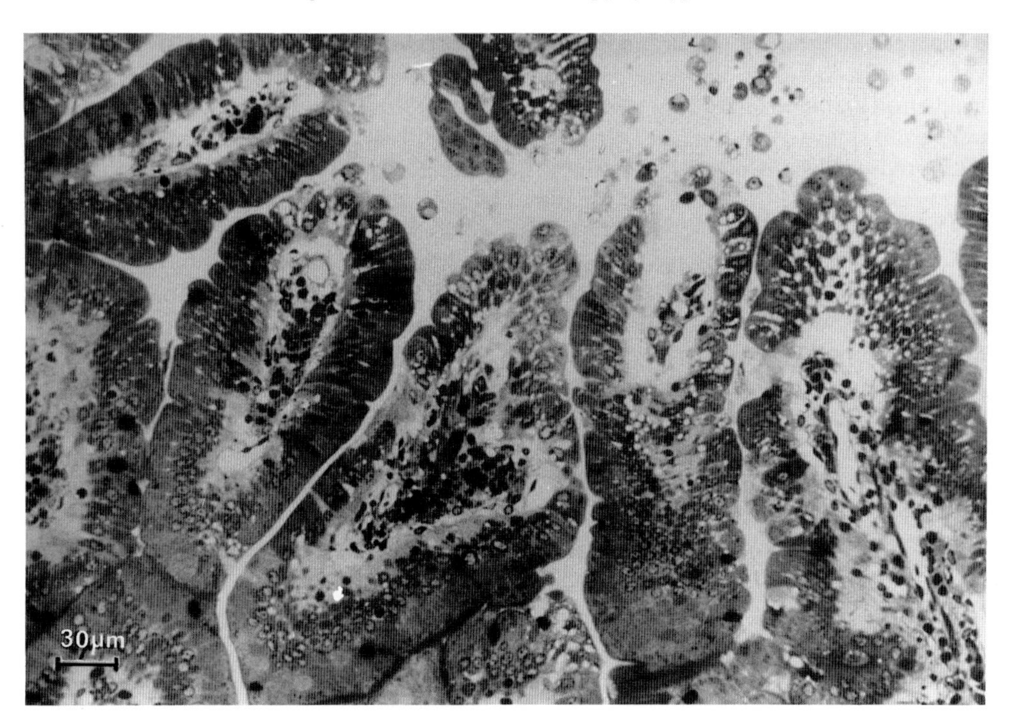

Fig. 21. Small intestine of mouse (jejunum) immediately after heat (30 min at 42° C). *v*, villi; *Kr*, crypts; *Sm*, first submucosa; *BK*, capillary; *DL*, lumen (x 165). Note the villi with cell desquamation. (Issa, 1985)

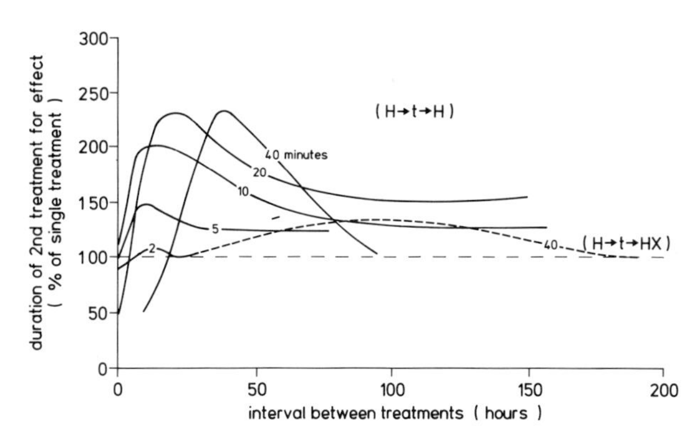

Fig. 22. Resistance to heat alone (—) or thermal enhancement of radiation damage (---) in mouse ear tested at various times after heat at 43.5° C. (Redrawn from Law et al. 1979)

mouse intestine and rat cartilage (Field 1978). In mouse intestine it has been observed for crypt loss that the TER reaches a maximal level with temperatures below 42° C. This phenomenon expresses a limit of interaction between ionizing radiation and heat (Hume 1985). In the higher temperature range it becomes difficult to separate between the radiosensitizing effect of heat and the effect of heat alone.

The thermal enhancement ratio of radiation damage is maximal when ionizing radiation and heat are applied simultaneously. As observed with cell killing the interaction between the two modalities decreases when heating and radiation are given with time intervals. The decrease in interaction is apparently faster when heat is given out after irradiation and it becomes zero with an interval of 4–6 h (Hume 1985). Recent studies on mouse tail skin suggest that the thermosensitivity of tissue is increased for a long period after irradiation. The higher sensitivity was expressed at times when the thermal response occurs as well as at the time when the radiation response is expressed (Hume and Myers 1984). In mouse tail skin the interaction took place when heat was given out 3 days after irradiation but the interaction was found at an interval of 6–10 days. It has been suggested that this time dependence may be influenced by cell turnover and repopulation in the tissue (Law and Ahier 1982). It is possible that these "recall" effects of radiation damage are observed during even longer periods in slowly dividing tissues. In several studies it has been demonstrated that thermosensitivity is increased when radiation damage has been expressed (Hume 1985).

The mechanisms by which effects develop in tissues are certainly much more complicated than in isolated cells. The influence of damage to the stroma including vasculature has to be considered. Radiation damage to the vasculature can alter the thermosensitivity considerably (Vaupel and Kallinowski, this volume). For the clinical application these situations are very important when retreatment of tissues is necessary.

Responses of Tumours to Heat

Single heat reatments can cure experimental animal tumours. Westermark (1927) observed such an effect with the Jensen sarcoma. Similar results were obtained by Goetze and Schmidt (1931). Crile (1962) found control of some spontaneous tumours in dogs. This author also treated mouse tumours (Crile 1963) and found that an temperature increase of 1° C above 42° C allowed a decrease of the heating time to half the value in order to obtain the same effect. This is in agreement with the data on cell killing which have been reported earlier. However, very great differences have been observed for the heat sensitivity of different tumours (Hahn 1982). Apparently also thermotolerance is induced in tumours by heat treatment, although the situation is not that clear as has been observed with cells in vitro.

Besides the cellular effects physiological and biochemical effects play a significant role when tumours are heated. These effects are described elsewhere in this volume. In this connection changes in blood flow, pH and nutrient supply are important. On one hand glucose infusion decreases the pH (von Ardenne 1982; Jähde and Rajewsky 1982) and therefore makes the tumour more thermosensitive; on the other hand it enhances glucose concentrations and can have the opposite effect through this condition.

Radiosensitization by Heat in Tumours

Overgaard (1935) showed that the X-ray dose for tumour eradication could be reduced when the tumour was heated at the same time or shortly after irradiation. Similar data have been obtained by a number of investigators since then (Hahn 1982). For the cure of tumours it is essential that the effect on tumours is higher than in normal tissues. Quantitatively this is frequently expressed as the therapeutic gain factor (TGF), which is defined as the ratio of the TER for the tumour over the TER for the normal tissue. The TGF was extensively studied by Hill and Denekamp (1979) for six transplantable mouse tumours. The tumours were of spontaneous origin in different tissues; they differed in growth rate and degree of differentiation. The TERs for the tumours varied from 1.1 to 1.8. The TGFs were larger when the heat was given after the irradiation. In some cases the TGF became larger when the interval between irradiation and heating was several hours (Overgaard 1981). Stewart and Denekamp (1980) found less therapeutic gain after fractionated treatments. On the other hand Emami et al. (1984) reported for a murine fibrosarcoma that the maximal control rate was achieved when irradiation and heat were given closely to each other on each session.

A very important question, which has been discussed frequently in this connection, is the sequence and interval between the two modalities. As hyperthermia alters a number of physiological conditions in the way that the radioresistance of the tumours might increase, agreement grows that irradiation should take place before the heating. In this connection the fraction of hypoxic cells in tumours has been studied. Song et al. (1982) showed that heating prior to irradiation increased the fraction of hypoxic cells. Similar data were obtained by Urano and Kahn (1983). If the interval becomes larger than 6, no or very little interaction apparently occurs and X-rays and heat act as independent modalities. The consequences have been discussed (Overgaard 1978).

In general it appears more favourable to use hyperthermia as a radiosensitizing agent with a short interval following irradiation (Streffer et al. 1983). Very extensive studies were performed by Dewhirst et al. (1984) with dogs and cats. Heating followed 10 min after X-irradiation with 4.6 Gy/fraction. The tumour temperature should be 44° C for 30 min. This combined treatment was compared with X-irradiation alone. The rate of complete tumour regression (CR) was higher after the combined treatment than after X-rays alone. The CR rate decreased with increasing tumour volume but the difference between the combined treatment and X-rays alone increased in this direction. The tumour response was optimal in those tumours for which an almost homogeneous temperature pattern was observed. The CR rate was low in those tumours in which regions existed which were not heated to a sufficient temperature range (Dewhirst et al. 1984). Similar data were obtained by Gilette (1985). The dose response curve was steeper for the combined treatment than after X-rays alone. This finding demonstrates a more homogeneous response after the combination with heating.

Development of thermotolerance was also observed in some tumours (Gilette 1985). Nielsen (1984) showed that the amount of radiosensitization was apparently also reduced in mouse mammary tumour when a conditioning dose of heating was given before the combined treatment. The effect of the conditioning heat treatment was maximal with an interval of 16 h. With an interval of 120 h no effect of a preceding heating occurred. However, again the degree of thermotolerance was smaller for radiosensitization that for treatments by heat alone (Dewey 1984). A high variability occurs for different tumours.

In general it appears that the use of heat as a radiosensitizing agent is the best modality. Heat should be applied as soon as possible after irradiation. A further advantage is that under these conditions the good localization of the radiation dose can be used as a localizing medium for the heat-induced radiosensitization. Heat should be given as often as possible.

Acknowledgement. These studies were supported by the Bundesminister für Forschung und Technologie, Bonn. We thank Mrs. J. Müller for the helpful preparation of the manuscript.

References

Adam G, Neumann H, Hinkelbein W, Weth R, Engelhardt R (1983) Metabolic changes in hyperthermia with chemotherapy. In: Engelhardt R (ed) Proceedings of the 13th international congress of chemotherapy, Vienna, Session 12.10, part 273, pp 37–40
Alper T (1979) Cellular radiobiology. Cambridge University Press, Cambridge
Altman D, Gerber GB, Ikada S (1970) Radiation biochemistry. Academic, New York
Anderson RL, Minton KW, Li GC, Hahn GM (1981) Temperature induced homeoviscous adaptation of Chinese hamster ovary cells. Biochem Biophys Acta 641: 334–348
Anghilari LJ, Crone-Escanye MC, Marchal C, Robert J (1984) Plasma membrane changes during hyperthermia: probable role of ionic modification in tumor cell death. In: Overgaard J (ed) Hyperthermic oncology, vol 1. Taylor and Francis, London, pp 49–52
Ashburner M, Bonner JJ (1979) The induction of gene activity in Drosophila by heat shock. Cell 17: 241–254
Atkinson ER (1977) Hyperthermia dose definition. J Bioengin 1: 487–492
Bass H, Moore JL, Coakely WT (1978) Lethality in mammalian cells due to hyperthermia under oxic and hypoxic conditions. Int J Radiat Biol 33: 57–67
Belt JA, Thomas JA, Buchsbaum RN, Racker E (1979) Inhibition of lactate transport and glycolysis in Ehrlich ascites tumor cells by bioflavonoids. Biochemistry 18: 3506–3511
Ben-Hur E, Riklis E (1979) Enhancement of thermal killing by polyamines. IV. Effects of heat sensitivity and spermine on protein synthesis and ornithine decarboxylase. Cancer Biochem Biphys 4: 25–31
Ben-Hur E, Bronk VB, Elkind MM (1972) Thermally enhanced radiosensitivity of cultured Chinese hamster cells. Nature (New Biol) 238: 209–211
Ben-Hur E, Elkind MM, Bronk BV (1974) Thermally enhanced radioresponse of cultured Chinese hamster cells: inhibition of rapair of sublethal damage and enhancement of lethal damage. Radiat Res 58: 38–51
Bhuyan BK, Day KJ, Edgerton CE, Ogunbase O (1977) Sensitivity of different cell lines and of different phases in the cell cycle to hyperthermia. Cancer Res 37: 3780–3784
Bowler K, Duncan CJ, Gladwell RT, Davison TF (1973) Cellular heat injury. Comp Biochem Physiol (A) 45: 441–450
Breasted JH (1930) The Edwin Smith surgical papyrus. In: Licht S (ed) Therapeutic heat and cold, 2nd edn, Waverly, Baltimore, p 196
Breipohl W, van Beuningen D, Ummels M, Streffer C, Schönfelder B (1983) Effect of hyperthermia on the intestinal mucosa of mice. Verh Anat Ges 77: 567–569
Burdon RH (1985) Heat shock proteins. In: Overgaard J (ed) Hyperthermic oncology, vol II. Taylor and Francis, London, pp 223–230
Burdon RH, Slater A, McMahon M, Cato ACB (1982) Hyperthermia and heat shock proteins of HeLa cells. Br J Cancer 45: 953–963
Burdon RH, Kerr SM, Cutmore CMM, Munro J, Gill V (1984) Hyperthermia, $Na^{+}K^{+}$ATPase and lactic acid production in some human tumour cells. Br J Cancer 49: 437–445
Burger F, Engelbrecht FW (1967) Changes in blood composition in experimental heat stroke. S African Med J 41: 718–721
Busch W (1866) Über den Einfluß welche heftigere Erysipeln zuweilig auf organisierte Neubildungen ausüben. Vrh Naturhist Preuss Rhein Westphal 23: 28–30
Calderwood StK, Hahn GM (1983) Thermal sensitivity and resistance of insulin-receptor binding. Biochim Biophys Acta 756: 1–8
Cavaliere R, Ciocatto EC, Giovanella BC, Heidelberger C, Johnson RO, Moricca G, Rossi-Fanelli A (1967) Selective heat sensitivity of cancer cells (biochemical and clinical studies). Cancer 20: 1351–1381
Chen TT, Heidelberger C (1969) Quantitative studies on the malignant transformation of mouse prostate cells by carcinogenic hydrocarbons in vitro. Int J Cancer 4: 166–178
Clark EP, Lett JT (1978) Possible mechanisms for hyperthermic inactivation of the rejoining of X-ray induced DNA strand breaks. In: 2–4 June 1977. Streffer C, van Beuningen D, Dietzel F, Röttinger E, Robinson JE, Scherer E, Seeber S, Trott K-R (eds) Cancer therapy by hyperthermia and radiation, Proceedings of the Second International Symposium, Essen, Germany. Urban and Schwarzenberg, Baltimore, pp 144–145

Coley WB (1893) The treatment of malignant tumors by repeated inoculations of erysipelas, with a report of ten original cases. Am J Med Sci 105: 488-511

Collins FG, Mitros FA, Skibba JL (1980) Effect of palmitate on hepatic biosynthetic functions at hyperthermic temperatures. Metabolism 29: 524-531

Connor WG, Gerner EW, Miller RC, Boone MLM (1977) Prospects for hypethermia in human cancer therapy: part II. Radiology 123: 497-503

Corry PM, Robinson S, Getz S (1977) Hyperthermic effects on DNA repair mechanisms. Radiology 123: 475-482

Coss RA, Dewey WC, Bamburg JR (1982) Effects of hyperthermia on dividing Chinese hamster ovary cells and on microtubules in vitro. Cancer Res 42: 1059-1071

Cress AE, Culver PS, Moon ThE, Gerner EW (1982) Correlation between amounts of cellular membrane components and sensitivity to hyperthermia in a variety of mammalian cell lines in cultures. Cancer Res 42: 1716-1721

Dennis WH, Yatvin MB (1981) Correlation of hyperthermic sensitivity and membrane microviscosity in E. coli K1060. Int J of Radiat Biol 39: 265-271

Dethlefsen LA, Dewey WC (eds) (1982) Third international symposium: cancer therapy by hyperthermia, drugs, and radiation. Natl Cancer Inst Monogr 61

Dewey WC (1984) Interaction of heat with radiation and chemotherapy. Cancer Res [Suppl] 44: 4714s-4720s

Dewey WC, Esch JL (1982) Transient thermal tolerance: cell killing and polymerase activities. Radiat Res 92: 611-614

Dewey WC, Holohan EV (1984) Hyperthermia - basic biology. In: Rosenblum ML, Wilson CB (eds) Progress in experimental tumor research: brain tumor therapy, vol 28, Karger, Basel, pp 198-219

Dewey WC, Westra A, Miller HH (1971) Heat-induced lethality and chromosomal damage in synchronized Chinese hamster cells treated with 5-bromodeoxyuridine. Int J Radiat Biol 20: 505-520

Dewey WC, Hopwood LE, Sapareto SA, Gerweck LE (1977) Cellular responses to combinations of hyperthermia and radiation. Radiology 123: 464-477

Dewey WC, Sapareto SA, Betten DA (1978) Hyperthermic radiosensitization of synchronous Chinese hamster cells: relationship between lethality and chromosomal aberrations. Radiat Res 76: 48-59

Dewey WC, Freeman ML, Raaphorst GP, Clark EP, Wong RS, Highfield DP, Spiro JS, Tomasovic SP, Denman DL, Coss RA (1980) Cell biology of hyperthermia and radiation. In: Meyn RE, Withers HR (eds) Radiation biology in cancer research. Raven, New York, pp 589-623

Dickson J, Calderwood StK (1979) Effect of hyperglycemia and hyperthermia on the pH, glycolysis and respiration of the Yoshida sarcoma in vivo. J Natl Cancer Inst 63: 1371-1381

Dietzel F (1975) Tumor and Temperatur. Urban and Schwarzenberg, Munich

Dikomey E (1978) Repair of DNA strand breaks in Chinese hamster ovary cells at 37 degrees or at 42 degrees C. In: Streffer C, van Beuningen D, Dietzel F, Röttinger E, Robinson JE, Scherer E, Seeber S, Trott K-R (eds) Cancer therapy by hyperthermia and radiation. Proceedings of the Second International Symposium, Essen, Germany, 2-4 June 1977. Urban and Schwarzenberg, Munich, pp 146-149

Dikomey E (1982) Effect of hyperthermia at 42° C and 45° C on repair of radiation-induced DNA strand breaks in CHO cells. Int J Radiat Biol 41: 603-614

Dube DK, Seal G, Loeb LA (1977) Differential heat sensitivity of mammalian DNA polymerase. Biochem Biophys Res Commun 76: 483-487

Durand RE (1978) Potentiation of radiation lethality by hyperthermia in a tumor model: effects of sequence, degree and duration of heating. Int J Radiat Oncol Biol Phys 4: 401-406

Eickhoff J, Dikomey E (1984) Development and decay of acutely induced thermotolerance in CHO cells by different heat shocks at various external pH values. In: Overgaard J (ed) Hyperthermic oncology, vol 1. Taylor and Francis, London, pp 91-94

Elkind MM, Sutton H, Moses WB (1967) Sublethal and lethal radiation damage. Nature 214: 1088-1092

Field SB (1978) The response of normal tissue to hyperthermia alone or in combination with X-rays. In: Streffer C, van Beuningen D, Dietzel F, Röttinger E, Robinson JE, Scherer E, Seeber S, Trott K-R (eds) Cancer therapy by hyperthermia and radiation. Proceedings of the Second International Symposium, Essen, Germany, 2-4 June 1977. Urban and Schwarzenberg, Baltimore, pp 37-48

Field SB, Hume S, Law MP, Morris C, Meyers R (1976) Some effects of combined hyperthermia and ionizing radiation on normal tissues. In: Proceedings of the international symposium on radiobiology research needed for the improvement of radiotherapy. IAEC, Vienna
Francesconi R, Mayer M (1979) Heat- and excercise-induced hyperthermia: Effects on high-energy phosphate. Aviat Space Environ Med 50: 799-802
Frankel HM, Ferrante FL (1966) Effects of pCO_2 on appearance of increased lactate during hyperthermia. Am J Physiology 210: 1269-1272
Frascella D, Frankel HM (1969) Liver pyridine nucleotides, lactate, and pyruvate in hyperthermic rats. Am J Physiology 217: 207-209
Freeman ML, Dewey WC, Hopwood LE (1977) Effect of pH on hyperthermic cell survival. J Natl Cancer Inst 58: 1837-1839
George KC, Singh BB (1982) Synergism of chlorpromazine and hyperthermia in two mouse solid tumours. Br J Cancer 45: 309-313
George KC, Singh BB (1985) Hyperthermic response of a mouse fibrosarcoma as modified by phenothiazine drug. Br J Cancer 51: 737-738
Gerner EW (1984) Definition of thermal dose. In: Overgaard J (ed) Hyperthermic oncology, vol II. Taylor and Francis, London, pp 245-251
Gerner EW (1984) Biological isoeffect relationships and dose for temperature induced cytotoxicity. In: Overgaard J (ed) Hyperthermic oncology, vol II. Taylor and Francis, London, pp 253-262
Gerner EW, Schneider MJ (1975) Induced thermal resistance in HeLa cells. Nature 256: 500-502
Gerner EW, Leith JT (1977) Interaction of hyperthermia with radiation of different linear energy transfer. Int J Radiat Biol 31: 238-288
Gerner EW, Connor WG, Boone MLM, Doss JD, Mayer EG, Miller RG (1975) The potential of localized heating as an adjunct to radiation therapy. Radiology 116: 433-489
Gerner EW, Leith JT, Boone MLM (1976) Mammalian cell survival response following irradiation with 4 MeV X-rays or accelerated helium ions combined with hyperthermia. Radiology 119: 715-720
Gerner EW, Holmes PW, McCullough JA (1979) Influence of growth state on several thermal response of EMT-6/Az tumor cells in vitro. Cancer Res 39: 981-986
Gerweck LE (1977) Modification of cell lethality at elevated temperatures: the pH effect. Radiat Res 70: 224-235
Gerweck LE (1978) Influence of microenvironmental condition on sensitivity to hyperthermia or radiation for cancer therapy. In: Caldwell W, Durand R (eds) Proceedings of the symposium on clinical prospects of hypoxic cell sensitizers and hyperthermia. University of Wisconsin, Madison
Gerweck LE (1982) Effect of microenvironmental factors on the response of cells to single and fractionated heat treatments. Natl Cancer Inst Monogr 61: 19-25
Gerweck LE (1984) Environmental and vascular effect. In: Overgaard J (ed) Hyperthermic oncology, vol II. Taylor and Francis, London, pp 253-262
Gerweck LE, Richards B (1981) Influence of pH on the thermal sensitivity of cultured human glioblastoma cells. Cancer Res 41: 845-849
Gerweck LE, Bascomb F (1982) Influence of hypoxia on the development of thermotolerance. Radiat Res 90: 356-361
Gerweck LE, Delaney TF (1984) Persistence of thermotolerance in slowly proliferating plateau phase cells. Radiat Res 97: 365-372
Gerweck LE, Epstein LF (1985) Cell proliferation, protein turnover, and the decay of thermotolerance in CHO cells. Radiat Res (to be published)
Gerweck LE, Nygaard TG, Burlett M (1979) Response of cells to hyperthermia under acute and chronic hypoxic conditions. Cancer Res 39: 966-972
Gerweck LE, Richards B, Michaels HB (1982) Influence of low pH on the development and decay of 42° C thermotolerance in CHO cells. Int J of Radiat Oncol Biol Phys 8: 1935-1941
Giovanella BC, Morgan AC, Stehlin JA, Williams LJ (1973) Selective lethal effect of supranormal temperatures on mouse sarcoma cells. Cancer Res 33: 2568-2578
Giovanella BC, Stehlin JS, Morgan AC (1976) Selective lethal effects of supranormal temperatures on human neoplastic cells. Cancer Res 36: 3944-3950
Goldin EM, Leeper DB (1981) The effect of low pH on thermotolerance induction using fractionated 45° C hyperthermia. Radiat Res 85: 472-479
Hahn GM (1974) Metabolic aspects of the role of hyperthermia in mammalian cell inactivation and their possible relevance to cancer treatment. Cancer Res 34: 3117-3123

Hahn GM (1980) Comparison of the malignant potential of 10T1/2 cells and transformants with their survival responses to hyperthermia and to amphotericin B. Cancer Res 40: 3763-3767
Hahn GM (1982) Hyperthermia and cancer. Plenum, New York
Hall E (1978) Radiobiology for the radiologist. Harper and Row, Hagestown
Hand JW, Walker H, Hornsey S, Field SB (1979) Effect of hyperthermia on the mouse testis and its response to X-rays, as assayed by weight loss. Int J Radiat Biol 35: 521-528
Harisiadis L, Hall EJ, Kraljevic U, Borek C (1975) Hyperthermia: biological studies at the cellular level. Radiology 117: 447-452
Havemann J (1983) Influence of a prior heat treatment on the enhancement by hyperthermia of X-ray-induced inactivation of cultured mammalian cells. Int J Radiat Biol 43: 267-280
Havemann J (1983) Influence of pH and thermotolerance on the enhancement of X-ray induced inactivation of cultured mammalian cells by hyperthermia. Int J Radiat Biol 43: 281-289
Havemann J, Hahn GM (1981) The role of energy in hyperthermia-induced mammalian cell inactivation: a study of the effects of glucose starvation and an uncoupler of oxidative phosphorylation. J Cell Physiol 107: 237-241
Heddle JA (1973) A rapid in vivo test for chromosomal damage. Mutation Res 18: 187-190
Hengstebeck S (1983) Untersuchungen zum Intermediärstoffwechsel in der Leber und in einem Adenocarcinom der Maus nach Hyperthermie. Dissertation, Universität-Gesamthochschule Essen
Henle KJ (1982) Thermotolerance in the murine jejenum. J Natl Cancer Int 68: 1033-1036
Henle KJ, Leeper DB (1976) Interaction of hyperthermia and radiation in CHO cells: recovery kinetics. Radiation Res 66: 505-518
Henle KJ, Leeper DB (1979) Effects of hyperthermia (45° C) on macromolecular synthesis in Chinese hamster ovary cells. Cancer Res 39: 2665-2674
Henle KJ, Bitner AF, Dethlefsen LA (1979) Induction of thermotolerance by multiple heat fractions in Chinese hamster ovary cells. Cancer Res 39: 2486-2491
Henle KJ, Peck JW, Higashikubo R (1983) Protection against heat-induced cell killing in polyols in vitro. Cancer Res 43: 1624-1627
Henle KJ, Nagle WA, Moss AJ, Herman TS (1984) Cellular ATP content of heated Chinese hamster ovary cells. Radiat Res 97: 630-633
Henriques FC, Jr (1947) Studies on thermal injury. Arch Pathol 43: 489-502
Holohan EV, Highfield DP, Dewey WC (1982) Induction during G_1 of heat radiosensitization in Chinese hamster ovary cells following single and fractionated heat doses. Natl Cancer Inst Monogr 61: 123-125
Holohan EV, Highfield DP, Holohan PK, Dewey WC (1984) Hyperthermic killing and hyperthermic radiosensitization in Chinese hamster ovary cells: effects of pH and thermal tolerance. Radiat Res 97: 108-131
Hume SP (1984) Experimental studies of normal tissue response to hyperthermia given alone or combined with radiation. In: Overgaard J (ed) Hyperthermic oncology, vol II. Taylor and Francis, London, pp 53-70
Hume SP (1985) Experimental studies of normal tissue response to hyperthermia given alone or combined with radiation. In: Overgaard J (ed) Hyperthermic oncology 1984, vol II. Taylor and Francis, pp 53-70
Hume SP, Marigold JCL (1980) Transient, heat induced, thermal resistance in the small intestine of mouse. Radiat Res 82: 526-535
Hume SP, Rogers MA, Field SB (1978) Two qualitatively different effects of hyperthermia on acid phosphatase staining in mouse spleen, dependent on the severity of the treatment. Int J Radiat Biol 34: 401-409
Hume SP, Marigold JCL, Field SB (1979) The effects of local hyperthermia on the small intestine of mouse. Br J Radiol 52: 657-662
Hume SP, Marigold JC, Michalowski A (1983) The effect of local hyperthermia on non proliferative, compared with proliferative, epithelial cells of the mouse intestinal mucosa. Radiat Res 94: 252-262
Hume SP, Myers R (1984) An unexpected effect of hyperthermia in the expression of X-ray damage in mouse skin. Radiat Res 97: 186-199
Issa MM (1985) Hyperthermie an Dünndarm der Maus. Eine licht- und elektronenmikroskopische Untersuchung. J. naugural dissertation, Universität Essen

Issel RD, Bournier S, Youngman R cited in: Leeper DB (1985) Molecular and cellular mechanisms of hyperthermia alone or combined with other modalities. In: Overgaard J (ed) Hyperthermic oncology, vol II. Taylor and Francis, London, pp 9–40
Jähde E, Rajewsky MF (1982) Sensitization of clonogenic malignant cells to hyperthermia by glucose-mediated, tumour-selective pH reduction. J Cancer Res Clin Oncol 104: 23–30
Jorritsma JBM, Konings AWT (1983) Inhibition of radiation-induced strand breaks by hyperthermia and its relationship to cell survival after hyperthermia alone. Int J Radiat Biol 43: 505–516
Jorritsma JBM, Kampinga HH, Konings AWT (1984) Role of DNA polymerase in the mechanisms of damage by heat and heat plus radiation in mammalian cells. In: Overgaard J (ed) Hyperthermic oncology 1984. Taylor and Francis, London, pp 61–64
Joshi DS, Jung H (1979) Thermotolerance and sensitization induced in CHO cells by fractionated hyperthermic treatments at 38°–45° C. Eur J Cancer 15: 345–350
Jung H (1982) Interaction of thermotolerance and thermosensitization induced in CHO cells by combined hyperthermic treatments at 40° and 43° C. Radiat Res 91: 433–446
Jung H, Kölling H (1980) Induction of thermotolerance and sensitization in CHO cells by combined hyperthermic treatments at 40° and 43° C. Eur J Cancer 16: 1523–1528
Kal HB, Hahn GM (1976) Kinetic responses of murine sarcoma cells to radiation and hyperthermia in vivo and in vitro. Cancer Res 36: 1923–1929
Kal HB, Hatfield M, Hahn GM (1975) Cell cycle progression of murine sarcoma cells after X-irradiation or heat shock. Radiology 117: 215–217
Kase K, Hahn GM (1975) Differential heat response of normal and transformed human cells in tissue culture. Nature 255: 228–230
Kiefer J, Kraft-Weyrather W, Hlawica M (1976) Cellular radiation effects and hyperthermia influence of exposure temperature on survival of diploid yeast irradiated under oxygenated and hypoxic conditions. Int J Radiat Biol 30: 293–300
Kim SH, Kim JH, Hahn EW (1975 a) The radiosensitization of hypoxic tumor cells by hyperthermia. Radiology 114: 727–728
Kim SH, Kim JH, Hahn EW (1975 b) Enhanced killing of hypoxic tumor cells by hyperthermia. Br J Radiol 48: 872–874
Kim SH, Kim JH, Hahn EW (1976) The enhanced killing of irradiated HeLa cells in synchronous culture by hyperthermia. Radiat Res 66: 337–345
Kim SH, Kim JH, Hahn EW (1978) Selective potentiation of hyperthermia killing of hypoxic cells by 5-thio-D-glucose. Cancer Res 38: 2935–2938
Kim SH, Kim JH, Hahn EW, Ensign NA (1980) Selective killing of glucose and oxygen-deprived HeLa cells by hyperthermia. Cancer Res 40: 3459–3462
Kim JH, Kim SH, Alfieri A, Young CW (1984) Quercetin, an inhibitor of lactate transport and a hyperthermic sensitizer of HeLa cells. Cancer Res 44: 102–106
Konings AWT, Penninga P (1983) Role of reduced glutathione in cellular heat sensitivity and thermotolerance. Strahlentherapie 159: 377–378
Konings AWT, Penninga P (1984) Role of reduced glutathione protein thiols, and pentose phosphate pathway in heat sensitivity and thermotolerance. Proc 4th Int Symp Hyperthermia Oncology. Aarhus, Denmark, 2–6 July 1984, pp 115–118
Landry J, Chretien P (1983) Relationship between hyperthermia induced heat shock proteins and thermotolerance in Morris hepatoma cells. Can J Biochem Cell Biol 61: 428–437
Langendorff H, Langendorff M (1943) Über die Wirkung einer mit Ultrakurzwelle kombinierte Röntgenbehandlung auf das Ehrlich-Karzinom der Maus. Strahlentherapie 72: 211–219
Law MP, Ahier RG (1982) Long-term thermal sensitivity of previously irradiated skin. Br J Radiol 55: 913–915
Law MP, Ahier RG, Field SB (1978) The response of the mouse ear to heat applied alone or combined with X-rays. Br J Radiol 51: 132–138
Law MP, Ahier RG, Field SB (1979) The effect of prior heat treatment on the thermal enhancement of radiation damage in the mouse ear. Br J Radiol 52: 315–321
Leeper DB (1985) Molecular and cellular mechanisms of hyperthermia alone or combined with other modalities. In: Overgaard J (ed) Hyperthermic oncology 1984. Taylor and Francis, London, pp 9–40
Leith JT, Miller RC, Gerner EW, Boone MLM (1977) Hyperthermic potentiation. Biological aspects and applications to radiation therapy. Cancer 39: 766–779

Lepock JR (1982) Involvement of membranes in cellular responses to hyperthermia. Radiat Res 92: 433-438

Lepock JR, Massicotte-Nolan P, Ruled GS, Kruuv J (1981) Lack of correlation between hyperthermic cell killing, thermotolerance, and membrane lipid fluidity. Radiat Res 87: 300-313

Lepock JR, Cheng KH, Al-Qysi H, Kruuv J (1983) Thermotropic lipid and protein transitions in Chinese hamster lung cell membranes: relationship to hyperthermic cell killing. Can J Biochem Cell Biol 61: 421-427

Li GC, Hahn GM (1980a) A proposed operational model of thermotolerance based on effects of nutrients and the initial treatment temperature. Cancer Res 40: 4501-4508

Li GC, Hahn GM (1980b) Adaptation to different growth temperatures modifies some mammalian cell survival responses. Exp Cell Res 128: 475-485

Li GC, Shiu EC, Hahn GM (1980) Similarities in cellular inactivation by hyperthermia or by ethanol. Radiat Res 82: 257-268

Li GC, Petersen NS, Mitchell HK (1982a) Induced thermal tolerance and heat shock protein synthesis in Chinese hamster ovary cells. Int J Radiat Oncol Biol Phys 8: 63-67

Li GC, Fisher GA, Hahn GM (1982b) Induction of thermotolerance and evidence for a well-defined thermotropic cooperative process. Radiat Res 89: 361-368

Li GC, Shrieve DC, Werb A (1982c) Correlations between synthesis of heat-shock proteins and development of tolerance to heat and to Adriamycin and Chinese hamster fibroblasts: heat shock and other inducers. In: Schlesinger MJ, Ashburner M, Tissieres (eds) Heat shock. Cold Spring Harbor, New York

Lin PS, Turi A, Kwock L, Lu RC (1982) Hyperthermia effect on microtubule organization. Natl Cancer Inst Monogr 61: 57-60

Loshek DD, Orr JS, Solomonidis E (1977a) Interaction of hyperthermia and radiation: the survival surface. Br J Radiol 50: 893-901

Loshek DD, Orr JS, Solomonidis E (1977b) Interaction of hyperthermia and radiation: temperature coefficient of interaction. Br J Radiol 50: 902-907

Loshek DD, Orr JS, Solomonidis E (1981) Interaction of hyperthermia and radiation: radiation quality. Br J Radiol 54: 40-47

Loven DP, Leeper DB, Oberley LW (1985) Superoxide dissmutase levels in Chinese hamster ovary cells and ovarian carcinoma cells after hyperthermia or exposure to cycloheximide. Cancer Res 45: 3029-3033

Lücke-Huhle C, Dertinger H (1977) Kinetic response of an in vitro "tumor model" (V99 spheroids) to 42° C hyperthermia. Eur J Cancer 13: 23-28

Lunec J, Cresswell SR (1983) Heat-induced thermotolerance expressed in the energy metabolism of mammalian cells. Radiat Res 93: 588-597

Lunec J, Hesslewood JP, Parker R, Leaper S (1981) Hyperthermic enhancement of radiation cell killing in HeLa S3 cells and its effect on the production and repair of DNA strand breaks. Radiat Res 85: 116-125

Magun BE, Fennie ChW (1981) Effects of hyperthermia on binding, internalization, and degradation of epidermal growth factor. Radiat Res 86: 133-146

Martinez A, Fajardo LF, Kernahan P, Prionas S, Hahn GM (1980) The effects of radio frequency heating on normal fat and muscular tissues: a histologically based tissue injury grading system. Presented at the third international symposium: cancer therapy by hyperthermia, drugs and radiation, Fort Collins, Co., June 22-26

Martinez AA, Meshorer A, Meyer JL, Hahn GM, Fajardo LF, Prionas SD (1983) Thermal sensitivity and thermotolerance in normal porcine tissues. Cancer Res 43: 2072-2075

Massicotte-Nolan P, Glofcheski DJ, Kruuv J, Lepock JR (1981) Relationship between hyperthermic cell killing and protein denaturation by alcohols. Radiat Res 87: 284-299

McCormick W, Penman SH (1969) Regulation of protein synthesis in HeLa cells: translation at elevated temperatures. J Mol Biol 39: 315-333

Mehdi SQ, Recktenwald DJ, Smith LM, Li GC, Armour EP, Hahn GM (1984) Effect of hyperthermia on murine cell surface histocompatibility antigens. Cancer Res 44: 3394-3397

Meyer KR, Hopwood LE, Gillette EL (1979) The thermal response of mouse adenocarcinoma cells at low pH. Eur J Cancer 15: 1219-1222

Milligan AJ, Metz JA, Leeper DB (1984) Effect of inestinal hyperthermia in the Chinese hamster. Int J Radiat Oncol Biol Phys 10: 259-263

Mills MD, Meyn RE (1983) Hyperthermic potentiation on unrejoined DNA strand breaks following irradiation. Radiat Res 95: 327–338
Mirtsch Sch, Streffer C, van Beuningen D, Rebmann A (1984) ATP metabolism in human melanoma cells after treatment with hyperthermia (42° C). In: Overgaard J (ed) Hyperthermic oncology 1984. Taylor and Francis, London, pp 19–22
Mitchell JB, Russo A, Kinsella TJ, Glatstein E (1983) Glutathione elevation during thermotolerance induction and thermosensitization by glutathione depletion. Cancer Res 43: 987–991
Mondovi B, Strom R, Rotilio G et al. (1969) The biochemical mechanism of selective heat sensitivity of cancer cells. I. Studies on cellular respiration. Eur J Cancer 5: 129–136
Mondovi B, Finazzi-Agro A, Rotilio G, Strom R, Moricca G, Rossi-Fanelli A (1969) The biochemical mechanism of selective heat sensitivity of cancer cells. II. Studies on nucleic acids and protein synthesis. Eur J Cancer 5: 137–146
Monson TP, Henle KJ, Moss AJ, Nagle WA (1984) Experimental test of the polyol hypothesis: the effect of aldose reductase inhibitors on thermotolerance development and measurements of intracellular sugar and polyol content in thermotolerant CHO cells. (Abstract) Proc 32nd annu mtg radiation research soc, Orlando, Florida, p 52
Morris CC, Myers R, Field SB (1977) The response of the rat tail to hyperthermia. Br J Radiol 50: 576
Moritz A, Henriques FC (1947) Studies of thermal injury: II. The relative importance of time and surface temperature in the causation of cutaneous burns. Am J Pathol 23: 695–720
Nagle WA, Moss AJ, Jr (1983) Inhibitors of poly (ADP-ribose) synthetase enhance the cytotoxicity of 42° C and 45° C hyperthermia in cultured Chinese hamster cells. Int J Radiat Biol 44: 475–481
Nagle WA, Moss AJ, Baker ML (1982) Increased lethatlity at 42° C for hypoxic Chinese hamster cells heated under conditions of energy depreviation. Natl Cancer Inst Monogr 61: 107–110
Nielsen OS, Overgaard J (1982) Influence of time and temperature on the kinetics of thermotolerance in L1A2 cells in vitro. Cancer Res 42: 4190–4196
Nielsen OS (1984) Fractionated Hyperthermia and Thermotolerance. Danish Medical Bulletin Vol. 31
Ngo FOH, Han A, Utsumi H, Elkind MM (1977) Comparative radiobiology of fast neutrons: relevance to radiotherapy and basic studies. Int J Radiat Oncol Biol Phys 3: 187–193
Ohyama H, Yamada T (1980) Reduction of rat thymocyte interphase death by hyperthermia. Radiat Res 82: 342–351
Omar RA, Lanks KW (1984) Heat shock protein synthesis and cell survival in clones of normal and SV40-transformed mouse embryo cells. Cancer Res 44: 3976–3982
Overgaard J (1976) Ultrastructure of a murine mammary carcinoma exposed to hyperthermia in vivo. Cancer Res 36: 983–995
Overgaard J (1977) Effect of hyperthermia on malignant cells in vivo: a review and hypothesis. Cancer 39: 2637–2646
Overgaard J (1985) History and heritage – an introduction. In: Overgaard J (ed) Hyperthermic oncology. Taylor and Francis, London, pp 3–8
Overgaard J (1985) Hyperthermic oncology. Taylor and Francis, London
Overgaard K (1934) Über Wärmetherapie bösartiger Tumoren. Acta Radiol (Ther) (Stockh) 15: 89–99
Overgaard K, Overgaard J (1972) Investigations on the possibility of a thermic tumour therapy: II. Action of combined heat-roentgen treatment on a transplanted mouse mammary carcinoma. Eur J Cancer 8: 573–575
Overgaard K, Overgaard J (1974) Radiation sensisitzing effect of heat. Acta Radiol (Ther) (Stockh) 13: 501–511
Palzer R, Heidelberger C (1973) Influence of drugs and synchrony on the hyperthermic killing of HeLa cells. Cancer Res 33: 422–427
Panniers R, Henshaw EC (1984) Mechanism of inhibition of polypeptide chain initiation in heat-shocked Ehrlich ascites tumour cells. Eur J Biochem 140: 209–214
Pincus G, Fischer A (1931) The growth and death of tissue cultures exposed to supranormal temperatures. J Exp Med 54: 323–332
Power J, Harris J (1977) Response of extremely hypoxic cells to hyperthermia: survival and oxygen enhancement ratios for exponential and plateau-phase cultures. Radiology 123: 767–770

Privalov PL (1979) Stability of proteins. In: Anfinsen CB, Edsall JT, Richards EM (eds) Advances in protein chemistry, vol 33. Academic New York, pp 167-241

Radford JR (1983) Effects of hyperthermia on the repair of X-ray-induced DNA double strand breaks in mouse L cells. Int J Rad Biol 43: 551-557

Rao B, Hopwood LE (1985) Effect of hypoxia on recovery from damage induced by heat and radiation in plateau-phase cells. Radiat Res 101: 312-325

Reeves O (1972) Mechanism of acquired resistance to acute heat shock in cultured mammalian cells. J Cell Physiol 79: 157-159

Reeves O (1982) Mechanism of acquired resistance to acute heat shock in cultured mammalian cells. J Cell Physiol 79: 157-159

Reinhold HS, Wike-Hooley JL, van den Berg AP, and van den Berg-Blok A (1984) Environmental factors, blood flow and microcirculation. In: Overgaard J (ed) Hyperthermic oncology, vol II. Francis and Taylor, London, pp 41-52

Reinhold HS, Wike-Hooley JL, van den Berg AP, and van den Berg-Blok A (1985) Environmental factors, blood flow and microcirculation. In: Overgaard J (ed) Hyperthermic oncology 1984, vol II. Taylor and Francis, London, pp 41-52

Reiter T, Penman S (1983) Prompt heat shock proteins: translationally regulated synthesis of new proteins associated with the nuclear matrix-intermediate filaments as an early response to heat shock. Proc Natl Assoc Sci USA 80: 4737-4741

Rice LC, Urano M, Maher J (1982) The kinetics of thermotolerance in the mouse foot. Radiat Res 89: 291-297

Robinson JE, Wizenberg MJ (1974) Thermal sensitivity and the effect of elevated temperatures on the radiation sensitivity of Chinese hamster cells. Acta Radiol (Ther) (Stockh) 13: 241-249

Robinson JE, Wizenberg MJ, McCready W, Scheltema J (1974b) Combined hyperthermia and radiation suggest an alternative to heavy particle therapy for reduced oxygen enhancement ratios. Nature 251: 521-522

Rofstad EK, Wahl A, Tveit KM, Monge OR, Brustad T (1985) Survival curves after X-ray and heat treatments for melanoma cells derived directly from surgical specimens of tumours in man. Radiother Oncol 4: 33-44

Roti Roti J (1982) Heat-induced cell death and radiosensitization: molecular mechanisms. Natl Cancer Inst Monogr 61: 3-9

Ruifrok ACC, Kanon B, Hulstaart CE, Konings AWT (1984) Permeability change of cells treated with hyperthermia alone and in combination with X-irradiation. In: Overgaard J (ed) Hyperthermic oncology, vol I. Taylor and Francis, London, pp 65-68

Russo A, Mitchell JB, McPherson S (1984) The effects of glutathione depletion on thermotolerance and heat stress protein synthesis. Br J Cancer 49: 753-758

Sapareto S, Hopwood L, Dewey W, Raju M, Gray J (1978) Effects of hyperthermia on survival and progression of Chinese hamster ovary cells. Cancer Res 38: 393-400

Schamhart DHJ, van Walraven HS, Weigant FAC, Linnemans WAM, van Rijn J, van den Berg J, van Wijk R (1984) Thermotolerance in cultured hepatoma cells: cell viability, cell morphology, protein synthesis, and heat shock proteins. Radiat Res 98: 89-95

Schlag H, Lücke-Huhle C (1976) Cytokinetic studies on the effect of hyperthermia on Chinese hamster lung cells. Eur J Cancer 12: 827-831

Schlesinger MJ, Ashburner M, Tissieres A (eds) (1982) Heat shock: from bacteria to man. Cold Spring Harbor Laboratory, New York

Schlesinger MJ, Aliperti G, Kelley PM (1982) The response of cells to heat shock. Trends Biochem Sci 7: 222-225

Schubert B, Streffer C, Tamulevicius P (1982) Glucose metabolism in mice during and after whole-body hyperthermia. Natl Cancer Inst Monogr 61: 203-205

Schulman N, Hall E (1974) Hyperthermia: its effect on proliferative and plateau phase cell cultures. Radiology 113: 207-209

Sciandra JJ, Subjeck JR (1984) Heat shock proteins and protection of proliferation and translation in mammalian cells. Cancer Res 44: 5188-5194

Sciandra JJ, Gerweck LE (1986) Thermotolerance in cells. In: Watmough DJ, Ross W (eds) Hyperthermia-Clinical and Scientific Aspects. Blackie and Son Ltd., Glasgow, pp 99-120

Shall S (1984) ADR-ribose in DNA repair: a new component of DNA excision repair. IN: Lett J (ed) Advances in radiation biology, vol II. Academic, Orlando, pp 1-69

Shenoy MA, Singh BB (1985) Temperature dependent modification of radiosensitivity following hypoxic cytocidal action of delorpromazine. Rad Envir Bio Phys 24: 113–117
Simard R, Bernhard W (1967) A heat-sensitive cellular function located in the nucleolus. J Cell Biology 34: 61–76
Skibba JL, Collins FG (1978) Effect of temperature on biochemical functions in the isolated perfused rat liver. J Surg Res 24: 435–441
Song CW, Clement SS, Levitt SH (1977) Cytotoxic and radiosensitizing effects of 5-thio-D-glucose hypoxic cells. Radiology 123: 201–205
Song CW, Kang MS, Rhee JG, Levitt S (1980) Vascular damage and delayed cell death in tumours after hyperthermia. Br J Cancer 41: 309–312
Spiro IJ, Denman DL, Dewey WC (1982) Effect of hyperthermia on CHO DNA polymerase- and *β*. Radiat Res 89: 134–139
Stevenson MA, Minton KW, Hahn GM (1981) Survival and concanavalin-A- induced capping in CHO fibroblasts after exposure to hyperthermia, ethanol, and X-irradiation. Radiat Res 86: 467–478
Streffer C (1963) Reaktivität und Struktur von Aminosäuren und Proteinen (Cystein und *β*-Galaktosidase). Dissertation, University Freiburg
Streffer C (1969) Strahlen-Biochemie. Springer, Berlin Heidelberg New York
Streffer C (1982) Aspects of biochemical effects by hyperthermia. Natl Cancer Inst Monogr 61: 11–16
Streffer C (1985) Mechanism of heat injury. In: Overgaard J (ed) Hyperthermic oncology 1984. Taylor and Francis, London, pp 213–222
Streffer C (1985) Metabolic changes during and after hyperthermia. Int J Hyperthermia 1: 305–319
Streffer C, van Beuningen D (1985) Zelluläre Strahlenbiologie und Strahlenpathologie (Ganz- und Teilkörperbestrahlung). In: Diethelm L, Heuck F, Olsson O, Strnad F, Vieten H, Zuppinger A (eds) Handbuch der medizinischen Radiologie, vol 20. Springer, Berlin Heidelberg New York Tokyo, pp 1–39
Streffer C, van Beuningen D, Elias S (1977) Comparative effects of tritiated water and thymidine on the preimplanted mouse embryos in vitro. Curr Topics Radiat Res Quart 12: 182–193
Streffer C, van Beuningen D, Dietzel F, Röttinger E, Robinson JE, Scherer E, Seeber S, Trott K-R (1978) Cancer therapy by hyperthermia and radiation. Urban and Schwarzenberg, Baltimore
Streffer C, van Beuningen D, Zamboglou N (1979) Cell killing by hyperthermia and radiation in cancer therapy. In: Abe M, Sakamoto K, Phillips TL (eds) Treatment of radioresistant cancers. Elsevier/North Holland Biomedical, Amsterdam, pp 55–70
Streffer C, Hengstebeck S, Tamulevicius P (1981) Glucose metabolism in mouse tumor and liver with and without hyperthermia. Henry Ford Hosp Med J 29: 41–44
Streffer C, Tamulevicius P, Schmidt K (1983) Poly (ADPR) synthetase activity in melanoma cells after hyperthermia and radiation. Radiat Res 94: 589 (Abstract)
Streffer C, van Beuningen D, Bertholdt G, Zamboglou N (1983) Some aspects of radiosensitization by hyperthermia: neutrons and X-rays. In: Kano E (ed) Fundamentals of cancer therapy by hyperthermia, radiation and chemicals. MAG Bros, Tokyo, pp 121–134
Streffer C, van Beuningen D, Uma Devi P (1984) Radiosensitization by hyperthermia in human melanoma cells: single and fractionated treatments. Cancer Treat Rev 11: 179–185
Strom R, Crifo C, Rossi-Fanelli A, Mondovi B (1977) Biochemical aspects of heat sensitivity of tumor cells. In: Rossi-Fanelli A, Cavaliere R, Mondovi B, Morrica G (eds) Selective heat sensitivity of cancer cells. Springer, Berlin Heidelberg New York, pp 7–35
Subjek JR, Sciandra JJ, Johnson RJ (1982) Heat shock proteins: a comparison of induction kinetics. Br J Radiol 55: 579–584
Suit HD, Shwayder M (1974) Hyperthermia: potential as an anti-tumor agent. Cancer 34: 122–129
Tamulevicius P, Streffer C (1983) Does hyperthermia produce increased lysosomal enzyme activity? Int J Radiat Biol 43: 321–327
Tamulevicius P, Schmidt K, Streffer C (1984) The effects of X-irradiation, hyperthermia and combined modality treatment on poly (ADPR) synthetase activity in human melanoma cells. Radiat Res 100: 65–77

Tamulevicius P, Würzinger U, Luscher G, Streffer C (1984) Lipid metabolism in mouse liver and adenocarcinoma following hyperthermia. In: Overgaard J (ed) Hyperthermic oncology 1984. Taylor and Francis, London, pp 23-26
Terasima T, Tolmach LJ (1963 a) Variations in several responses of HeLa cells to X-irradiation during the division cycle. Biophys J 3: 11-33
Terasima T, Tolmach LJ (1963 b) X-ray sensitivity and DNA synthesis in synchronously dividing populations of HeLa cells. Science 140: 490-492
Tomasovic SP, Steck PA, Heitzman D (1983) Heat stress proteins and thermal resistance in rat mammary cells. Radiat Res 95: 399-413
Tomasovic SP, Rosenblatt PL, Johnston DA, Tang K, Lee PSY (1984) Heterogeneity in induced heat resistance and its relation to synthesis of stress proteins in rat tumor cell clones. Cancer Res 44: 5850-5856
van Beuningen D (1983) Hyperthermie als cytotoxisches und strahlensensibilisierendes Agens: zelluläre Effekte - eine Übersicht. Strahlentherapie 159: 60-66
van Beuningen D, Molls M, Schulz S, Streffer C (1978) Effects of irradiation and hyperthermia on the development of preimplanted mouse embryos in vitro. In: Streffer C et al. (eds) Cancer therapy by hyperthermia and radiation. Urban and Schwarzenberg, Baltimore, pp 151-153
van Beuningen D, Issa M, Breipohl W, Streffer C, Raumwolf M (1983) Light- and electron-microscopical investigations on the effect of hyperthermia on the small intestine (Abstract). Strahlentherapie 159: 367
van Beuningen D, Streffer D, Spalthoff C (to be published) Effects of hyperthermia on glucose, pyruvate, and lactate metabolism in human melanoma cell cultures. Int J Hyperthermia
van Beuningen D, Streffer C, Pelzer T (1985) Radiosensitization of exponential and plateau phase cells. Strahlentherapie 161: 552
van Rijn J, van den Berg J, Schamhart DHJ, van Wijk R (1984) Effect of thermotolerance on thermal radiosensitization in hepatoma cells. Radiat Res 97: 318-328
Vaupel P, Müller-Klieser W, Otte J, Manz R, Kallinowski F (1983) Durchblutung, Sauerstoffversorgung des Gewebes und pH-Verteilung in malignen Tumoren nach Hyperthermie. Pathophysiologische Grundlagen und Einfluß verschiedener Hyperthermiedosen. Strahlentherapie 159: 73-81
Vaupel P, Benzing H, Egelhof E, Müller-Klieser W, Müller-Schauenburg (1983) The effect of various thermal doses on the regional tumor blood flow measured by heat clearance. Strahlentherapie 159: 384 (Abstract)
Verma SP, Wallach DFH (1976) Erythrocyte membranes undergo cooperative, pH-sensitive state transitions in the physiological temperature range: evidence from Raman spectroscopy. Proc Natl Acad Sci USA 73: 3558-3561
von Ardenne M (1971) The cancer multi-step therapy concept. Panminerva Med 13: 509-519
von Ardenne M (1975) Prinzipien und Konzept 1974 der "Krebs-Mehrschritt-Therapie". Radiobiol Radiother 16: 99-119
von Ardenne M (1978) On a new physical principle for selective local hyperthermia of tumor tissue. In: Streffer C, van Beuningen D, Dietzel F, Röttinger E, Robinson JE, Scherer E, Seeber S, Trott K-R (eds) Cancer therapy by hyperthermia and radiation. Proceedings of the Second International Symposium, Essen, 2-4 June 1977. Urban and Schwarzenberg, Baltimore München, pp 96-104
von Ardenne M (1980) Hyperthermia and cancer therapy. Adv Pharmacol Chemother 10: 137-138
von Ardenne M, Reitnauer P (1976) Verstärkung der mit Glukoseinfusion erzielbaren Tumorübersäuerung in vivo durch NAD. Arch Geschwulstforsch 30: 319-330
von Ardenne M, Chaplain R, Reitnauer P (1969) Selektive Krebszellenschädigung durch eine Attackenkombination mit Übersäuerung Hyperthermie, Vitamin A, Dimethylsulfoxid und weiteren die Freisetzung lysosomaler Enzyme fördernden Agenzien. Arch Geschwulstforsch 33: 331-344
Wallach D (1977) Basic mechanisms in tumor thermotherapy. J Mol Med 2: 381-403
Wallach DHF (1978) Action of hyperthermia and ionizing radiation on plasma membranes: In: Streffer C, van Beuningen D, Dietzel F, Röttinger E, Robinson JE, Scherer E, Seeber S, Trott K-R (eds) Cancer therapy by hyperthermia and radiation. Urban and Schwarzenberg, Baltimore, pp 19-28
Wallenfels K, Streffer C (1964) Chemische Reaktivität von Proteinen. In: 14. Colloquium der Gesellschaft für physiologische Chemie in Mosbach/Baden. Springer, Berlin Göttingen Heidelberg, pp 6-40
Wallenfels K, Streffer C (1966) Das Dissoziationsverhalten von Cystein und verwandten SH-Verbindungen. Biochem Z 346: 119-132

Warburg O, Wind F, Negelein E (1926) Über den Stoffwechsel von Tumoren im Körper. Klin Wschr 5: 829-834
Warocquier R, Scherrer K (1969) RNA metabolism in mammalian cells at elevated temperature. Eur J Biochem 10: 362-370
Warters RL, Roti Roti JL (1982) Hyperthermia and the cell nucleus. Radiat Res 92: 458-462
Warters RL, Stone OL (1983) Effects of hyperthermia on DNA replication in HeLa cells. Radiat Res 93: 71-84
Warters RL, Stone OL (1983) Histone protein and DNA synthesis by HeLa cells and thermal shock. Radiat Res 96: 646
Weber G (1983) Biochemical strategy of cancer cells and the design of chemotherapy: GHA Glowes memorial lecture. Cancer Res 43: 3466-3492
Westermark F (1898) Über die Behandlung des ulcerierenden Cervixcarcinoms mittels konstanter Wärme. Zentralbl Gynakol 22: 1335
Westermark N (1927) The effect of heat on rat tumors. Skand Arch Physiol 52: 257-322
Wiegant F, Karelaars A, Blok F, Linnemanns W (1984) Effects of extra cellular Ca^{2+} concentrations upon hyperthermia induced cell death. In: Overgaard J (ed) Hyperthermic oncology, vol I. Taylor and Francis, London, pp 3-6
Wike-Hooley JL, Faithfull NS, van der Zee J, van den Berg AP (1983) Liver damage and extraction of indocyamine green under whole body hyperthermia. Eur J Appl Physiol 51: 269-279
Williamson DH, Lund P, Krebs HA (1967) The redox state of free nicotinamide adenine dinucleotide in the cytoplasm and mitochondria of rat liver. Biochem J 103: 514-527
Wizenberg M, Robinson JE (1975) Proceedings of the international symposium on cancer therapy by hyperthermia and radiation. American College of Radiology Press, Baltimore
Wondergem J, Havemann J (1983) The response of previously by irradiated mouse skin to heat alone or combined with irradiation: influence of thermotolerance. Int J Radiat Oncol 44: 539-552
Wong RSL, Dewey WC (1982) Molecular studies on the hyperthermic inhibition of DNA synthesis in Chinese hamster ovary cells. Radiat Res 92: 370-395
Yatvin MB (1977) The influence of membrane lipid composition and procaine on hyperthermic death of cells. Int J Radiat Biol 32: 513-521
Yatvin MB, Cree TC, Elson CE, Gipp JJ, Tegmo I-M, Vorpahl JW (1982) Probing the relationship of membrane "fluidity" to heat killing of cells. Radiat Res 89: 644-646
Yatvin MB, Abuirmeileh NM, Vorpahl JW, Elson CE (1983) Biological optimization of hyperthermia: modification of tumor membrane lipids. Eur J Cancer 19: 657-663
Yatvin M-B, Vorpahl JW, Gould MN, Lyte M (1983) The effects of membrane modification and hyperthermia on the survival of P-388 and V-79 cells. Eur J Cancer 19: 1247-1253
Yi PN (1983) Hyperthermia-induced intracellular ionic level changes in tumor cells. Radiat Res 93: 534-544

Physiological Effects of Hyperthermia[*, **]

P. Vaupel and F. Kallinowski

Abteilung für Angewandte Physiologie, Universität Mainz, Saarstrasse 21, 6500 Mainz, FRG

Introduction

Hyperthermia as a modality for the treatment of malignant tumors, either alone or in combination with radiation or anticancer drugs, is rapidly becoming a clinical reality. Three different mechanisms of action have provided the rationale for considering the use of hyperthermia as an antitumor agent. At moderate hyperthermia ($T = 40° - 42.5°$ C), heat can increase cell killing in a synergistic way following exposure of a tumor to ionizing radiation. This *radiosensitization* is probably based on, among other things, the inhibited repair of radiation-induced DNA lesions. Elevated tissue temperatures at 40°-42.5° C also sensitize tumor cells to certain chemotherapeutic drugs, particularly to alkylating agents *(chemosensitization)*. In this context it has been shown that the action of bleomycin, Adriamycin, and *cis*-platinum is also enhanced by heat treatment (see detailed literature data in: Bicher and Bruley 1982; Dethlefsen and Dewey 1982; Dietzel 1975; Hahn 1982; Hornback 1984; Jain and Gullino 1980; Nussbaum 1982; Overgaard 1984/85; Storm 1983; Streffer et al. 1978).

At temperatures higher than 42.5° C, hyperthermia acts as a *cytotoxic agent* since mammalian cells die after heating in a time-temperature and cell cycle dependent manner. According to our present knowledge, this cell-killing effect of hyperthermia is based on multiple mechanisms leading to heat-induced lesions. One of the main cellular targets for hyperthermia is the cell membrane. Whereas changes of membrane lipids may not be critical during heat treatment, there is considerable evidence that hyperthermia affects certain membrane proteins (Arancia et al. 1986). Conformational changes of membrane proteins are followed by an instability of the phospholipid bilayer and by an altered permeability for cations through the cell membrane. After cytotoxic heat doses an increased K^+ efflux as well as an enhanced Ca^{2+} and H^+ influx have been reported recently (Anghileri et al. 1985; Ruifrok et al. 1985). Hyperthermia also affects the number and distribution of protein surface receptors for hormones or growth factors. Most probably changes are induced not only in the plasmalemma but also in the intracellular membrane structures and in the cytoskeleton. The disaggregation of microtubules and microfilaments may result from conformational changes of proteins and from imbalances of the intra-extracellular ion distribution. In addition, hyperthermia inhibits DNA replication as well as RNA and protein synthesis. Furthermore, heat usually leads to changes in physicochemical equilibria and to alterations of chemical reactions including metabolic processes. During moder-

* Dedicated to Prof. Dr. med. Dr. rer. nat G. Thews on his 60th birthday.

** Supported by the Bundesministerium für Forschung und Technologie (Grant 01 VF 034) and by the Deutsche Forschungsgemeinschaft (Grants Va 57/2-1, Va 57/2-2).

ate hyperthermia an increased rate of metabolic reactions is often found, followed by a dysregulation of the metabolism upon heating. As a consequence of the disturbed substrate turnover, the energy production is inhibited in most cells. This ATP depletion is concomitant with a disturbance in the Krebs cycle, while the glycolytic pathway remains almost unaffected for a longer time. Whether or not the intracellular ATP starvation plays a cardinal role during heat-induced cell killing is as yet unclear and thus needs to be further investigated (Bicher and Bruley 1982; Dethlefsen and Dewey 1982; Dietzel 1975; Hahn 1982; Hornback 1984; Jain and Gullino 1980; Nussbaum 1982; Overgaard 1984/85; Storm 1983; Streffer et al. 1978).

The cytocidal mechanisms of hyperthermia are effective under both in vitro and in vivo conditions and kill tumor cells preferentially in the S-phase. In solid tumors in vivo there are additional physiological (or pathophysiological) changes upon heating that can distinctly modulate the cytotoxic capacity of hyperthermia via modifications of the cellular microenvironment. The microenvironment of tumor cells is characterized by hypoxia or even anoxia, acidity and nutrient and energy deprivation (Vaupel and Mueller-Klieser 1983b; Vaupel et al. 1981). This is in contrast to normal tissues where the nutrient supply meets the requirements of the parenchymal cells under physiological conditions. The specific features characterizing the interstitial milieu of most tumors are generated by an inadequate and inhomogeneously distributed nutritive blood flow (Peterson 1979; Vaupel 1982; Vaupel and Hammersen 1983) and are known to sensitize cancer cells to treatment with hyperthermia (von Ardenne and Reitnauer 1968; Freeman et al. 1977; Gerweck and Richards 1981; Gerweck et al. 1974; Overgaard and Bichel 1977; Suit and Gerweck 1979). Gerweck et al. (1984) have demonstrated recently that sensitivity of heat treatment under nutrient-deprived conditions correlates with intracellular ATP levels at the time of treatment rather than with nutrient concentrations per se. Furthermore, low intratumor pH values may also inhibit the induction of thermotolerance, which can impede the effective use of hyperthermia, at least under in vitro conditions (Dikomey et al. 1981; Goldin and Leeper 1981; Nielson and Overgaard 1979).

The preferential effect of heat on tumor cells may also be attributable to blood flow in tumors which is mostly poor and sluggish when compared with that of normal tissues. Furthermore, the perfusion rate deteriorates with increasing size of a tumor (Vaupel 1974, 1977, 1979, 1982b). At a given energy input, temperature rises distinctly more in tumors than in normal tissues since in tumors the efficacy of heat dissipation by convection is reduced. The magnitude and the regional distribution of blood flow through tumors thus play a cardinal role during heat treatment through modifications of the micromilieu of the tumor cells and of the energy input required to achieve a therapeutic tissue temperature level. For this reason, besides the modulations of the microenvironment upon hyperthermia, considerations concerning heat-induced changes of tumor blood flow are discussed in detail when the physiological effects of tissue heating are compiled in the following chapters (for reviews dealing with the physiological effects of hyperthermia see also Emami and Song 1984; Mueller-Klieser and Vaupel 1984; Reinhold et al. 1984; Song 1982; Vaupel et al. 1983d). Immunological mechanisms involved in tumor cell-killing under in vivo conditions are not dealt with in this review.

Effect of Hyperthermia on Blood Flow Rates and Microcirculation in Normal and Neoplastic Tissues

Normal Tissue Blood Flow

Normal tissues consistently respond to hyperthermia by a remarkable increase in blood flow as long as tissue temperature levels of 45° C, exposure times of 30–60 min and heating up rates of 0.7° C/min are not exceeded (Dewhirst et al. 1984a; Dickson and Calderwood 1980; Dudar and Jain 1984; Jain and Ward-Hartley 1984; Milligan et al. 1983; Peck and Gibbs 1983; Rappaport and Song 1983; Sekins et al. 1980; Song 1978; Song et al. 1980b, 1980d; Stewart and Begg 1983). After termination of heating, the perfusion rate usually returns to normal levels within 1–2 h.

Skin and muscle are the most extensively studied normal tissues when physiological and pathophysiological effects of external heating are considered. This is understandable since mostly these tissues are involved during therapeutic heating of superficially seated tumors using an external heat source. Under thermoneutral and resting conditions the specific perfusion of skin and skeletal muscle is 3–5 ml/100 g per minute (Thews and Vaupel 1985). Heating at 45° C causes a pronounced flow increase up to a factor of 20 in skin, and a factor of 10 in muscle. This increase in blood flow in normal tissues is due to thermoregulatory mechanisms and is accompanied by a similar enhancement of heat dissipation by convection, counteracting a deleterious heat load of these tissues.

The magnitude of the flow increase in normal tissues upon heating seems to vary considerably depending on the species studied, on the tissue investigated, on the heating technique, and on the experimental protocol. According to Rappaport and Song (1983), blood flow in the skin and muscle adjacent to a tumor is about twice that in normal skin and muscle. This may be due to inflammatory reactions at the invasion front of a tumor. Upon heating at 43° C for 60 min, blood flow in skin and muscle adjacent to tumors increased even more than that in the respective tissue far away from the tumor.

Reports describing a mild to marked and reversible increase in normal tissue blood flow are consistent with observations that the energy input for maintaining the tissue temperature at a certain level is several times higher in normal tissues than in tumors. For example, the energy input required to keep the tissue temperature in the subcutis at 44° C is four to five times higher than in s. c. tumors (Mueller-Klieser and Vaupel 1984).

The heat-induced flow increase in normal tissues is caused by a reactive vasodilation, followed by a significant drop in the vascular resistance and a distinct increase in functional vascular volumes. Whether the vascular permeability at 43° C is enhanced in normal tissues (Emami and Song 1984) or whether it remains unaltered (Young et al. 1985) is not yet clear.

Tumor Blood Flow During and After Hyperthermia

The first quantitative data on tumor blood flow considering both tissue temperature level and exposure time are derived from rodent tumor systems (Scheid 1961; Vaupel et al. 1976). Thereafter many investigations have been performed on different tumor types, utilizing various heating techniques and protocols or devices for thermometry. Whereas data on hyperthermia-induced changes in blood flow derived from animal tumor systems are abundant, blood flow investigations in human tumors during hyperthermic treatment are scarce (see Tables 1-3). Most of the studies have shown that the tumor perfusion rate is decreased as a result of hyperthermic treatment. This flow decline is occasionally preceded by an initial, transient flow improvement when mild hyperthermia levels and/or short exposure times are applied (see Fig. 1).

Table 1. Response of tumor blood flow *(TBF)* to localized hyperthermia in mice

Hyperthermia level (° C)	Exposure time (min)	Tumor	Flow response	Reference
42-45	35	S_2 sarcoma	Temperature- and time-dependent flow stoppage	Scheid 1961
41	-	Mammary carcinomas	TBF increase up to 41° C, a decline thereafter	Johnson 1978
Continuous increase up to 45° C	-		TBF increase at T ≤ 41° C, flow collapse above 41° C	Bicher et al. 1982
42/44	30		No significant flow changes during heating, flow decrease after heating	Robinson et al. 1982
41.5/44	ca. 60		Immediate TBF increase after heating for 3 min, at longer exposure times TBF decreases	Peck and Gibbs 1983
43.5	30		No significant changes during heating, TBF decrease after heating	Song et al. 1980c
42.5	60	SA FA fibrosarcoma	Transient TBF increase and return to pre-treatment level during heating, TBF decrease after heating	Stewart and Begg 1983
40-45	15-75	Glioma	TBF increase at low thermal doses, flow decline at high thermal doses	Sutton 1980
42/45	40	S-180 tumor	42° C: TBF decline; flow restored 16 h after heating 45° C: marked decrease in TBF	Tanaka et al. 1984
43.5	60	RIF-1 tumor	Initial TBF increase, flow decrease after heating	Song et al. 1984

Table 2. Response of tumor blood flow *(TBF)* to localized hyperthermia in rats

Hyperthermia level (° C)	Exposure time (min)	Tumor	Flow response	Reference
39.5/42	30	DS carcinosarcoma	39.5° C: TBF increases 42.0° C: TBF decreases	Vaupel et al. 1976, 1980
39.5-44	30		TBF increase at 39.5° C, TBF decreases above 42° C	Vaupel et al. 1982a
43	100		Significant flow drop with heat alone in only a few cases	von Ardenne and Reitnauer 1982

Table 2 *(continued)*

Hyperthermia level (° C)	Exposure time (min)	Tumor	Flow response	Reference
41.3	60	BA 1112 rhabdomyosarcoma	Significant flow reduction	Endrich et al. 1979
41-44	30-60		TBF reduction dependent on temperature and exposure time	Emami et al. 1980
42/42.5	160/226		Flow stoppage in 50% of the tumors	Reinhold and van den Berg-Blok 1981
43	180		Red blood cell velocity increases slightly within the first 30 min of treatment, followed by a decrease and eventually a complete stop	Reinhold and van den Berg-Blok 1983
42-43.5	180		Stoppage of microcirculation dependent on temperature and exposure time	van den Berg-Blok and Reinhold 1984
40.3-43	30-60	Walker 256 tumor	No significant flow change	Gullino 1980, Gullino et al. 1978
42-45	30-60		43° C: TBF remains unchanged 45° C: TBF increases slightly in very small tumors; TBF in large tumors decreased after heating	Song 1984; Song et al. 1980a, c, d
42	60-180	Yoshida sarcoma	TBF reduction dependent on exposure time	Dickson and Calderwood 1980
40-46	20-60		TBF reduction dependent on exposure time and temperature	Kallinowski et al. 1984, Vaupel et al. 1983d
43.5	60	Mammary adenocarcinomas 13762 A	TBF decreases after heating	Rappaport and Song 1983
39/42/44	3-60	SMT-2A	39° C: no change 42° C: TBF decreases after heating for 10 min, then returns to control level 44° C: TBF decreases after heating for 3 min, then returns to initial level	Shrivastav et al. 1983
Continuous increase up to 45° C		R 3230	Vascular stasis at 42.5° C	Dewhirst et al. 1984b
41-45	120	Hepatoma AH 100 B	41° C: TBF increase during heating 43° C: initial TBF increase, gradual decrease thereafter 45° C: continuous TBF decrease	Karino et al. 1984

Table 3. Response of tumor blood flow *(TBF)* to localized hyperthermia in various species (other than mice and rats)

Hyperthermia level (° C)	Exposure time (min)	Tumor (host)	Flow response	Reference
40	> 10	TV tumor (dog)	TBF increase	Voorhees and Babbs 1982
44	40	Mast cell tumor (dog)	No TBF change on first treatment	Milligan and Panjehpour 1983
42.5	10/60	Melanoma A-MEL-3 (hamster)	Fifty percent reduction in TBF after 15 min heating	Endrich et al. 1984
40-52	60	VX2 carcinoma (rabbit)	Rise in red blood cell velocity at temperatures < 41° C, at higher temperatures vascular stasis can occur	Dudar 1982; Dudar and Jain 1984
40-42	30/60	Different tumors (human)	TBF increased in 7/9 tumors during heating and decreased in 8/12 patients after heating	Olch et al. 1983
41-44	40	Different tumors (human)	TBF increased during heating	Waterman et al. 1982
41.5-42[a]	240	Melanoma, colon carcinoma (human)	Melanoma: initial flow increase, TBF decline thereafter Colon carcinoma: flow decrease	Karino et al. 1984

[a] Whole body hyperthermia.

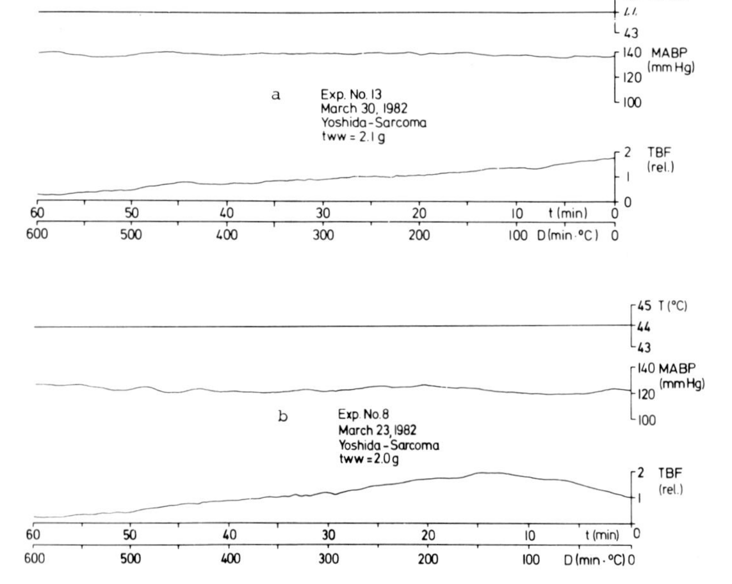

Fig. 1 a, b. Original tracing of the tumor tissue temperature *(T)*, the mean arterial blood pressure *(MABP)*, and the tumor blood flow in relative units *(TBF)* during localized ultrasound hyperthermia. Tumor temperature before heating was approximately 34° C; *tww*, tumor wet wt. **a** Continuous decrease in TBF with increasing thermal dose *(D)*; **b** biphasic flow pattern as a function of thermal dose. (Mueller-Klieser and Vaupel 1984)

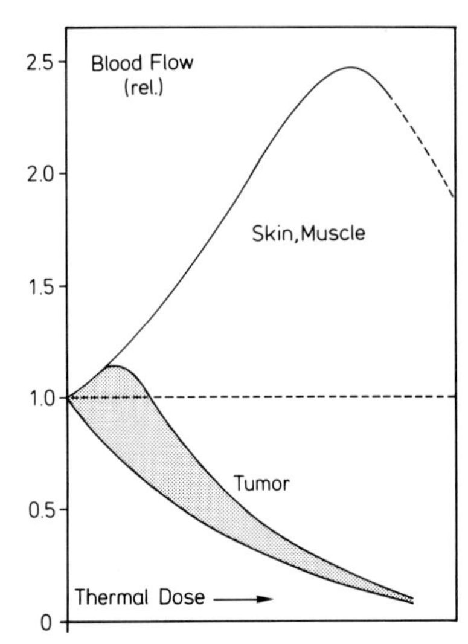

Fig. 2. Relative changes in blood flow of skin, muscle, and experimental tumors at different thermal doses. Heat-induced blood flow changes mostly observed in tumors (continuous flow decline versus transient flow increase followed by a flow reduction) are marked by the two curves enclosing a *shaded area* which includes the flow variations that may occur upon heat treatment. (Vaupel et al. 1986)

The data obtained during the past decade usually exhibit heat-induced changes in tumor blood flow that are considerably different from those obtained in most normal tissues (see Fig. 2). Despite a certain conformity in heat-induced tumor blood flow changes, there are, however, great differences in the exposure times and hyperthermic levels required to achieve this effect. Furthermore, it is obvious that some tumors undergo a collapse of their blood supply already *during* heating, whereas the breakdown of the perfusion in other tumors is evidenced only *after* heating. These differences are of great relevance, especially when sequencing and timing have to be considered during multifraction heat therapy or during combination treatment with radiation or anticancer drugs. This is due to the fact that tumor blood flow is of paramount importance for radiosensitivity, for intratumor pharmacokinetics, as well as for pharmacodynamics.

The observed shut-down of tumor blood flow is frequently irreversible (particularly after high thermal doses). On the other hand, blood flow can be restored 1–3 days after heat treatment depending on the tumor type investigated.

Considering the relevant literature on murine tumors it is striking that there seem to be only two significant exceptions from the generally found blood flow pattern upon tumor hyperthermia: (1) in tissue-isolated tumor preparations connected only to one artery and one vein, no significant flow changes at 40° -43° C for 30–60 min occurred (Gullino 1980; Gullino et al. 1978), or an initial flow increase of approximately 39% was followed by only a minimal decline of about 12% at 44° C for 20–30 min (Vaupel et al. 1982a) and (2) in Walker 256 carcinosarcomas no significant alterations in blood flow can be observed as long as tissue temperatures of 45° C are not established for 1 h. This finding implies that the circulation in Walker tumors cannot be shut down significantly without disturbing the blood perfusion in the normal tissue surrounding the tumor at the same time.

The very few clinical data available today are derived from sporadic observations rather than from systematic studies. They do not allow conclusive statements to be made concerning tumor blood flow changes upon hyperthermia in patients or comparisons between animal and human tumors. Whereas a few human tumors apparently are adequately perfused so that they cannot be treated efficiently (Olch et al. 1983), other investigations are indicative of a blood flow decrease in patient tumors comparable to that obtained in animal neoplasms (see Table 3).

Since sweeping conclusions from these casuistic pilot studies cannot be drawn, we have tried to close the gap between isotransplanted rodent and human tumors in situ by investigating the heat-induced blood flow changes of xenotransplanted human breast carcinomas in T-cell-deficient, nude *rnu/rnu* rats (Vaupel et al. 1986). From these systematic studies there is clear evidence that in both medullary and anaplastic mammary tumors the blood circulation distinctly decreases if mean tissue temperatures of 42° C and 44° C are applied for 20-60 min (see Fig. 3). At comparable heat doses, the observed shut-down of the blood flow, however, is somewhat less pronounced as compared with isotransplanted Yoshida sarcomas. In some of the xenotransplanted tumors investigated a (nonsignificant) increase in blood flow initially occurs before the perfusion rate steadily decreases as the treatment continues.

A complete stoppage of tumor blood flow during hyperthermia with the thermal doses applied was observed in a few cases only. As is the case with isotransplanted tumors, blood flow in xenotransplanted breast carcinomas exhibits cardinal intertumor differences as well as pronounced intratumor heterogeneities which are intensified when the tumor grows to larger sizes. A prospective view of the biological attitudes of an individual tumor thus is distinctly impeded. Besides these tumor-to-tumor variations and intratumor heterogeneities the scattering of the results obtained by the different investigators may also be explained by a nonconformity of other relevant factors. Among those, differing starting temperatures before heating and differing heating-up rates, the use of different tumor types, implantation and growth sites, and different growth stages have to be mentioned. Furthermore, different techniques for blood flow measurement, for temperature monitoring, and for tissue heating may partially be responsible for the variability in the flow pattern obtained. In addition, one has to be aware that blood flow changes in periph-

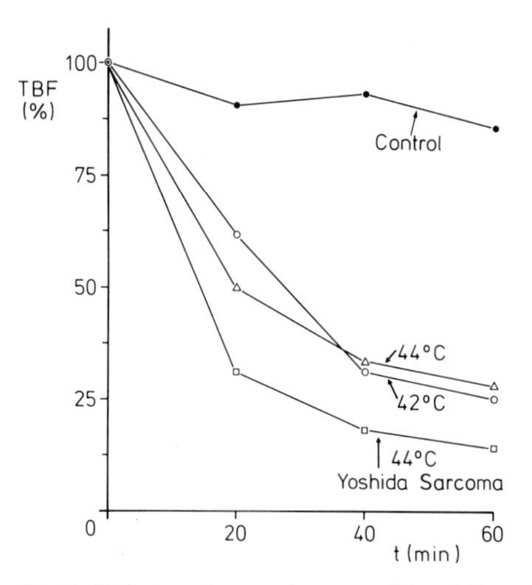

Fig. 3. Relative changes in tumor blood flow *(TBF)* of xenotransplanted human breast carcinomas (*rnu/rnu* rats) as a function of exposure time (20-60 min) and tissue hyperthermia level (42° C, 44° C, ultrasound heating). For comparison, the respective flow decline of isotransplanted Yoshida sarcomas in SD rats is also given. Sham-heated breast tumors served as controls. (Vaupel et al. 1986)

eral regions of a tumor may differ significantly from those observed in central tissue areas at a given thermal dose (Karino et al. 1984).

The nonconsistent initial flow increase observed at low thermal doses is usually more pronounced in smaller tumors than in larger ones. This may indicate that the flow rise at low thermal doses can be at least partially explained by residual normal host tissue vessels incorporated into the tumors. Unlike tumor microvessels arising from neovascularization, these host vessels may still exhibit some physiological (or "regulatory") responsiveness to temperature changes. Therefore, the reaction of blood flow to elevated tissue temperatures may be determined to some extent by the existence (or absence) and by the proportion of "physiologically" responsive vessels still present in the tumor. Since this proportion may vary among different tumors to a large extent, a great variability in the flow response to hyperthermia has to be expected. It is quite unlikely that the transient flow increase during low thermal doses is due to a reaction of the newly formed tumor microvessels since the latter constitute a more or less passive vascular bed devoid of any contractile elements in the vessel wall (Vaupel and Gabbert 1986; Vaupel et al. 1983d). The slight increase in tumor blood flow during heating at low thermal doses (and before the vasculature is damaged) may also be explained by an arousal reaction increasing the cardiac output of the host and thus raising the portion of the cardiac output reaching the tumor tissue (Peck and Gibbs 1983). This interpretation is consistent with our own observations that at the beginning of ultrasound or microwave heating a transient increase in the mean arterial blood pressure can often be observed.

The inhibition of tumor perfusion with increasing thermal dose is consistent with histopathological observations in heated tumors (Eddy 1980; Eddy et al. 1982; Emami et al. 1980, 1981; Endrich et al. 1984; Song 1984; Vaupel et al. 1983b, d). According to these investigations, pronounced vasodilation and marked congestion (stasis) occur. Endothelial swelling with projection of the endothelial cell nuclei into the vessel lumen and a disintegration of the endothelial cell lining can be observed. Ruptures of the microvessel walls lead to a massive plasma leakage or even hemorrhage into the interstitial space. Concomitant findings at higher thermal doses are red blood cell and platelet aggregations, so that tumor microvessels are densely packed with blood cells forming sludges. At sites with incomplete endothelial cell lining thrombus formation is described. During heating, leukocytes tend to stick to the wall of vessels with low shear stresses, i.e., in venules. The mechanisms described, together with high blood viscosity and interstitial edema due to plasma leakage, lead to a dramatic elevation of the flow resistance in tumor microvessels. In melanomas the elevated flow resistance in tumor capillaries and venules also leads to an extreme vasoconstriction in larger preexisting arterioles at the tumor periphery (Endrich et al. 1984) excluding the tumor tissue from the general circulation for at least 3 days.

An increase in red blood cell rigidity has been proposed to be one of the factors responsible for vascular plugging and breakdown of tumor blood flow upon hyperthermia (Vaupel et al. 1980; von Ardenne and Reitnauer 1980). Temperatures exceeding 42° C combined with an intensified acidosis, the latter being frequently found under these hyperthermic conditions (see Chap. 4), yield a tremendous increase in the number of crenated red blood cells. Under these conditions, approximately 80% of the red blood cells (RBCs) have changed their physiological cell shape, and shrinkage with wrinkling of the cell membrane can be observed (Vaupel et al. 1980). These observations are in good agreement with earlier findings describing a loss of RBC flexibility at temperatures above 40° C and at lowered pH values (Bieri and Wallach 1975; Verma and Wallach 1976). Recent observations also confirm a decrease in RBC deformability at 43° C and pH = 6.5 as compared with data obtained at 37° C and pH = 7.0 (Barnikol and Burkhard 1985). According to Schmid-Schönbein et al. (1984), at least subpopulations of RBCs may be stiffened under these conditions and may lead to vascular stasis at low perfusion pressures, i. e., at low arteriovenous pressure gradients. However, since the transit or passage times of erythrocytes in treated tumors are totally unknown, the significance of pH- and hyperthermia-induced RBC stiffening for the breakdown of tumor blood flow during treatment cannot be assessed at present.

The hyperthermia-induced shutdown of tumor blood flow may also be enhanced by the (nonenzymatic) intravascular formation of fibrinogenin gel at low tissue pH-values which can take place despite heparinization (Copley 1980; Copley and King 1984). Another relevant mechanism leading to a drop in tumor blood flow during heat treatment may be the pronounced reactive hyperemia of the adjacent normal tissues surrounding a tumor. The vasodilation in the tissues in the neighborhood of a tumor can lead to a steal phenomenon and thus to diversion of blood from the tumor, which can be so pronounced that a great portion of tumor hypoperfusion during heating may be explained by this mechanism. The role of a steal phenomenon in this context is substantiated by the fact that during hyperthermic in situ perfusion of tissue-isolated tumors without significant heating of the surrounding tissues there is no substantial reduction of blood flow to be observed (Vaupel et al. 1980).

The mechanisms which lead to vascular stasis in tumors during hyperthermia are most probably quite complex (see Fig. 4). As mentioned earlier, the results of the studies available are indicative of definite changes in the tumor microcirculation that are reversible at low thermal doses and irreversible at higher temperatures when appropriate exposure times are applied. In the latter case coagulation necrosis is observed following treatment at therapeutic temperatures. The histopathological changes in the tumor microvasculature upon hyperthermia are identical in human and animal tumors so that the mechanisms which lead to a breakdown of blood flow appear to be quite similar (for a review see Overgaard 1983).

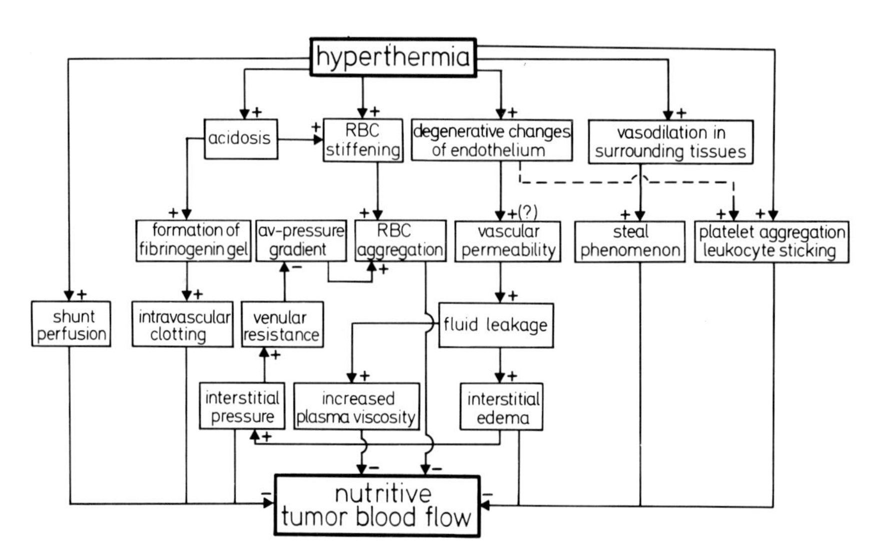

Fig. 4. Possible mechanisms involved in hyperthermia-induced breakdown of nutritive blood flow in malignant tumors

As a general finding, interstitial edema develops in the tumors during heat treatment as a consequence of a pronounced plasma leakage from the tumor microvessels into the interstitial space and due to the absence of a functioning lymphatic drainage system within the tumor tissue. A rise in the fluid pressure within the interstitial compartment obligatorily follows, leading to a further increase in the vascular resistance, especially in the venules and thus to a further impairment of tumor blood flow (development of a vicious circle!). Immediately after 44° C hyperthermia for 60 min, in subcutaneous Yoshida sarcomas the gain in tumor volume due to interstitial edema is 15% on average (tumor wt. before heating: 3–4 g). Twenty-four hours after heating there is still a mean increment in tumor volume of approximately 2%–3%. Hyperthermia-induced interstitial edemas are even more pronounced in smaller tumors and are persistent for a longer period.

For further quantification of the effect of hyperthermia on tumor perfusion, the time to achieve a 50% reduction of blood flow is plotted as a function of tissue temperature level in Fig. 5. Our results and those from the relevant literature as indicated by different symbols (Dudar 1982; Eddy et al. 1982; Emami et al. 1980; Endrich et al. 1984; Peck and Gibbs 1983; Reinhold and van den Berg-Blok 1983; Sutton 1980; van den Berg-Blok and Reinhold 1984; Vaupel and Mueller-Klieser 1983a; Vaupel et al. 1983a, d 1986, imply that on average heat exposure times for inducing a biological isoeffect can be halved for every 2.2° C temperature increment in the range between 40° and 42.5°. For temperatures above 42.5° C the respective heat exposure times can be halved for every 1.7° C of temperature increment. This is in contrast to the generally accepted finding that a 1° C increase in temperature requires a twofold decrease in time for the same effect above 42.5° C and a three- to fourfold decrease in time for an isoeffect below 42.5° C; for a review see Sapareto and Dewey (1984). In support of the generally accepted finding there is only one tumor model (rhabdomyosarcoma grown as a "sandwich" preparation) which exhibits the latter phenomenon when time-temperature relationships to achieve a 50% stoppage of microcirculation are investigated (solid line in Fig. 5; van den Berg-Blok and Reinhold 1984). When these results are expressed as a log-linear relationship, a slope value of 0.46/°C is obtained, which does not differ significantly from the "*t*/2 for every degree Celsius" rule.

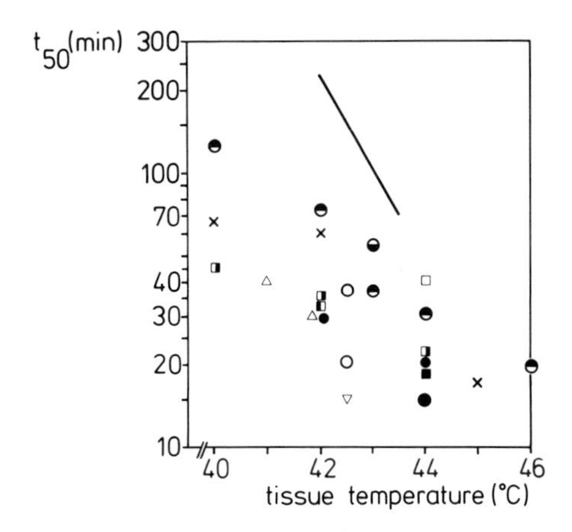

Fig. 5. Exposure time (t_{50}) required to achieve a 50% flow inhibition in solid tumors as a function of tissue temperature during localized hyperthermia. Relevant literature data are indicated by *different symbols;* the *line* represents the 50% stoppage time of the rhabdomyosarcoma BA1112. (van den Berg-Blok and Reinhold 1984)

The relationship between tissue temperature and exposure time can be mathematically described by:

$$t_1/t_2 = R^{(T_2 - T_1)}$$

where t=treatment time, T=tissue temperature, and R represents a factor that can be calculated as a function of the activation energy (calories/mole) and the absolute temperature (° K) from the Arrhenius plot. This formula has been derived from many studies both in vivo and in vitro aimed at relating time and temperature ("heat dose") to produce a given level of damage with a single heat treatment (Field and Morris 1984). The values reported range from 0.4–0.8 above 42.5° C, with 0.5 being the most frequent value. Below 42.5° C in general the R value is approximately a factor of 2 smaller than that above 42.5° C ($R \approx 0.25$). From the data of Fig. 5 mean R values of 0.77 can be calculated for temperatures above 42.5° C. Below 42.5° C mean R values of 0.69 are obtained (the R value for the straight line is calculated to be 0.46).

The time to achieve a total flow stoppage (t_{st}) in heated tumors is plotted as a function of tissue temperature level in Fig. 6. Here again, there is evidence that an exponential relationship exists between hyperthermia level and exposure time in the temperature range depicted (the different symbols indicate relevant literature data: Dudar and Jain 1984; Eddy 1980; Emami et al. 1980; Reinhold and van den Berg-Blok 1983; Scheid 1961).

Modifications of heat-induced flow changes can be obtained by combining hyperthermia with vascoactive drugs, irradiation, anticancer compounds, hyperglycemia and inhibitors of glucose metabolism. *Vasodilators* (such as hydralazine, Voorhees and Babbs 1982) may decrease the blood flow in tumors whereas the perfusion rate is increased in the adjacent normal tissue (steal phenomenon!). This result demonstrates that adjuvant treatment with vasodilators may be a promising procedure to increase the temperature difference between tumors and surrounding normal tissues during local heat therapy. *Irradiation* also altered the response of tumor microcirculation to heat. The most efficient combination was 2000 rad followed after 1 h by 42° C for 30 min. Longer time intervals between irradiation and hyperthermia as well as radiation given after heat treatment were less effective (Eddy and Chmielewski 1982; Eddy et al. 1982). An enhanced shut-down of tumor microcirculation was also obtained during *sequential heat treatment* (similar to human hyperthermia protocols) when compared with effects after a single hyperthermia session (Eddy and Chmielewski 1982; Robert et al. 1982). A greater depression of tumor blood flow also occurs by a combination of hyperthermia and *hyperglycemia* (von Ar-

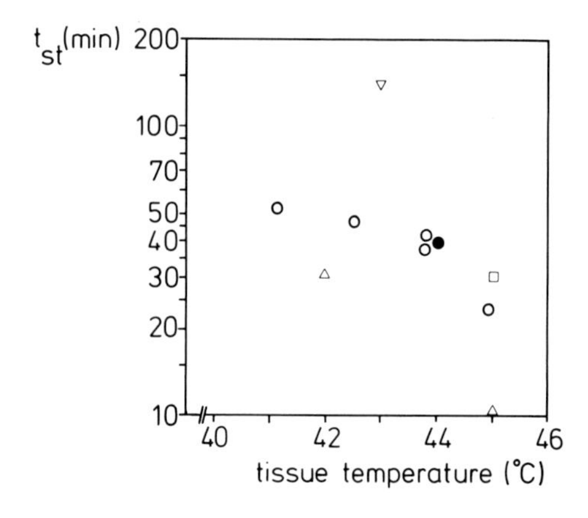

Fig. 6. Exposure time (t_{st}) required to achieve a total flow stoppage in malignant tumors as a function of tissue temperature upon localized hyperthermia. Data points are taken from relevant literature data

denne and Reitnauer 1982; Dickson and Calderwood 1980; Reinhold and van den Berg-Blok 1980).

The mechanisms for this amplification of the heat-induced flow decline may be an increase in the blood viscosity (Dickson and Calderwood 1980) and/or an enhanced rigidification of red blood cells due to an intensified tissue acidification after stimulation of glycolysis (von Ardenne and Reitnauer 1982). The heat-induced disruption of the microcirculation in rat tumors could also be fortified by treating the host animals with misonidazole and 5-thio-D-glucose, both *inhibitors of the glucose utilization.*

The combined effects of heat and *chemotherapy* on tumor blood flow remain unclear since relevant data are scarce. According to one report studying the impact of heat/Adriamycin combinations the latter drug did not influence tumor blood flow to a greater extent than heat given alone (Eddy and Chmielewski 1982).

In conclusion, the data presented provide evidence for a paramount role of tumor blood flow in hyperthermic treatment. Most tumors show a tremendous flow reduction if appropriate tissue temperature levels and heat exposure times are chosen. Thus, differences in blood perfusion rates between tumors and normal tissues already existing before heating can be considerably magnified upon heat application. Tumor perfusion is less able to dissipate heat and, therefore, tumors acquire higher temperatures than the surrounding normal tissues at a certain energy input. The inhibition of blood flow at high thermal doses may lead to pathophysiological changes in the tumors that:

1. Can increase the cytocidal effect of hyperthermia
2. Can modulate pharmacodynamics of anticancer drugs (via modifications of the cellular microenvironment)
3. Can further intensify the inefficient drug delivery in many tumors (changes in intratumor pharmacokinetics)
4. Can make a tumor more hypoxic, thus reducing its overall radiosensitivity
5. Require rational sequencing and timing (i.e., treatment schedule) when heat is used in combination with chemotherapeutic agents and irradiation

The inconsistently found transient flow increase in some tumors upon heating at low thermal doses can enhance the efficiency of irradiation and the delivery of antiproliferative drugs and through modifications of the tumor micromilieu may modulate pharmacodynamics.

Temperature Distributions in Tumors

"Good thermal dosimetry is essential to good hyperthermia research" (Robinson et al. 1978) and, accordingly, intratumor temperature monitoring is an obligatory prerequisite for heat treatment. Tumor blood flow and tissue temperature are to some extent interrelated since (1) blood flow can act as a mechanism of heat transfer and (2) heat can modify the tissue perfusion rate (see pp. 73 ff.). As a rule, with rising flow rate, the percentage of heat transfer by convection increases up to a plateau, the level of which is dependent on the given basic conditions.

Temperature Distributions in Tumors During Normothermia

Using tissue-isolated preparations it has been shown by Gullino et al. (1982) that conductive heat transfer in tumors prevails over convective heat dissipation. In contrast to this, measurements of temperature distributions in peripheral s. c. tumors, i. e., in a tumor model which is more commonly used in tumor biology, and considering the impact of flow-related tissue temperature variations, there is clear evidence that convection is the prevailing heat transfer mechanism (Dave et al. 1984). In this model, changes in tumor blood flow were always followed by immediate alterations in tumor temperature: halving or doubling the flow rate produced concomitant tissue temperature changes of approximately 1° C. The results of Dave et al. (1984) are further indicative of heterogeneous temperature distributions within s. c. tumors (with variations between room and rectal temperature) paralleled or caused by pronounced inhomogeneities of the tumor microcirculation. Temperature gradients of up to 2° C within a short distance apart were measured in central areas of the tumors, the differences between tumor core and peripheral tissue areas being even more pronounced. Contrary to previous reports by Gullino et al. (1978, 1982) stating the tumor temperature to be higher than that of the subcutis, the results of Dave et al. (1984) clearly show that the tumor temperature can be distinctly lower than the subcutaneous temperature (see Fig. 7). This discrepancy regarding the role of convection and conduction is likely to be due to the fact that different tumor models were used. Therefore, it has to be stressed that statements regarding the temperature distribution in normal and tumor tissues should be made only in reference to the model used. Furthermore, it is also evident that conclusions concerning the proportion of individual heat transfer mechanisms can be drawn only on considering the actual tumor perfusion rate. The latter in turn is dependent on various predefined morphological and functional attributes such as the tumor cell line used, the growth site, the tumor size, the perfusion pressure, and the use of different anesthetics.

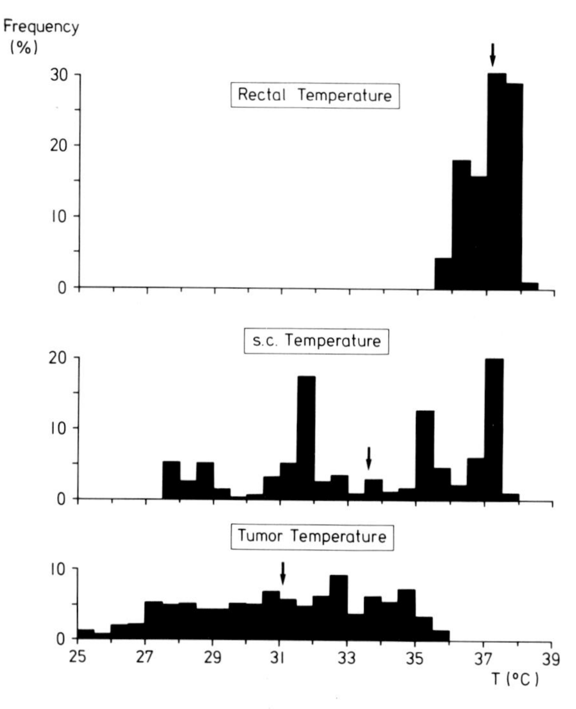

Fig. 7. Frequency distributions of rectal *(upper panel)*, of subcutaneous *(central panel)*, and of s. c. tumor *(lower panel)* temperatures. *Arrows* indicate the respective mean values (Dave et al. 1984)

Temperature Distributions in Neoplastic Tissues During Hyperthermia

Hyperthermia has little hope of progressing as a clinical modality without accurate assessment of the temperature distributions in the tumors treated (Fessenden et al. 1984). At the present time only direct, invasive temperature-measuring devices are possible, posing severe limitations in clinical use. Uncertainty in the temperature probe placement, in the intensity variation in the near field of ultrasound transducers, in thermal smearing errors due to a good thermal conductivity of thermocouple probes, and in the possibility of reradification (especially in electromagnetic fields) still impedes an accurate correlation of the treatment response with adequacy of heating.

Table 4. Temperature nonuniformity in tumors during waterbath immersion

Bath temperature	Tumor (species)	Tumor temperature (° C)	Reference
42.8° C	Sarcoma F (mouse)	Foot: 42.4-42.8 Tail: 41.3-42.8 Leg: 41.3-42.8 Chest: 41.9-42.8	Hill and Denekamp 1982
42.8° C	Different tumors (mouse)	41.1-42.8	Hill et al. 1980
43.1° C	Mammary carcinoma (mouse)	42.2-43.0	O'Hara et al. 1985b
42.9° C	Mammary carcinoma (mouse)	41.6-42.9	Robinson et al. 1978
43.0° C	Mammary carcinoma (mouse)	Flank: 42.2 Foot: 42.9 Leg: 42.5	Gibbs et al. 1981
42.5° C	Fibrosarcoma (mouse)	42.3	Stewart and Denekamp 1978
42.8° C	Different tumors (mouse)	42.5	Hill and Denekamp 1979
43.5° C	Mammary carcinoma (mouse)	43.3	Overgaard 1980
42.7° C	Yoshida sarcoma (rat)	ca. 42.2	Dickson and Suzangar 1974

When comparing different heating modalities usually used in hyperthermia research and in patients, temperature mapping in tumors reveals that a homogeneous temperature distribution cannot be obtained in any of these modalities (see Tables 4 and 5). The most uniform temperature distributions have been reported with waterbath immersion although differences between liquid bolus and tumor temperature as great as 1.7° C have been measured (Hill and Denekamp 1982; Hill et al. 1980). Electromagnetic heating using radiofrequencies and microwaves as well as ultrasound, which probably show the greatest potential for hyperthermia treatment in patients, give a somewhat larger variation in temperature within the heated volume (see Table 5). Using sophisticated devices with improved thermometric techniques, temperature variations within a treated tumor should not exceed 2° C when heating is performed by electromagnetic fields. The same holds true for ultrasound heating when the advantages offered by this technique are utilized. When temperature mappings were performed in experimental rodent tumors during ultrasound

Table 5. Temperature nonuniformity in tumors during localized hyperthermia

Heating method	Tumor (species)	Intratumor temperature variations (ΔT)	Reference
Diathermic lamp	Walker 256 tumor (rat)	2.2° C	Jain 1980
	Walker 256 tumor (rat)	Up to 4.5° C	Gullino et al. 1982
Electromagnetic field heating	R1 H-tumor (rat)	2° C	Zywietz et al. 1986
	Mammary carcinoma (mouse)	Up to 2° C	Hetzel et al. 1984
	Different tumors (human)	ca. 1° C	Kim and Hahn 1979
	Different tumors (human)	Up to 7° C	Bagshaw et al. 1984
	Different tumors (human)	Up to 4° C	Fessenden et al. 1984
	Different tumors (human)	Up to 4° C	Nakajima et al. 1984
Ultrasound heating	Different tumors (human)	Up to 3° C	Bagshaw et al. 1984
	Different tumors (human)	Up to 6° C	Fessenden et al. 1984
	Spontaneous tumors (cats + dogs)	Up to 2° C	Marmor et al. 1978
	Mammary carcinosarcoma (mouse)	ca. 1° C	Marmor et al. 1979
	Yoshida sarcoma (rat)	2° C	Blendstrup et al. 1985

heating there is clear evidence that the temperature distribution is somewhat "homogenized" (Blendstrup et al. 1985). Whereas in unheated s.c. tumors temperature variations of up to 5° C can be detected, in heated tumors these variations are only up to 2° C (see Fig. 8). In contrast to Gullino et al. (1982), an exaggeration of the nonuniformity in temperature was never observed during ultrasound hyperthermia. However, we did not succeed in eliminating these temperature variations completely under hyperthermic conditions. Furthermore, the temperature data are indicative that a single measurement of tumor temperature is not representative of the temperature profile in the tumor. This emphasizes the importance of knowing the temperature distribution within a tumor instead of reading a "mean" temperature. Using two microthermocouples which were drawn stepwise through the tumors by means of a mechanical microdrive (the tracks of the microthermocouples are presented in the schematic illustration of the setup in Fig. 8), it could be shown that during ultrasound-induced hyperthermia tumors were often somewhat cooler in the periphery than in the center. Furthermore, higher temperatures were found in tumor areas close to the ultrasonic transducer than in more distant parts of a tumor.

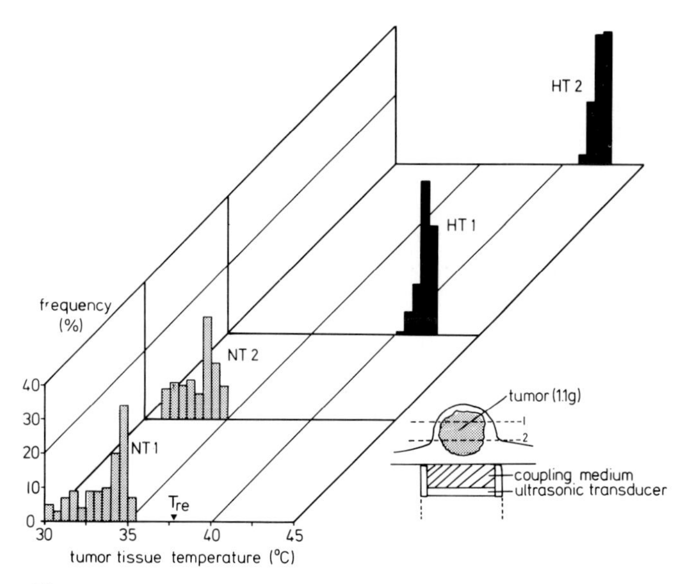

Fig. 8. Frequency distributions of intratumor temperatures during control conditions (*lower histograms,* NT1/NT2) and during 42° C hyperthermia *(HT1/HT2).* The tracks of the microthermocouples No. 1 and No. 2 are presented in the schematic illustration of the setup for ultrasound heating. T_{re}, rectal temperature of the host animal. (Blendstrup et al. 1985)

Hyperthermia-Induced Changes in Nutrient and Oxygen Consumption Rates in Tumors

Nutrient and Oxygen Availabilities to Tumors During Hyperthermia

The paramount parameter governing nutrient and oxygen availabilities to solid tumors is the nutritive blood flow to the cancer cells, provided that the concentration of the respective substances is kept constant in the arterial blood. In the case of the substrates essential for tumor growth, a constant arterial concentration has to be taken for granted. For oxygen, the arterial concentration is dependent on the temperature of the blood, since both the binding of O_2 to hemoglobin and the physical solubility of O_2 in the blood are temperature-dependent processes. Nevertheless, the O_2 delivery to the tumor cells is mostly determined by the efficiency of tumor blood flow as is the case with the relevant substrates. The improved O_2 release from the blood passing through a tumor to the cells (due to a right shift of the O_2 dissociation curve as a consequence of localized tissue heating and of the Bohr effect due to an intensified tissue acidosis, see pp. 95 ff.) may also be of minor importance for the actual O_2 uptake by the cancer cells. As a consequence, all changes in nutritive blood flow during hyperthermia are obligatorily paralleled by similar changes in the availabilities to the cells. This is unequivocally shown in Fig. 9. Here, tumor blood flow (TBF) and O_2 availability are depicted as a function of the tissue temperature level using tissue-isolated tumor preparations. The same pattern holds for all substrates if the respective concentrations in the arterial blood do not change significantly during hyperthermia. At higher thermal doses a breakdown of the nutrient supply to the cancer cells has to be expected whereas at lower thermal doses in some tumors an improved nutrient supply may be expected if a transient flow increase occurs.

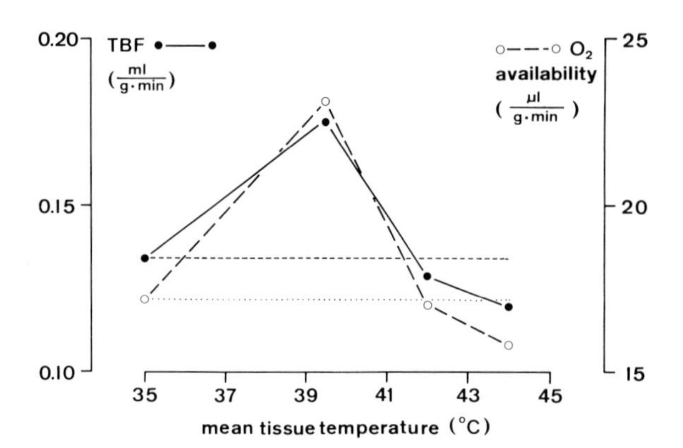

Fig. 9. Blood flow *(TBF)* and oxygen availabilities in tissue-isolated tumor preparations as a function of tumor temperature

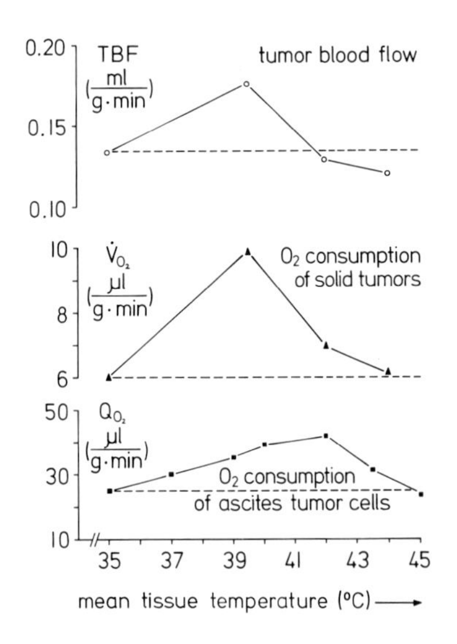

Fig. 10. Blood flow *(TBF)*, in vivo O_2 consumption rate ($\dot{V}_{O_2}$) of tissue-isolated tumor preparations, and in vitro O_2 consumption rate (Q_{O_2}) of ascites tumor cells (DS carcinosarcoma) as a function of temperature

Oxygen and Glucose Uptake Rates by Tumors During Hyperthermia

In general the actual O_2 and nutrient uptake by a tissue is determined by the respective availabilities, the diffusional flux, and the metabolic requirements of the cells. This holds true for both normal and neoplastic tissues during normothermic and hyperthermic conditions. Whereas under in vitro conditions without any supply limitations the capacity of the cells to consume oxygen is the limiting factor, the O_2 availability is the paramount parameter for tumors in vivo. This is explicitly shown in Fig. 10: At moderate hyperthermia ($T=39.5°$ C, applied for approximately 30 min) a characteristic maximum is found with a mean oxygen consumption rate ($\dot{V}_{O_2}$) of 10 µl/g per minute. At control level ($T=35°$ C) the O_2 consumption rate is 6 µl/g per minute. Similar data are obtained for 42° C and 44° C (Mueller-Klieser et al. 1984). When comparing the temperature-dependent pattern of the O_2 consumption rate with that of tumor blood flow, there is clear indication that the O_2 availability (i.e., TBF × arterial O_2 concentration) governs the actual O_2 consumption rate of tumors *in vivo*. This means that $\dot{V}_{O_2}$ changes during hyperthermia only occur when tumor heating is accompanied by changes in nutritive blood flow. If perfusion changes are

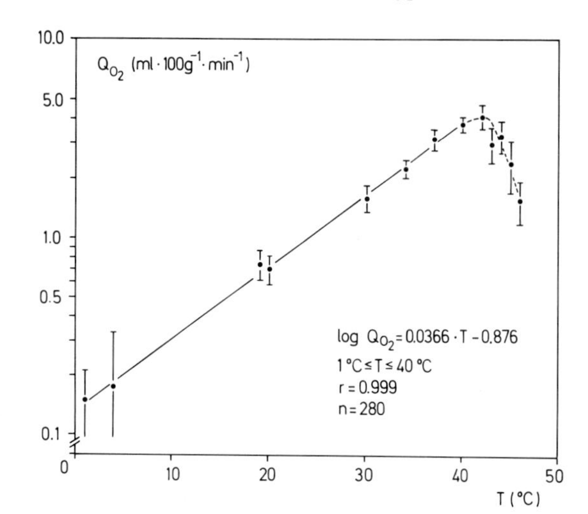

Fig. 11. Temperature dependency of the O_2 consumption rate (Q_{O_2}) of isolated ascites cells suspended in native ascites fluid (DS carcinosarcoma, Mueller-Klieser and Vaupel 1983a)

not apparent during hyperthermia, significant alterations in the O_2 (or nutrient) uptake rates cannot be expected (Gullino 1980).

In contrast to this, the O_2 consumption rate of ascites cells of DS carcinosarcoma (suspended in native ascitic fluid) exhibits a distinctly different dependence on temperature. Upon heating, isolated tumor cells increase their O_2 consumption (Q_{O_2}) to maximum values at 42° C (see Fig. 10). If the O_2 consumption rate is limited only by the cellular capacity to consume O_2 and not by the O_2 delivery, it is increased to 42 μl/g per minute at 42° C in comparison to 25 μl/g per minute at 35° C. Heating up to 45° C decreases the O_2 consumption to values of 24 μl/g per minute.

The temperature dependency of the O_2 consumption rate of isolated ascites cells (DS carcinosarcoma) is depicted in more detail in Fig. 11. Within a wide temperature range there is an exponential increase of the cellular respiration rate with a maximum at 42° C and a decline at temperatures exceeding 42° C (Mueller-Klieser and Vaupel 1983a).

Using different tumor cell lines in vitro and taking an optimal O_2 supply into account, it is a common finding that higher thermal doses can inhibit the respiration rate of tumor cells (see Table 6). The temperature level and the exposure time to achieve this depression may be cell type specific (and certainly somewhat dependent on the kind of suspension medium used). Normal tissues seem to be less heat sensitive (Cavaliere et al. 1969; Westermark 1927).

Predictions concerning the actual tumor tissue oxygenation on the basis of both total tumor blood flow and the capacity of cancer cells to consume oxygen in vitro at different hyperthermia levels are not possible since, inter alia, changes in the O_2 hemoglobin binding, in the O_2 solubility, and in the amount of blood passing the tumor through arteriovenous shunts have also to be considered. Among these mechanisms, alterations in the shunt perfusion seem to play the most important role (Dewhirst et al. 1984a, b).

Impact of Hyperthermia on the Cellular Microenvironment in Tumors

As already mentioned, in most tumors an inadequate blood flow not only yields a quasi-selective heating of the tumor mass in comparison with the surrounding normal tissues but also generates a certain cellular microenvironment which can sensitize cancer cells to

Table 6. Changes in cellular oxygen consumption rates during hyperthermic conditions in vitro

Cell line (host)	Technique	Results	Reference
Jensen-sarcoma (rat)	Warburg manometry	44.5° C: respiration rate drops after 60–90 min 45.0° C: respiration rate drops after 40–60 min 45.5° C: respiration rate drops after 45–60 min 47.0° C: respiration rate drops after 25 min 49.0° C: respiration rate drops after 6 min (liver and spleen tissues are less heat sensitive)	Westermark 1927
Flexner-Jobling carcinoma (rat)		45.0° C: respiration rate decreases after 60 min 46.5° C: respiration rate decreases after 25 min 46.8° C: respiration rate decreases after 15 min 47.0° C: respiration rate decreases after 20 min	Westermark 1927
VX2 carcinoma (rabbit)		Maximum respiration at 40° C Inhibition of respiration at higher temperatures	Dickson and Muckle 1972
Yoshida sarcoma (rat)		Maximum respiration at 40° C Inhibition of respiration at higher temperatures	Dickson and Calderwood 1980, Dickson and Suzanger 1974
Breast carcinoma (rat)		Maximum respiration at 40° C Inhibition of respiration at higher temperatures	Dickson and Shah 1972
Novikoff hepatoma (rat)		Lower O_2 consumption at 43° C than at 38° C	Mondovi et al. 1969
Novikoff hepatoma (rat)		Lower O_2 consumption at 42° C than at 38° C ($t > 1$ h)	Cavaliere et al. 1967
Hepatoma 5123 (rat)		No significant temperature effect (similar results with normal liver cells)	Cavaliere et al. 1967
Ehrlich ascites carcinoma (mouse)		Lower O_2 consumption at 44° C than at 38° C ($t > 3$ h)	Cavaliere et al. 1967
Ehrlich ascites carcinoma (mouse)		Irreversible stoppage of respiration at temperatures > 42.5° C ($t > 33$ min)	von Ardenne and Krüger 1966
Ehrlich ascites carcinoma (mouse)		Depression of O_2 consumption at 44° C (for 30 min)	Rapoport et al. 1971
V79 spheroids	Clark electrode	Depression of respiration rate at temperatures ≥ 41° C (after an initial enhancement)	Durand 1978
DS carcino-sarcoma ascites cells	Colorimetric method	Maximum O_2 consumption at 42° C, inhibition at higher temperatures	Mueller-Klieser et al. 1983
EMT6/Ro spheroids		Maximum O_2 consumption at 39° C, inhibition at higher temperatures	Mueller-Klieser et al. 1983

heat treatment. In general, this microenvironment (=intercellular or interstitial milieu) is characterized by inhomogeneously distributed hypoxic and acidic tissue areas as well as nutrient and energy deprivation which already occur in early growth stages. Hyperthermia can greatly modulate this microenvironment and, therefore, the heat sensitivity of the cancer cells may be changed upon treatment.

Tumor Tissue Oxygenation upon Hyperthermia

As a rule, tissue oxygenation is the resultant of the oxygen availability to the parenchymal cells and the actual respiration rate of the cells. In most tumors oxygenation is quite poor, exhibiting pronounced intertumor and intratumor variations (Vaupel 1977, 1986; Vaupel et al. 1981; Wendling et al. 1984). As a measure of tissue oxygenation, the oxygen partial pressure (pO_2) distribution in microareas of a tumor and/or the hemoglobin saturation (HbO_2) of individual red blood cells within tumor microvessels (Ø < 12 μm) have been utilized in the past (see Table 7).

Table 7. Changes in tumor tissue oxygenation upon heating

Cell line (host)	Technique	Hyperthermia level (° C)	Exposure time (min)	Results	Reference
Mammary carcinoma (mouse)	pO_2 measurements with O_2-sensitive electrodes	Continuous temperature increase up to 45° C		Improvement up to 40° C, pO_2 drop at higher temperatures	Bicher and Vaupel 1980; Bicher et al. 1980; Vaupel et al. 1982a
Mammary carcinoma (mouse)		40		Slight increase in tumor oxygenation during treatment	O'Hara et al. 1985a
S 180-tumor (mouse)		42/45	30	42° C: pO_2 temporarily decreased and returned to the initial level 18 h later 45° C: pO_2 drop	Tanaka et al. 1984
Melanoma A-Mel-3 (hamster)		40/42.5	15	42.5° C: severe hypoxia 40° C: improvement	Endrich et al. 1984
Epithelioma (human)		(unknown)	10(+5)	pO_2 decreases	Cater and Silver 1960
Different tumors (human)		42–45	60	Improvement at low thermal doses, pO_2 decreases at higher temperatures	Bicher and Mitagvaria 1984
DS carcinosarcoma (rat)	Measurement of HbO_2 saturation in red blood cells within tumor microvessels	40/43/45	30	40° C: improvement 43–45° C: deterioration in tissue oxygenation	Vaupel et al. 1982b, c
DS carcinosarcoma (rat)		40/43/45	30/60	Improvement at thermal doses below 250° C·min; at higher thermal doses tissue oxygenation worsens	Mueller-Klieser et al. 1984; Vaupel et al. 1983d

If the oxygen partial pressure distribution is measured for characterization of the tumor tissue oxygenation during elevated temperatures, there is clear evidence from the relevant literature that low thermal doses can improve tissue oxygenation provided that tumor microcirculation exhibits an improvement under these conditions (to simplify matters, in the following heat dose is expressed in terms of time spent at the elevated temperature × the change in temperature, i.e., HD [° C · min] = $\Delta T \cdot t$; it appears that heat dose expressed in this way is sufficient to describe many biological or pathophysiological aspects during heat treatment (Hill and Denekamp 1982; Vaupel et al. 1983d).

At higher thermal doses tissue oxygenation deteriorates, reflecting the breakdown of tumor microcirculation under these conditions (see Fig. 12; Bicher and Mitagvaria 1984; Bicher and Vaupel 1980; Bicher et al. 1980; O'Hara et al. 1985a; Tanaka et al. 1984; Vaupel et al. 1982a). Metabolic effects, i.e., changes in the capacity of the tumor cells to consume O_2, seem to have a subordinate influence on the oxygenation status. Since blood flow in normal tissues is impaired only at distinctly higher thermal doses, oxygenation is decreased solely at very high thermal doses. Despite this clear interrelationship between oxygenation status and heat dose, apparent from a series of animal and patient data, up to now unequivocal correlations between the microvascular changes and tumor treatment responses have not been developed.

For characterization of the oxygenation status of solid tumors, the oxyhemoglobin saturation (HbO_2) of individual red blood cells within tumor microvessels has been measured using a cryophotometric micromethod. This method has been proved in previous investigations to be valid for quantifying the oxygenation both in rodent tumors (Manz et al. 1983; Mueller-Klieser and Vaupel 1983b; Mueller-Klieser et al. 1980; Vaupel et al. 1979) and in human tumors (Mueller-Klieser et al. 1981; Vaupel 1986; Vaupel et al. 1983c; Wendling et al. 1984). Using this method only red blood cells within microvessels with di-

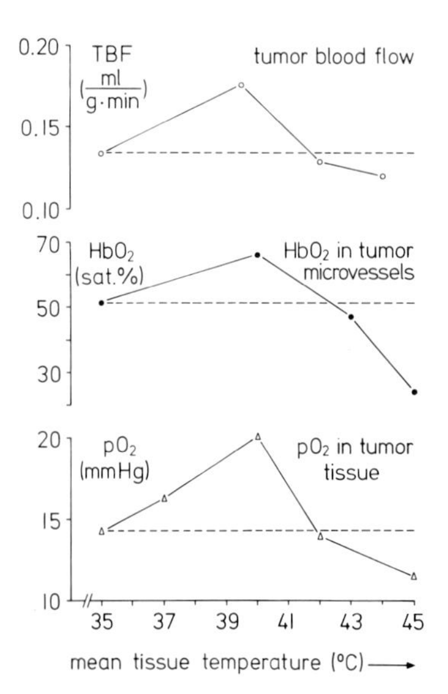

Fig. 12. Blood flow *(TBF)* of tissue-isolated tumor preparations and tumor tissue oxygenation as a function of tumor temperature. For characterization of the tissue oxygenation the mean oxyhemoglobin saturation *(HbO_2)* of individual red blood cells in tumor microvessels ($\varnothing < 12\,\mu m$) and the mean intratumor O_2 partial pressure *(pO_2)* is given

ameters less than 12 μm have been investigated since this vessels size ensures that nutritive blood vessels have been considered only (Vogel 1965).

The cumulative frequency distribution curves of the measured oxyhemoglobin saturation of individual red blood cells in tumors as a measure of the efficiency of O_2 supply at three different temperature levels (40° C, 43° C, and 45° C) and two different exposure times are plotted in Figs. 13–15. In untreated control tumors (mean tissue temperature approximately 35° C) the mean HbO_2 saturation is 51 sat. %. Upon heating for *30 min*, the oxygenation of the tumor tissue significantly improved at 40° C as compared with control conditions. In contradiction to this, after 43° C hyperthermia the tumor oxygenation was significantly lower and reached a mean saturation value of 47 sat. % (see Fig. 12). A further temperature rise to 45° C caused the oxygenation to drop drastically due to a severe restriction of nutritive blood flow. The mean oxyhemoglobin saturation value under these conditions was 24 sat. % (Mueller-Klieser et al. 1984; Vaupel et al. 1982b, c).

For investigating the influence of the exposure time on tumor oxygenation during heat application, HbO_2 saturations in DS carcinosarcoma were also measured after microwave-induced hyperthermia maintained for *60 min*. These experiments led to results that differ considerably from those obtained after a heating period of 30 min. A marked drop of the mean HbO_2 values can be registered even at 40° C (see Fig. 13). Elevating the tumor temperature to 43° C and 45° C is followed by a further deterioration of the O_2 supply to the cancer cells with mean HbO_2 saturations of 17.5 and 19 sat. %, respectively (see Figs. 14 and 15 (Mueller-Klieser et al. 1984; Vaupel et al. 1983d).

Considering all the data on HbO_2 distributions in tumors and taking into account different hyperthermia levels and exposure times, there is clear indication that low thermal doses (<250° C · min) on average can lead to an improvement in the tumor tissue oxygenation as compared with physiological control conditions. A maximum oxygenation is obtained at thermal doses of 150–200° C · min. The application of higher thermal doses leads

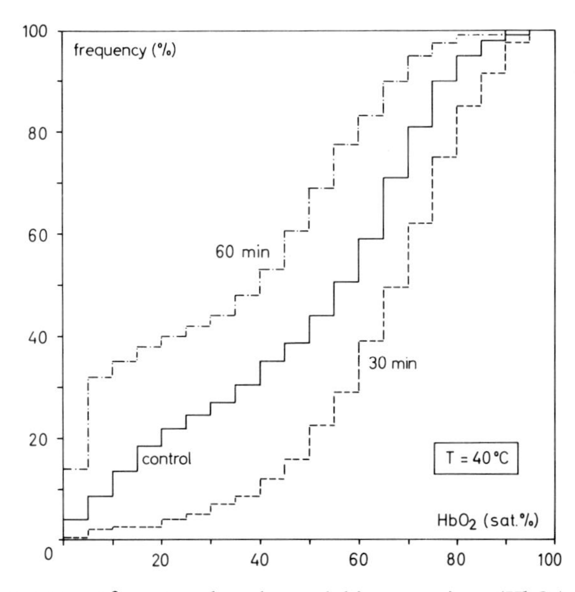

Fig. 13. Cumulative frequency distribution curves of measured oxyhemoglobin saturations *(HbO_2)* of individual red blood cells within tumor microvessels during control conditions (approx. 35° C) and upon localized ultrasound heating at 40° C for 30 and 60 min

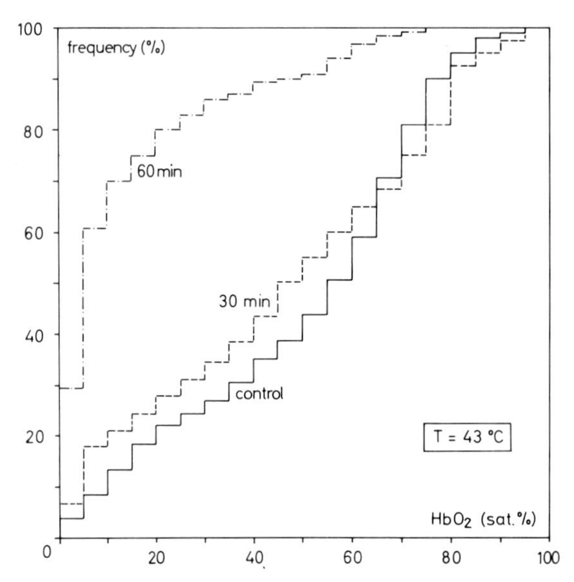

Fig. 14. Cumulative frequency distribution curves of measured oxyhemoglobin saturation values *(HbO_2)* of individual red blood cells within tumor microvessels during control conditions (ca. 35° C) and upon localized ultrasound heating at 43° C for 30 and 60 min

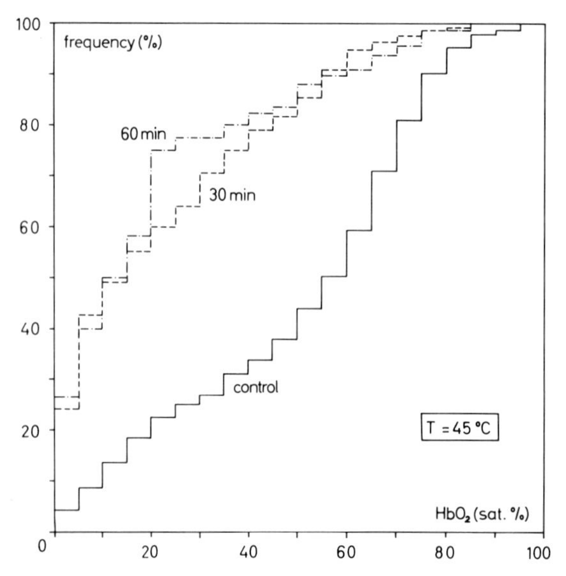

Fig. 15. Cumulative frequency distribution curves of HbO_2 data within tumor microvessels during sham-heating *(control)* and upon localized ultrasound hyperthermia at 45° C for 30 and 60 min

to a continuous decline of the tissue oxygenation. A 50% reduction in tissue oxygenation is elicited after application of heat doses of about 400° C · min. As was the case with tumor blood flow measurements, in this series of experiments (Vaupel et al. 1983d) we observed no indications of a total stoppage of microcirculation even if thermal doses >600° C · min were applied to the tumor tissue.

The changes in tumor tissue oxygenation upon hyperthermia are of great relevance, especially when sequencing and timing have to be considered during multifraction heat therapy or during combination treatment with irradiation or anticancer drugs. In general the consequences are similar to those already mentioned when the heat-induced alterations in tumor blood flow were discussed in detail (see pp. 73 ff.).

Impact of Localized Hyperthermia on the pH Distribution in Malignant Tumors

Tumors in vivo convert large amounts of glucose into nonessential amino acids (which are incorporated into tumor proteins) and into the pentose compound of nucleic acids (for review see Shapot 1980). Besides glucose oxidation, cancer cells intensely split glucose to lactic acid, i.e., glucose also contributes to the energy metabolism of rapidly growing tumor cells in vivo. Although there are no longer any reasons to ascribe to the aerobic glycolysis any specificity in malignant growth, the increased capacity for glycolysis still remains characteristic of tumors.

Frequently, the amount of lactic acid actually produced is higher than that expected to be formed when considering the actual amount of glucose consumed. This discrepancy is probably due to the fact that glucose is not the only source of lactate formation in tumor tissues. As other important sources, glutamine and several other amino acids have to be considered which also contribute to the energy metabolism of malignant tumors (Eigenbrodt et al. 1985).

Due to the elevated rate of lactic acid production and its subsequent inadequate removal, a severe tissue acidosis is evident in malignant tumors (for reviews see Vaupel et al. 1981; Wike-Hooley et al. 1984). Within different microareas of the same tumor distinct heterogeneities in the pH distribution occur. This is especially evident in partially necrotic tumors where tissue pH values even higher than the arterial blood pH can be observed in necrotic regions (Vaupel et al. 1981, 1983d).

It has already been mentioned that tissue acidosis can enhance the therapeutic efficiency of elevated tissue temperatures. This is due to the fact that:

1. Lowered tumor pH values increase the cytocidal capacity.
2. Low pH values inhibit the repair of thermal damage.
3. Tissue acidosis may inhibit the development of thermotolerance at least in animal tumors.

Therefore, it may also be possible that changes in the pH distribution during heat treatment can modify the thermal sensitivity of tumor cells. In order to elucidate this question, in a first series of experiments the impact of localized microwave hyperthermia (2.45 GHz) on the tissue pH distribution is investigated in C3H mouse mammary adenocarcinoma. The measurements are performed in microareas of the tissue using pH microelectrodes with tip diameters of approximately 1 μm. pH readings during localized heating reveal that tissue temperatures exceeding 42° C are followed by a pronounced pH drop in tumors with a negligible amount of necrosis. Tissue pH values measured immediately after heating (43° C for 60 min) are lowered by an average of 0.5 ± 0.2 pH units as compared with preheating data ($2P < 0.005$; see Fig. 16). In large tumors with extensive necrosis the mean tissue pH value is distinctly higher (pH = 7.21) than in smaller tumors (pH = 6.75). Significant changes of the mean pH values during hyperthermia are not obvious in these larger mammary carcinomas (Vaupel 1982a).

In a second series of experiments the tissue pH distribution is measured in Yoshida sarcomas implanted subcutaneously into the hind foot dorsum of Sprague-Dawley rats.

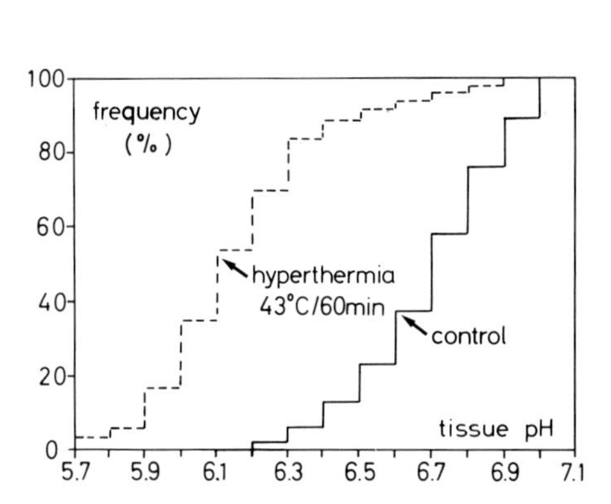

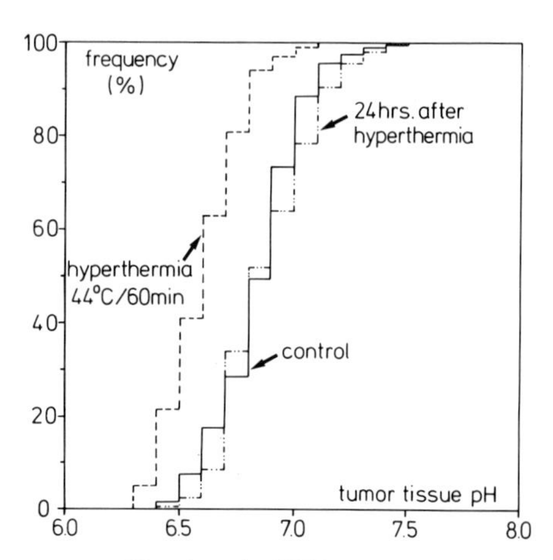

Fig. 16 *(left)*. Cumulative fequency distributions of intratumor pH values in C3H mouse mammary adenocarcinomas before *(control)* and after 1 h of 43° C hyperthermia (microwave heating)

Fig. 17 *(right)*. Cumulative frequency distributions of intratumor pH values in Yoshida sarcomas (wet weight, 2–3 g) before (control, *solid line*), immediately after *(broken line)*, and 24 h after *(dotted line)* localized ultrasound heating at 44° C for 60 min

Hyperthermia is induced by ultrasound heating (1.7 MHz) using a feedback control system (Vaupel et al. 1982c, 1983d). Measurements of pH values are carried out before, 1 h, and 24 h after tissue heating (mean tissue temperature, 44° C; exposure time, 60 min) by means of miniaturized needle pH electrodes. About 45–50 locations throughout each tumor are included and cumulative pH frequency distribution curves for each group are compiled (see Fig. 17).

In tumors with wet weights between 2 and 3 g the tissue pH ranges from 6.40 to 7.60 with an average pH of 6.89 under control conditions (solid line in Fig. 17). At 1 h after heating, the pH distribution significantly shifts to lower values with an average pH of 6.65 (broken line in Fig. 17), i.e., upon hyperthermia employing thermal doses of 540° C · min the mean tumor pH drops by about 0.25 pH units. From these results again it can be concluded that the intratumor pH is usually lower than that of normal tissues and hyperthermia further decreases tumor pH. At 24 h after heating the tumor pH distribution has returned to values obtained during control conditions. The average tissue pH 1 day after heating is 6.93 (dotted line in Fig. 17); pH values higher than 7.2 are more frequent as compared with control conditions. Probably this is due to enlarged areas with necrotic debris following heat application. In partially necrotic tumors with mean wet weights of approximately 4 g, the effect of hyperthermia on tumor pH is similar, but less pronounced, whereas in smaller tumors (mean wet weight, 1.45 g) no significant changes in the tumor tissue pH distribution are detectable. In the latter case, the application of localized ultrasound hyperthermia is often accompanied by the development of a pronounced interstitial edema, so that the tumors are enlarged considerably (see p. 81). Obviously, the increased extravasation of fluid and plasma proteins into the interstitial space of tumors can prevent a pH drop upon hyperthermia.

The findings of an intensified tissue acidosis upon localized hyperthermia is in agreement with results obtained with other tumor cell lines or heating techniques as long as temperatures higher than 42° C and exposure times of 30 min or longer are employed. Tissue temperatures of 42° C or even lower seem to have no significant influence on the tissue pH distribution in tumors. In one case a small pH increase at temperatures <42° C has been reported (see Table 8). Most probably this is due to an increased tissue perfusion and, thus, an improved drainage of acidic waste products.

Hyperthermia-induced pH changes are similar in different tissue layers of a tumor if appropriate heating techniques are utilized. The data in Table 9 indicate that there is a continuous drop of the mean pH values from the outer rim to central tissue layers. This gradient is roughly maintained upon hyperthermia, the absolute pH values being somewhat lower than those before heating at 44° C for 60 min.

Different heating up rates of the tumor tissue seem to induce different pH changes during 44° C hyperthermia sustained for 60 min (see Fig. 18). Whereas heating up rates <0.7° C/min are followed by a mean pH drop of approximately 0.4 pH units, after rapid

Table 8. Hyperthermia-induced changes in tumor pH

Cell line (host)	Hyperthermia level (° C)	Exposure time (min)	pH decrease (mean)	Reference
DS carcinosarcoma (rat)	42-43	150	0.42	von Ardenne and Reitnauer 1978
	42-43	130-195	0.47	von Ardenne and Reitnauer 1979
Yoshida sarcoma (rat)	42	60	∅	Dickson and Calderwood 1979
	44	60	0.24	Vaupel et al. 1983d
Walker 256 tumor	43	60	0.10-0.30	Song et al. 1980c
(rat)	46	60	0.15-0.40	Song et al. 1980c
C3H mammary carcinoma (mouse)	43	60	0.35-0.95	Vaupel 1982a
SCK mammary carcinoma	43.5	30	0.25-0.40	Song et al. 1980c
(mouse)	43.5	30	0.20	Rhee et al. 1984
16/C mammary carcinoma	47	15	0.10	Evanochko et al. 1983
(mouse)	47	30	0.50	
Dunn osteosarcoma	47	15	0.25	Ng et al. 1982
(mouse)	47	30	0.62	
Different tumors	<42	60	pH increase	Thistlethwaite et al. 1984
(human)	41.8	120	∅	Wike-Hooley et al. 1984b

Table 9. Hyperthermia-induced pH changes in different tissue layers of Yoshida sarcomas

	Peripheral tissue layers (0.5-3.5 mm)		Intermediate tissue layers (4.0-7.5 mm)		Central tissue layers (>7.5 mm)	
	NT	HT	NT	HT	NT	HT
pH range	6.52-7.70	5.57-7.42	6.48-7.40	5.97-7.05	6.42-7.30	6.23-7.19
Mean pH	6.95	6.82	6.86	6.71	6.80	6.70
Median pH	6.96	6.85	6.87	6.73	6.78	6.70
Modal class	6.90-7.00	6.80-6.90	6.90-7.00	6.70-6.80	6.70-6.80	6.60-6.70

NT, normothermia; *HT*, hyperthermia (ultrasound heating).

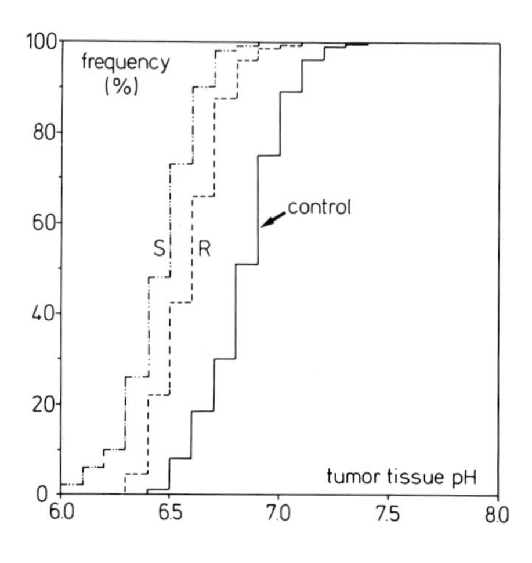

Fig. 18. Cumulative frequency distributions of intratumor pH values in Yoshida sarcomas (wet weight, 2–3 g) before (control, *solid line*) and directly after ultrasound heating at 44° C for 60 min. *S*, heating up rate <0.7° C/min ("slow heating up"); *R*, heating up rate >1.0° C/min ("rapid heating up")

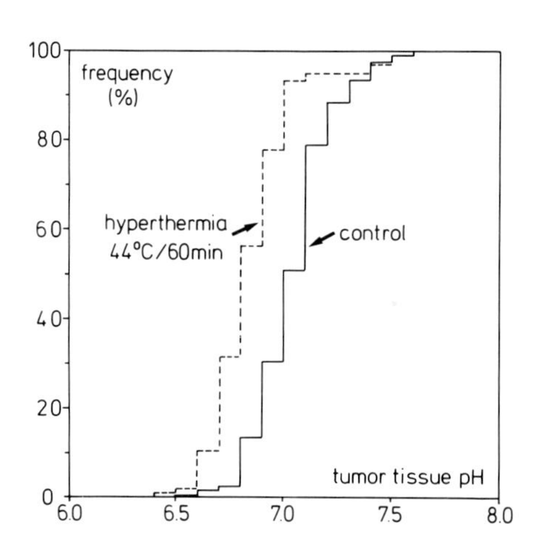

Fig. 19. Cumulative frequency distributions of intratumor pH values in xenotransplanted undifferentiated human breast carcinomas before (control, *solid line*) and immediately after localized ultrasound heating at 44° C for 60 min *(broken line)*

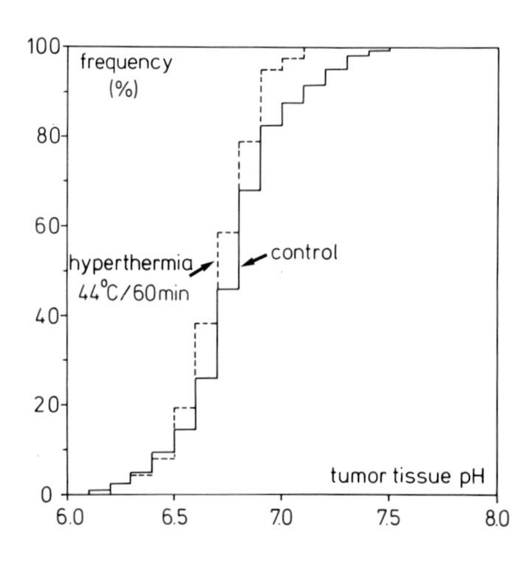

Fig. 20. Cumulative frequency distributions of intratumor pH values in xenotransplanted medullary human breast carcinomas before (control, *solid line*) and directly after localized ultrasound heating at 44° C for 60 min *(broken line)*

heating ($\Delta T/\Delta t > 1.0°$ C/min) the average pH drop is only 0.25 pH units. Up to now there have been no convincing experimental hints that this difference is caused by the somewhat greater heat dose applied during 44° C hyperthermia for 60 min and heating up rates $< 0.7°$ C/min as compared with rapidly heated tumors.

The heat-induced pH changes in xenotransplanted human breast carcinomas in T-cell-deficient nude rats are compiled in Figs. 19 and 20. Localized tissue hyperthermia at 44° C for 60 min caused a distinct pH drop in undifferentiated mammary carcinomas (Fig. 19), whereas identical heat doses were followed by only a modest pH change in medullary breast tumors (Fig. 20). This result provides evidence that, besides other parameters (e.g., heating up rate, heat dose, possible existence of a temperature threshold at 42° C), the use of different cell lines may produce differing pH changes upon heating.

As to the causes of the intensified tissue acidosis in malignant tumors during localized hyperthermia, several pathogenetic mechanisms have to be considered. The results of a pH drop in the cancer tissue upon heating are not surprising if one takes into account the familiar principle that temperature variations lead to changes in chemical equilibria of the intra- and extracellular buffer systems, which in turn are followed by an altered dissociation of H^+ ions and by a change in the Donnan equilibrium. As a rule, there is a shift to lower pH values if the temperature is elevated in any aqueous solution due to changes in ionization of solute and solvent. For simple solutions, the temperature-induced pH changes can be calculated by means of the van't Hoff equation. In complex solutions (such as in living tissues) the problem becomes more sophisticated. pH changes in blood with alterations in temperature were described as early as 1924 by Stadie and Martin and some 20 years later by Rosenthal (1948). According to Rahn et al (1974) in tissues of vertebrates the average change in *intracellular pH* with temperature ($\Delta pH/\Delta T$) is -0.016 pH units/1° C. Measuring the *extracellular pH* with pH-sensitive microelectrodes, the average change was -0.023 pH units/1° C (Harrison and Walker 1979).

Another relevant pathogenetic mechanism yielding an intensified tissue acidosis is based on the intensified ATP hydrolysis during tissue heating and on the fact that the glycolytic breakdown of glucose by the cancer cells obviously is not reduced by hyperthermia. The further degradation of pyruvate or of acetyl CoA in the Krebs cycle, however, is diminished upon hyperthermia (Streffer 1984). These heat-induced metabolic changes lead to an enhanced production of the acidic metabolites lactic acid (Ryu et al. 1982), β-hydroxybutyrate, and acetoacetate (Streffer 1982). The latter waste products in turn accumulate in the heated tumor tissue due to a progressive impairment of the venous drainage during and/or after heating. The heat-induced acidification of the tumor is further intensified by a distinct increase of the CO_2 partial pressures within the tissue (Vaupel et al. 1980; Vaupel et al. 1982a).

Investigating tumor cell sensitivity to hyperthermia as a function of extracellular and intracellular pH, Hofer and Mivechi (1980) showed that intracellular pH (pH_i) rather than extracellular pH (pH_e) changes are responsible for the phenomenon of heat sensitization at reduced environmental pH. This finding can support the observations that a lysosome-dependent cell injury is one of the paramount causative factors in the hyperthermic cell lesion (Overgaard 1976). Since most of the pH studies upon heat treatment were performed with pH-sensitive electrodes, the intratumor pH observed represents a "mixed" value of extracellular and intracellular pH.

Considering the relevant literature on proton transport and its regulation in tumor cells there is clear evidence that pH_i is somewhat correlated to pH_e. Hereby, the relationship between pH_i and pH_e is relatively complex, depending on the buffer system of the

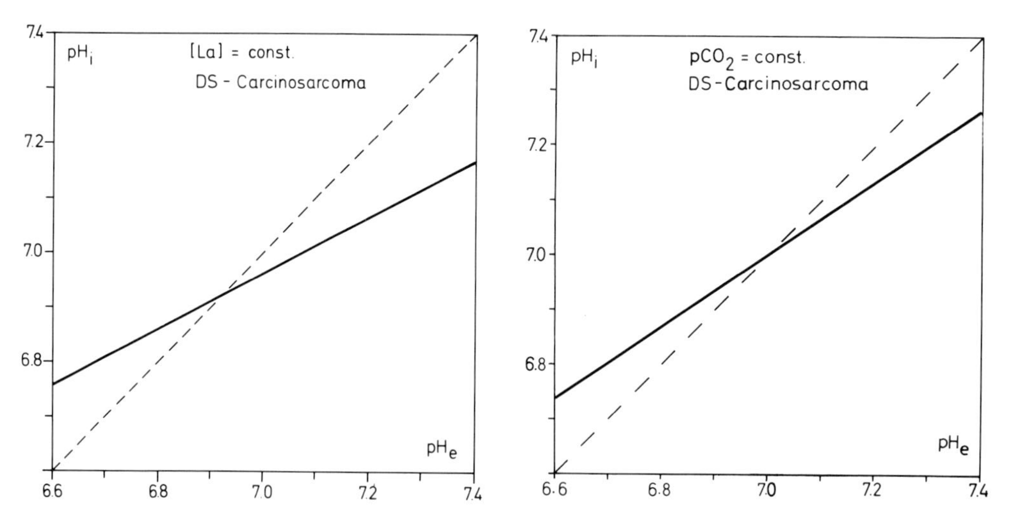

Fig. 21 *(left)*. Relationship of extracellular pH (*pH$_e$*, measurement with a capillary glass electrode) to intracellular pH (*pH$_i$*, DMO method) of DS carcinosarcoma ascites tumor cells suspended in the native ascites fluid. *pH$_e$* was varied by changing the CO_2 partial pressure of the gas mixture with which the cell suspension was equilibrated. *[La]*, lactate concentration. The *dashed line* indicates where $pH_i = pH_e$

Fig. 22 *(right)*. Relationship of extracellular pH *(pH$_e$)* to intracellular pH *(pH$_i$)* of DS carcinosarcoma ascites tumor cells. *pH$_e$* was varied by adding lactic acid to the suspension medium ($pCO_2 = CO_2$ partial pressure). The *dashed line* indicates where $pH_i = pH_e$

medium used, on the method used for acidification, on the initial pH, on the incubation temperature, and on the metabolic state of the suspended cells under investigation (Hofer and Mivechi 1980). According to our own studies on DS carcinosarcoma cells suspended in the native ascitic fluid, a linear relationship between pH_i and pH_e was obtained (Albers et al. 1981). When the tumor cells incubated in ascitic fluid were equilibrated with different CO_2 concentrations and no lactic acid was added, pH_i exceeded pH_e if the extracellular pH value is decreased below a critical level ($pH_e = 6.93$). As illustrated in Fig. 21, intracellular and extracellular pH are equal at 6.93. Conversely, when pH_e is greater than 6.93, pH_i is below pH_e. During lactic acid loading to the ascitic fluid at constant CO_2 tensions a similar relationship was obtained. Here, intracellular and extracellular pH were equal at 7.0 (see Fig. 22).

Using Ehrlich ascites tumor cells a linear relationship between pH_i and pH_e was also obtained by Poole et al. (1964, with phosphate buffer in the medium), Bowen and Levinson (1984), and Navon et al. (1977). Data derived from [^{31}P]NMR spectroscopy show that, at a $pH_e < 7.1$, the pH_i remains more alkaline than the pH_e · At $pH_e > 7.1$, pH_i and pH_e are equal (Gillies et al. 1982). Linear pH_e - pH_i relationships were also obtained for a series of other cell lines but only when certain buffers were used (Dickson and Oswald 1976; Hofer and Mivechi 1980; Schloerb et al. 1965).

From this compilation of experimental data the conclusion can be drawn that both an extracellular as well as an intracellular pH drop have to be expected when appropriate hyperthermia levels and exposure times are chosen. These pH changes are paralleled by a potassium efflux, thus achieving charge neutrality in the intracellular space (Yi 1979; Yi et al. 1983; 1985). Both alterations are detrimental for cell survival. Since the sodium level in the tumor cells increased by an amount much smaller than the potassium changes in these

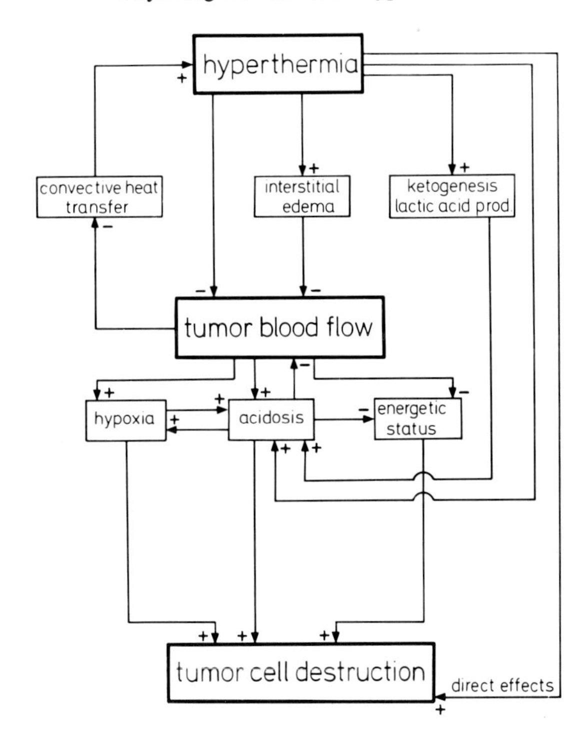

Fig. 23. Diagram showing the principles of heat-induced tumor cell destruction in vivo

experiments, hyperthermia produced an apparent deficit in the cellular osmolarity. The rate of K^+ loss from the intracellular space as well as the ultimate level reached is heat dose dependent (Ruifrock et al. 1985).

Conclusions

Due to a poor perfusion rate, the convective heat dissipation is distinctly reduced in most tumors, thus creating a "heat reservoir" with impaired heat clearance from the tissue. For this reason, at a given energy input temperature rises significantly more in malignant tumors than in normal tissues, allowing a quasi selective treatment of malignant neoplasms. In contrast to normal tissues, the microcirculation of malignant tumors, however, cannot tolerate elevated tissue temperatures for a longer period.

Whereas at lower thermal doses, an (at least transient) increase in blood flow is sometimes observed, a prolonged exposure at higher temperatures is generally followed by a shut-down of nutritive tumor blood flow. The pathomechanisms which lead to vascular stasis in tumors during hyperthermia are most probably quite complex. The deterioration of tumor blood flow results in a further impairment of convective heat dissipation (in the sense of a positive feedback system) and in this way facilitates tumor heating. Furthermore, the flow reduction brings about an intensified tissue acidosis, an enlargement of the hypoxic tissue areas, and an exhaustion of the energy supply (see Fig. 23). This means that the hyperthermia-induced disturbances in tumor microcirculation not only improve the conditions for selective tumor heating but further sensitize cancer cells to hyperthermic treatment via modifications of the intratumor milieu (again in the sense of a positive feedback system). Since the pathomechanisms described are mostly derived from experimental results on animal tumors, the clinical relevance, i.e., the applicability of these pathophysiological aspects to human tumors, has to be tested during clinical application of hyperthermia.

References

Albers C, van den Kerckhoff W, Vaupel P, Müller-Klieser W (1981) Effect of CO_2 and lactic acid on intracellular pH of ascites tumor cells. Respir Physiol 45: 273-285
Anghileri LJ, Marcha C, Crone-Escanyé MC, Robert J (1985) Effects of extracellular calcium on calcium transport during hyperthermia of tumor cells. Eur J Cancer Clin Oncol 21: 981-984
Arancia G, Malorni W, Mariutti G, Trovalusci P (1986) Effect of hyperthermia on the plasma membrane structure of Chinese hamster V79 fibroblasts: a quantitative freeze-fracture study. Radiat Res 106: 47-55
Bagshaw MA, Taylor MA, Knapp DS, Meyer JL, Samulski TV, Lee ER, Fessenden P (1984) Anatomical site-specific modalities for hyperthermia. Cancer Res (Suppl) 44: 4842s-4852s
Barnikol WKR, Burkhard O (1985) Die Abhängigkeit der Erythrozyten-Deformierbarkeit von der Osmolarität, dem pH-Wert, der Temperatur und der Proteinkonzentration. Funkt Biol Med 4: 55-60
Bicher HI, Bruley DF (1982) Hyperthermia. Plenum, New York
Bicher HI, Mitagvaria NP (1984) Changes in tumor tissue oxygenation during microwave hyperthermia-clinical relevance. In: Overgaard J (ed) Hyperthermic oncology 1984. Taylor and Francis, London, pp 169-172
Bicher HI, Vaupel PW (1980) Physiological mechanisms of localized microwave hyperthermia. In: Arcangeli G, Mauro F (eds) Hyperthermia in radiation oncology. Masson, Milan, pp 95-99
Bicher HI, Hetzel FW, Sandhu TS, Frinak S, Vaupel P, O'Hara MD, O'Brien T (1980) Effects of hyperthermia on normal and tumor microenvironment. Radiology 137: 523-530
Bicher HI, Sandhu TS, Vaupel P, Hetzel FW (1982) Effect of localized microwave hyperthermia on physiological responses. Natl Cancer Inst Monogr 61: 217-219
Bieri VG, Wallach DFH (1975) Variations of lipid-protein interactions in erythrocyte ghosts as a function of temperature and pH in physiological and non-physiological ranges. Biochem Biophys Acta 406: 415-423
Blendstrup K, Kluge M, Vaupel P (1985) Dynamic temperature mapping of tumors. Strahlentherapie 161: 525
Bowen JW, Levinson C (1984) H^+ transport and the regulation of intracellular pH in Ehrlich ascites tumor cells. J Membrane Biol 79: 7-18
Cater DB, Silver IA (1960) Quantitative measurements of oxygen tension in normal tissues and in the tumours of patients before and after radiotherapy. Acta Radiol 53: 233-256
Cavaliere R, Ciocatto EC, Giovanella BC, Heidelberger C, Johnson RO, Margottini M, Mondovi B, Moricca G, Rossi-Fanelli A (1967) Selective heat sensitivity of cancer cells. Cancer 20: 1351-1381
Copley AL (1980) Fibrinogen gel clotting, pH and cancer therapy. Thrombosis Res 18: 1-6
Copley AL, King RG (1984) A survey of surface hemorrheological experiments on the inhibition of fibrinogenin formation employing surface layers of fibrinogen systems with heparins and other substances. A contribution on antithrombogenic action. Thrombosis Res 35: 237-256
Dave S, Vaupel P, Mueller-Klieser W, Blendstrup K (1984) Temperature distribution in peripheral s.c. tumors in rats. In: Overgaard J (ed) Hyperthermic oncology 1984. Taylor and Francis, London, pp 503-506
Dethlefsen LA, Dewey WC (1982) Cancer therapy by hyperthermia, drugs, and radiation. National Cancer Institute Monograph 61
Dewhirst M, Gross JF, Sim D, Arnold P, Boyer D (1984a) The effect of rate of heating or cooling prior to heating on tumor and normal tissue microcirculatory blood flow. Biorheology 21: 539-558
Dewhirst M, Sim DA, Gross JF, Kundrat MA (1984b) Effect of heating rate on tumour and normal tissue microcirculatory function. In: Overgaard J (ed) Hyperthermic oncology 1984. Taylor and Francis, London, pp 177-180
Dickson JA, Calderwood SK (1979) Effects of hyperglycemia and hyperthermia on the pH, glycolysis, and respiration of the Yoshida sarcoma in vivo. J Natl Cancer Inst 63: 1371-1381

Dickson JA, Calderwood SK (1980) Temperature range and selective sensitivity of tumors to hyperthermia: a critical review. Ann NY Acad Sci 335: 180-205

Dickson JA, Muckle DS (1972) Total-body hyperthermia versus primary tumor hyperthermia in the treatment of the rabbit VX-2 carcinoma. Cancer Res 32: 1916-1923

Dickson JA, Oswald BE (1976) The sensitivity of a malignant cell line to hyperthermia (42° C) at low intracellular pH. Br J Cancer 34: 262-271

Dickson JA, Shah DM (1972) The effects of hyperthermia (42° C) on the biochemistry and growth of a malignant cell line. Eur J Cancer 8: 561-571

Dickson JA, Suzangar M (1974) In vitro - in vivo studies on the susceptibility of the solid Yoshida sarcoma to drugs and hyperthermia (42° C). Cancer Res 34: 1263-1274

Dietzel F (1975) Tumor und Temperatur. Urban und Schwarzenberg, Munich

Dikomey E, Eickhoff J, Jung H (1981) The effect of extracellular pH on heat-sensitivity and thermotolerance of CHO and R1H cells. Strahlentherapie 157: 617

Dudar TE (1982) Flow modifications in normal and neoplastic tissues during growth and hyperthermia. PhD thesis, Faculty of Carnegie Institute of Technology, Carnegie - Mellon University, Pittsburgh, PA

Dudar TE, Jain RK (1984) Differential response of normal and tumor microcirculation to hyperthermia. Cancer Res 44: 605-612

Durand RE (1978) Potentiation of radiation lethality by hyperthermia in a tumor model: effects of sequence, degree, and duration of heating. Int J Radiat Oncol Biol Phys 4: 401-405

Eddy HA (1980) Alterations in tumor microvasculature during hyperthermia. Radiology 137: 515-521

Eddy HA, Chmielewski G (1982) Effect of hyperthermia, radiation and adriamycin combinations on tumor vascular function. Int J Radiat Oncol Biol Phys 8: 1167-1175

Eddy HA, Sutherland RM, Chmielewski G (1982) Tumor microvascular response: hyperthermia and radiation combinations. Natl Cancer Inst Monogr 61: 225-229

Eigenbrodt E, Fister P, Reinacher M (1985) New perspectives on carbohydrate metabolism in tumor cells. In: Beitner R (ed) Regulation of carbohydrate metabolism, vol II. CRC, Boca Raton, pp 141-179

Emami B, Song CW (1984) Physiological mechanisms in hyperthermia: a review. Int J Radiat Oncol Biol Phys 10: 289-295

Emami B, Nussbaum GH, Ten Haken RK, Hughes WL (1980) Physiological effects of hyperthermia: response of capillary blood flow and structure to local tumor heating. Radiology 137: 805-809

Emami B, Nussbaum GH, Hahn N, Piro A, Dritschilo A, Quimby F (1981) Histopathological study on the effects of hyperthermia on microvasculature. Int J Radiat Oncol Biol Phys 7: 343-348

Endrich B, Zweifach BW, Reinhold HS, Intaglietta M (1979) Quantitative studies of microcirculatory function in malignant tissue: influence of temperature on microvascular hemodynamics during the early growth of the BA1112 rat sarcoma. Int J Radiat Oncol Biol Phys 5: 2021-2030

Endrich B, Voges J, Lehmann A (1984) The microcirculation of the amelanotic melanoma A-Mel-3 during hyperthermia. In: Overgaard J (ed) Hyperthermic oncology 1984. Taylor and Francis, London, pp 137-140

Evanochko WT, Ng TC, Lilly MB, Lawson AJ, Corbett TH, Durant JR, Glickson JD (1983) In vivo ^{31}P NMR study of the metabolism of murine mammary 16/C adenocarcinoma and its response to chemotherapy, x-radiation, and hyperthermia. Proc Natl Acad Sci USA 80: 334-338

Fessenden P, Lee ER, Samulski TV (1984) Direct temperature measurements. Cancer Res (Suppl) 44: 4799s-4804s

Field SB, Morris CC (1984) Application of the relationship between heating time and temperature for use as a measure of thermal dose. In: Overgaard J (ed) Hyperthermic oncology 1984. Taylor and Francis, London, pp 183-186

Freeman ML, Dewey WC, Hopwood LE (1977) Effect of pH on hyperthermic cell survival. J Natl Cancer Inst 58: 1837-1839

Gerweck LE, Richards B (1981) Influence of pH on the thermal sensitivity of cultured human glioblastoma cells. Cancer Res 41: 845-849

Gerweck LE, Gillette EL, Dewey WC (1974) Killing of Chinese hamster cells in vitro by heating under hypoxic and aerobic conditions. Eur J Cancer 10: 691-693

Gerweck LE, Dahlberg WK, Epstein LF, Shimm DS (1984) Influence of nutrient and energy deprivation on cellular response to single and fractionated heat treatments. Radiat Res 99: 573-581

Gibbs FA, Peck JW, Dethlefsen LA (1981) The importance of intratumor temperature uniformity in the study of radiosensitizing effects of hyperthermia in vivo. Radiat Res 87: 187-197

Gillies RJ, Ogino T, Shulman RG, Ward DC (1982) ^{31}P Nuclear magnetic resonance evidence for the regulation of intracellular pH by Ehrlich ascites tumor cells. J Cell Biol 95: 24-28

Goldin EM, Leeper DB (1981) The effect of reduced pH on the induction of thermotolerance. Radiology 141: 505-508

Gullino PM (1980) Influence of blood supply on thermal properties and metabolism of mammary carcinomas. Ann NY Acad Sci 335: 1-21

Gullino PM, Yi PN, Grantham FH (1978) Relationship between temperature and blood supply or consumption of oxygen and glucose by rat mammary carcinomas. J Natl Cancer Inst 60: 835-847

Gullino PM, Jain RK, Grantham FH (1982) Temperature gradients and local perfusion in a mammary carcinoma. J Natl Cancer Inst 68: 519-533

Hahn GM (1982) Hyperthermia and cancer. Plenum, New York

Harrison DK, Walker WF (1979) Micro-electrode measurement of skin pH in humans during ischemia, hypoxia and local hypothermia. J Physiol 291: 339-350

Hetzel FW, O'Hara MD, Frinak S (1984) Comparison of temperature distributions between microwave and waterbath heated murine tumors. In: Overgaard J (ed) Hyperthermic oncology. Taylor and Francis, London, pp 565-567

Hill SA, Denekamp J (1979) The response of six mouse tumours to combined heat and X rays: implications for therapy. Br J Radiol 52: 209-218

Hill SA, Denekamp J (1982) Site dependent response of tumours to combined heat and radiation. Br J Radiol 55: 905-912

Hill SA, Denekamp J, Travis EL (1980) Temperature nonuniformity in waterbath-heated tumours. In: Arcangeli G, Mauro F (eds) Hyperthermia in radiation oncology. Masson, Milan, pp 45-51

Hofer KG, Mivechi NF (1980) Tumor cell sensitivity to hyperthermia as a function of extracellular and intracellular pH. J Natl Cancer Inst 65: 621-625

Hornback NB (1984) Hyperthermia and cancer: human clinical trial experience. Vol I + II. CRC, Boca Raton

Jain RK (1980) Temperature distributions in normal and neoplastic tissues during normothermia and hyperthermia. Ann NY Acad Sci 335: 48-66

Jain RK, Gullino PM (1980) Thermal characteristics of tumors: applications in detection and treatment. NY Academy Sciences, New York

Jain RK, Ward-Hartley K (1984) Tumor blood flow-characterization, modifications, and role in hyperthermia. IEEE Trans Sonics Ultrasonics SU-31: 504-526

Johnson RJR (1978) Radiation and hyperthermia. In: Streffer C, van Beuningen D, Dietzel F, Röttinger E, Robinson JE, Scherer E, Seeber S, Trott KR (eds) Cancer therapy by hyperthermia and radiation. Urban und Schwarzenberg, Baltimore, pp 89-95

Kallinowski F, Vaupel P, Schaefer C, Benzing H, Mueller-Schauenburg W, Fortmeyer HP (1984) Hyperthermia-induced blood flow changes in human mammary carcinomas transplanted into nude *(rnu/rnu)* rats. In: Overgaard J (ed) Hyperthermic oncology 1984. Taylor and Francis, London, pp 133-136

Karino T, Koga S, Maeta M, Hamazoe R, Yamane T, Oda M (1984) Experimental and clinical studies on effects of hyperthermia on tumor blood flow. In: Overgaard J (ed) Hyperthermic oncology 1984. Taylor and Francis, London, pp 173-176

Kim JH, Hahn EW (1979) Clinical and biological studies of localized hyperthermia. Cancer Res 39: 2258-2261

Manz R, Otte J, Thews G, Vaupel P (1983) Relationship between size and oxygenation status of malignant tumors. Adv Exp Med Biol 159: 391-398

Marmor JB, Pounds D, Hahn N, Hahn GM (1978) Treating spontaneous tumors in dogs and cats by ultrasound-induced hyperthermia. Int Radiat Oncol Biol Phys 4: 967-973

Marmor JB, Hilerio FJ, Hahn GM (1979) Tumor eradication and cell survival after localized hyperthermia induced by ultrasound. Cancer Res 39: 2166-2171

Milligan AJ, Panjehpour M (1983) Canine normal and tumor tissue blood flow during fractionated hyperthermia. In: Broerse JJ, Barendsen GW, Kal HB, van der Kogel AJ (eds) Proc. 7th ICRR. Martinus Nijhoff Publ., Boston, The Hague, Nr D6-35

Milligan AJ, Conran PB, Ropar MA, McCulloch HA, Ahuja RK, Dobelbower RR (1983) Predictions of blood flow from thermal clearance during regional hyperthermia. Int J Radiat Oncol Biol Phys 9: 1335-1343
Mondovi B, Strom R, Rotilio G, Agro AF, Cavaliere R, Fanelli AR (1969) The biochemical mechanism of selective heat sensitivity of cancer cells. I. Studies on cellular respiration. Eur J Cancer 5: 129-136
Mueller-Klieser W, Vaupel P (1983a) Oxygen availability as the main determinant of O_2 consumption in tumors during hyperthermia. Proc 3rd Ann Meeting North American Hyperthermia Group, San Antonio, pp 38-39
Mueller-Klieser W, Vaupel P (1983b) Tumor oxygenation under normobaric and hyperbaric conditions. Br J Radiol 56: 559-564
Mueller-Klieser W, Vaupel P (1984) Effect of hyperthermia on tumor blood flow. Biorheology 21: 529-538
Mueller-Klieser W, Vaupel P, Manz R, Grunewald WA (1980) Intracapillary oxyhemoglobin saturation in malignant tumors with central or peripheral blood supply. Eur J Cancer 16: 195-201
Mueller-Klieser W, Vaupel P, Manz R, Schmidseder R (1981) Intracapillary oxyhemoglobin saturation of malignant tumors in humans. Int J Radiat Oncol Biol Phys 7: 1397-1404
Mueller-Klieser W, Vaupel P, Sutherland RM (1983) Impact of hyperthermia on the oxygen consumption of single tumor cells, of multicellular tumor spheroids, and of solid tumors. Strahlentherapie 159: 380-381
Mueller-Klieser W, Manz R, Otte J, Vaupel P (1984) Effect of localized hyperthermia on tumor blood flow and oxygenation. In: Francis E, Ring J, Phillips B (eds) Recent advances in medical thermology. Plenum, New York, pp 669-676
Nakajima T, Tsumura M, Onoyama Y (1984) Clinical experience with hyperthermia in cancer radiotherapy: special reference to in vivo thermometry. In: Modification of radiosensitivity in cancer treatment. Academic, Tokyo
Navon G, Ogawa S, Shulman RG, Yamane T (1977) ^{31}P Nuclear magnetic resonance studies of Ehrlich ascites tumor cells. Proc Natl Acad Sci 74: 87-91
Ng TC, Evanochko WT, Hiramoto RN, Ghanta VK, Lilly MB, Lawson AJ, Corbett TH, Durant JR, Glickson JD (1982) ^{31}P NMR spectroscopy of in vivo tumors. J Magnet Res 49: 271-286
Nielsen OS, Overgaard J (1979) Effect of extracellular pH on thermotolerance and recovery of hyperthermic damage in vitro. Cancer Res 39: 2772-2778
Nussbaum GH (1982) Physical aspects of hyperthermia. Medical physics Monogr No 8
O'Hara M, Hetzel FW, Avery K (1985a) Mild (40° C) microwave hyperthermia and tumor oxygenation. 33rd Annual Meeting Radiat Res Soc, Los Angeles, Abstr No. Fa-35
O'Hara M, Hetzel FW, Frinak S (1985b) Thermal distributions in a water bath heated mouse tumor. Int J Radiat Oncol Biol Phys 11: 817-822
Olch AJ, Kaiser LR, Silberman AW, Storm FK, Graham LS, Morton DL (1983) Blood flow in human tumors during hyperthermia therapy: demonstration of vasoregulation and an applicable physiological model. J Surg Oncol 23: 125-132
Overgaard J (1976) Influence of extracellular pH on the viability and morphology of tumor cells exposed to hyperthermia. J Natl Cancer Inst 56: 1243-1250
Overgaard J (1980) Simultaneous and sequential hyperthermia and radiation treatment of an experimental tumor and its surrounding normal tissue in vivo. Int J Radiat Oncol Biol Phys 6: 1507-1517
Overgaard J (1983) Histopathologic effects of hyperthermia. In: Storm FK (ed) Hyperthermia in cancer therapy. Hall Medical, Boston, pp 163-185
Overgaard J (1984/85) Hyperthermic oncology 1984, Vol I + II, Francis and Taylor, London
Overgaard J, Bichel P (1977) The influence of hypoxia and acidity on the hyperthermic response of malignant cells in vitro. Radiology 123: 511-514
Peck JW, Gibbs FA (1983) Capillary blood flow in murine tumors, feet, and intestines during localized hyperthermia. Radiat Res 96: 65-81
Peterson HI (1979) Tumor blood circulation: angiogenesis, vascular morphology and blood flow of experimental and human tumors. CRC, Boca Raton
Poole DT, Butler TC, Waddell WJ (1964) Intracellular pH of the Ehrlich ascites tumor cell. J Natl Cancer Inst 32: 939-946
Rahn H, Reeves RB, Howell BJ (1974) Intra- and extracellular pH as a function of body temperature. Proc Internat Union Physiol Sci 10: 56-57

Rapoport S, Nieradt-Hiebsch C, Thamm R (1971) Über den Einfluß der Hyperthermie auf die Verwertung von Substraten in Ehrlich-Aszites-Tumorzellen und Kaninchenretikulozyten. Acta Biol Med Germ 26: 483-500
Rappaport DS, Song CW (1983) Blood flow and intravascular volume of mammary adenocarcinoma 13726A and normal tissues of rat during and following hyperthermia. Int J Radiat Oncol Biol Phys 9: 539-547
Reinhold HS, van den Berg-Blok A (1980) Enhancement of thermal damage to "sandwich" tumours by additional treatment. In: Arcangeli G, Mauro F (eds) Hyperthermia in radiation oncology. Masson, Milan, pp 179-183
Reinhold HS, van den Berg-Blok A (1981) Enhancement of thermal damage to the microcirculation of „sandwich" tumours by additional treatment. Eur J Cancer Clin Oncol 17: 781-795
Reinhold HS, van den Berg-Blok AE (1983) Hyperthermia-induced alteration in erythrocyte velocity in tumors. Int J Microcirc Clin Exp 2: 285-295
Reinhold HS, Wike-Hooley JL, van den Berg AP, van den Berg-Blok A (1984) Environmental factors, blood flow and microcirculation. In: Overgaard J (ed) Hyperthermic oncology 1984, vol II, Taylor and Francis, London
Rhee JG, Kim TH, Levitt SH, Song CW (1984) Changes in acidity of mouse tumor by hyperthermia. Int J Radiat Oncol Biol Phys 10: 393-399
Robert J, Escanye JM, Marchal C, Thouvenot P (1982) Blood flow and temperature evolution of rhabdomyosarcoma-bearing mice during normal growth and during sequential hyperthermia treatment. In: Gautherie M, Albert E (eds) Biomedical thermology. AR Liss, New York, pp 85-95
Robinson JE, Harrison GH, McCready WA, Samarar GM (1978) Good thermal dosimetry is essential to good hyperthermia research. Br J Radiol 51: 532-534
Robinson JE, McCulloch D, McCready WA (1982) Blood perfusion of murine tumors at normal and hyperthermal temperatures. Natl Cancer Inst Monogr 61: 211-215
Rosenthal TB (1948) The effect of temperature on the pH of blood and plasma in vitro. J Biol Chem 173: 25-30
Ruifrock ACC, Kanon B, Konings AWT (1985) Correlation between cellular survival and potassium loss in mouse fibroblasts after hyperthermia alone and after a combined treatment with X rays. Radiat Res 101: 326-331
Ryu HL, Song CW, Kang MS, Levitt SH (1982) Changes in lactic acid content in tumors by hyperthermia. Proc 2nd Ann Meeting North American Hyperthermia Group, Salt Lake City
Sapareto SA, Dewey WC (1984) Thermal dose determination in cancer therapy. Int J Radiat Oncol Biol Phys 10: 787-800
Scheid P (1961) Funktionale Besonderheiten der Mikrozirkulation im Karzinom. Bibl Anat 1: 327-335
Schloerb PR, Blackburn GL, Grantham JJ, Mallard DS, Cage GK (1965) Intracellular pH and buffering capacity of the Walker-256 carcinoma. Surgery 58: 5-11
Schmid-Schönbein H, Singh M, Malotta H, Leschke D, Teitel P, Driessen G, Scheidt-Bleichert H (1984) Subpopulations of rigid red cells in hyperthermia and acidosis: effect on filterability in vitro and on nutritive capillary perfusion in the mesenteric microcirculation. Int J Microcirc Clin Exp 3: 497
Sekins M, Dundore D, Emery A, Lehmann J, McGrath P, Nelp W (1980) Muscle blood flow changes in response to 915 MHz diathermy with surface cooling as measured by Xe^{133} clearance. Arch Phys Med Rehabil 61: 105-113
Shapot VS (1980) Biochemical aspects of tumour growth. MIR, Moscow
Shrivastav S, Kaelin WG, Joines WT, Jirtle RL (1983) Microwave hyperthermia and its effect on tumor blood flow in rats. Cancer Res 43: 4665-4669
Song CW (1978) Effect of hyperthermia on vascular functions of normal tissues and experimental tumors. J Natl Cancer Inst 60: 711-713
Song CW (1982) Physiological factors in hyperthermia of tumors. In: Nussbaum GH (ed) Physical aspects of hyperthermia. Med Phys Monogr 8: 43-62
Song CW (1984) Effect of local hyperthermia on blood flow and microenvironment. Cancer Res (Suppl) 44: 4721s-4730s
Song CW, Rhee JG, Levitt SH (1980a) Blood flow in normal tissues and tumors during hyperthermia. J Natl Cancer Inst 64: 119-124

Song CW, Kang MS, Rhee JG, Levitt SH (1980b) Effect of hyperthermia on vascular function in normal and neoplastic tissues. Ann NY Acad Sci 335: 35-47
Song CW, Kang MS, Rhee JG, Levitt SH (1980c) The effect of hyperthermia on vascular function, pH, and cell survival. Radiology 137: 795-803
Song CW, Kang MS, Rhee JG, Levitt SH (1980d) Vascular damage and delayed cell death in tumours after hyperthermia. Br J Cancer 41: 309-312
Song CW, Pattan MS, Rhee JG, Schuman VL, Levitt SH (1984) Role of blood flow in the response of RIF-1 tumors to combined treatment of hyperthermia and radiotherapy. In: Overgaard J (ed) Hyperthermic oncology 1984, Taylor and Francis, London, pp 293-296
Stadie WC, Martin KA (1924) The thermodynamic relations of the oxygen and base combining properties of blood. J Biol Chem 60: 191-235
Stewart F, Begg A (1983) Blood flow changes in transplanted mouse tumours and skin after mild hyperthermia. Br J Radiol 56: 477-482
Stewart FA, Denekamp J (1978) The therapeutic advantage of combined heat and X rays on a mouse fibrosarcoma. Br J Radiol 51: 307-316
Storm FK (1983) Hyperthermia in cancer therapy. Hall, Boston
Streffer C (1982) Aspects of biochemical effects by hyperthermia. Natl Cancer Inst Monogr 61: 11-17
Streffer C (1984) Mechanism of heat injury. In: Overgaard J (ed) Hyperthermic oncology 1984, vol II, Taylor and Francis, London, pp 213-222
Streffer C, van Beuningen D, Dietzel F, Röttinger E, Robinson JE, Scherer E, Seeber S, Trott KH (1978) Cancer therapy by hyperthermia and radiation. Urban and Schwarzenberg, Baltimore
Suit HD, Gerweck LE (1979) Potential for hyperthermia and radiation therapy. Cancer Res 39: 2290-2298
Sutton CH (1980) Discussion. Ann NY Acad Sci 335: 35-47
Tanaka Y, Hasegawa T, Murata T (1984) Effect of irradiation and hyperthermia on vascular function in normal and tumor tissue. In: Overgaard J (ed) Hyperthermic oncology 1984, Taylor and Francis, London, pp 145-148
Thews G, Vaupel P (1985) Autonomic functions in human physiology. Springer, Berlin Heidelberg New York Tokyo
Thistlethwaite AJ, Leeper DB, Moylan DJ, Nerlinger RE (1984) pH distribution in human tumors. Proc 4th Ann Meeting North American Hyperthermia Group, Orlando
van den Berg-Blok AE, Reinhold HS (1984) Time-temperature relationship for hyperthermia induced stoppage of the microcirculation in tumors. Int J Radiat Oncol Biol Phys 10: 737-740
Vaupel P (1974) Atemgaswechsel und Glucosestoffwechsel von Implantationstumoren (DS-Carcinosarkom) in vivo. Funktionsanalyse Biolog Systeme 1: 1-138
Vaupel P (1977) Hypoxia in neoplastic tissue. Microvasc Res 13: 399-408
Vaupel P (1979) Oxygen supply to malignant tumors. In: Peterson HI (ed) Tumor blood circulation: angiogenesis, vascular morphology, and blood flow of experimental and human tumors. CRC, Boca Raton, pp 143-168
Vaupel P (1982a) Einfluß einer lokalisierten Mikrowellenhyperthermie auf die pH-Verteilung in bösartigen Tumoren. Strahlentherapie 158: 168-173
Vaupel P (1982b) Pathophysiologie der Durchblutung maligner Tumoren. Funktionsanalyse Biolog Systeme 8: 155-170
Vaupel P (1986) Durchblutung, Oxygenierung und pH-Verteilung in malignen Tumoren: biologische und therapeutische Aspekte. In: Bromm B, Lübbers DW (eds) Physiologie Aktuell I. Fischer, Stuttgart, pp 53-67
Vaupel P, Hammersen F (1983) Mikrozirkulation in malignen Tumoren. Karger, Basel
Vaupel P, Gabbert H (1986) Evidence for and against a tumor type-specific vascularity. Strahlentherapie Onkol 162: 633-638
Vaupel P, Mueller-Klieser W (1983a) Heat susceptibility of tumor blood flow. Proc 31st Ann Meeting Radiat Res Soc, San Antonio, pp 68-69
Vaupel P, Mueller-Klieser W (1983b) Interstitieller Raum und Mikromilieu in malignen Tumoren. Mikrozirk Forsch Klin 2: 78-90
Vaupel P, Ostheimer K, Thomé H (1976) Blood flow, vascular resistance, and oxygen consumption of malignant tumors during normothermia and hyperthermia. Ann Meeting Gesellschaft für Mikrozirkulation, Aachen, FRG (Microvasc Res 13: 272, 1977)

Vaupel P, Manz R, Mueller-Klieser W, Grunewald WA (1979) Intracapillary HbO_2 saturation in malignant tumors during normoxia and hyperoxia. Microvasc Res 17: 181-191
Vaupel P, Ostheimer K, Mueller-Klieser W (1980) Circulatory and metabolic responses of malignant tumors during localized hyperthermia. J Cancer Res Clin Oncol 98: 15-29
Vaupel P, Frinak S, Bicher HI (1981) Heterogeneous oxygen partial pressure and pH distribution in C3H mouse mammary adenocarcinoma. Cancer Res 41: 2008-2013
Vaupel P, Frinak S, Mueller-Klieser W, Bicher HI (1982a) Impact of localized hyperthermia on the cellular microenvironment in solid tumors. Natl Cancer Inst Monogr 61: 207-209
Vaupel P, Otte J, Manz R (1982b) Changes in tumor oxygenation after localized microwave hyperthermia. In: Gautherie M, Albert E (eds) Biomedical thermology, Liss, New York, pp 65-74
Vaupel P, Otte J, Manz R (1982c) Oxygenation of malignant tumors after localized microwave hyperthermia. Radiat Environment Biophys 20: 289-300
Vaupel P, Benzing H, Egelhof E, Mueller-Klieser W, Mueller-Schauenburg W (1983a) The effect of various thermal doses on the regional tumor blood flow measured by heat clearance. Strahlentherapie 159: 384
Vaupel P, Mueller-Klieser W, Gabbert H (1983b) Experimental evidence for a hyperthermia-induced breakdown of tumor blood flow during normoglycemia. J Cancer Res Clin Oncol 105: 303-304
Vaupel P, Mueller-Klieser W, Manz R, Wendling P, Strube HD, Schmidseder R (1983c) Heterogeneous oxygenation of malignant tumors in humans. Verhdlg Dt Krebsges 4: 153
Vaupel P, Mueller-Klieser W, Otte J, Manz R, Kallinowski F (1983d) Blood flow, tissue oxygenation, and pH-distribution in malignant tumors upon localized hyperthermia. Strahlentherapie 159: 73-81
Vaupel P, Kallinowski F, Kluge M (1986) Pathophysiologische Aspekte der Hyperthermiewirkung in malignen Tumoren: Durchblutungsänderungen in Xenotransplantaten menschlicher Mammacarcinome. In: Streffer C, Herbst M, Schwabe HW (eds) Lokale Hyperthermie, Deutscher Ärzte-Verlag, Cologne, pp 39-46
Verma SP, Wallach DFA (1976) Erythrocyte membranes undergo cooperative, pH-sensitive state transitions in the physiological temperature range: evidence from Raman spectroscopy. Proc Natl Acad Sci USA 73: 3558-3561
Vogel AW (1965) Intratumoral vascular changes with increased size of a mammary adenocarcinoma. New method and results. J Natl Cancer Inst 34: 571-578
von Ardenne M, Krüger W (1966) Messungen zur irreversiblen Schädigung der Atmung von Krebszellen durch Extremhyperthermie. Zschr Naturforsch 21b: 836-840
von Ardenne M, Reitnauer PG (1968) Selektive Krebszellenschädigung durch Proteindenaturation. Dtsch Gesundheitswesen 23: 1681-1685
von Ardenne M, Reitnauer PG (1978) Amplification of the selective tumor acidification by local hyperthermia. Naturwissenschaften 65: 159
von Ardenne M, Reitnauer PG (1979) Verstärkung der mit Glukoseinfusion erzielbaren Tumorübersäuerung durch lokale Hyperthermie. Res Exp Med (Berl) 175: 7-18
von Ardenne M, Reitnauer PG (1980) Selective occlusion of cancer tissue capillaries as the central mechanism of the cancer multistep therapy. Jpn J Clin Oncol 10: 31-48
von Ardenne M, Reitnauer PG (1982) Die manipulierte selektive Hemmung der Mikrozirkulation im Krebsgewebe. J Cancer Res Clin Oncol 103: 269-279
Voorhees WD, Babbs CF (1982) Hydralazine-enhanced selective heating of transmissible venereal tumor implants in dogs. Eur J Cancer Clin Oncol 18: 1027-1033
Waterman FM, Fazekas J, Nerlinger RE, Leeper DB (1982) Blood flow rates in human tumors during hyperthermia treatments as indicated by thermal washout. Proc 2nd Ann Meeting North American Hyperthermia Group, Salt Lake City, p E-6
Wendling P, Manz R, Thews G, Vaupel P (1984) Inhomogeneous oxygenation of rectal carcinomas in humans. A critical parameter for perioperative irradiation? Adv Exp Med Biol 180: 293-300
Westermark N (1927) The effect of heat upon rat-tumors. Scand Arch 52: 257-318
Wike-Hooley JL, Haveman J, Reinhold HS (1984a) The relevance of tumour pH to the treatment of malignant disease. Radiother Oncol 2: 343-366
Wike-Hooley JL, van der Zee J, van Rhoon GC, van den Berg AP, Reinhold HS (1984b) Human tumour pH changes following hyperthermia and radiation therapy. Eur J Cancer Clin Oncol 20: 619-623

Yi PN (1979) Cellular ion content changes during and after hyperthermia. Biochem Biophys Res Comm 91: 177-181

Yi PN, Chang CS, Tallen M, Bayer W, Ball S (1983) Hyperthermia-induced intracellular ionic level changes in tumor cells. Radiat Res 93: 534-544

Yi PN, Fenn JO, Jarrett JH (1985) Hyperthermia and osmoregulation. Proc 33rd Ann Meeting Radiat Res Soc, Los Angeles, Abstr Ac-20

Young MA, Gerlowski LE, Jain RK (1985) Effects of hyperthermia, glucose and galactose on normal and tumor microvascular permeability. Microvasc Res 29: 262

Zywietz F, Knöchel R, Kordts J (1986) Heating of a rhabdomyosarcoma of the rat by 2450 MHz microwaves - technical aspects and temperature distributions. Rec Res Cancer Res 101: 36-46

The Combination of Hyperthermia and Radiation: Clinical Investigations

M. Molls and E. Scherer

Strahlenklinik, Universitätsklinikum Essen, Hufelandstrasse 55, 4300 Essen, FRG

History and Heritage

Heat has been used through the ages as a therapeutic agent. Hippocrates (470-377 B.C.) stated that a patient who could not be cured by heat was actually incurable. He adapted the bathing customs of the Egyptians to ancient Greece and induced hyperthermia in patients suffering from paralysis. The Japanese of the seventeenth century performed hyperthermia by balneotherapy. They treated syphilis, arthritis, and gout by using hot water at temperatures between 45° and 53° C. Thus the body temperature was elevated to at least 39° C (Giles 1938).

The German physician Busch (1886) was the first to present a scientific report on the effect of hyperthermia in neoplastic tissue. He described the spontaneous disappearance of a facial sarcoma after a febrile period caused by erysipelas infection. Bruns (1888) reported a comparable case. Erysipelas infection led to complete regression of a recurrent malignant melanoma after a period of high fever. Coley (1893, 1894, 1911), a New York surgeon, induced fever by inoculation of erysipelas. He obtained disease-free survival from 1 to 7 years in 7 of 17 unresectable sarcomas and a permanent cure in 3 of 17 other inoperable carcinomas. He also extracted bacterial toxins which were injected in patients with malignant tumors. The treatment was effective when the temperature was elevated to 39°-40° C over a period of several days. It is difficult to judge whether heat alone or the combination of whole-body hyperthermia and unspecific immunotherapy exerted the favorable effect. Nauts (1985) has reviewed the different procedures by which systemic fever together with other undefined reactions has been induced.

Based on the observations of Coley (1893, 1894, 1911) local hyperthermic treatment in animal and human tumors was tested. The Swedish gynecologist Westermark (1898) treated nonoperable carcinomas of the cervix uteri. He reported on direct local heating by positioning a metallic coil filled with hot water in the vagina. The water was provided from a vessel heated by a gas flame. A mechanism regulating the strength of the flame controlled the water temperature automatically. The treatment lasted 48 h and the temperature of the metal loop was between 42° and 44° C. Westermark (1898) obtained a beneficial effect and observed a marked regression of large cervix carcinomas.

A selective effect of hyperthermia on tumors was also seen in experimental models. Lambert (1912) demonstrated the greater susceptibility of sarcoma cells to heat as compared with actively proliferating connective tissue cells. Westermark (1927) published a thesis on „The effect of heat upon rat tumors," which showed a selective effect of hyperthermia on tumor tissue at temperatures of 44°-45° C. During the subsequent decades the anticancer effect of heat was repeatedly shown under different experimental conditions (Overgaard 1934; Johnson 1940; Overgaard and Okkels 1940). The possibility of using

Recent Results in Cancer Research. Vol 104

heat as a radiosensitizer was suggested early in this century by Schmidt (1909). During the following years several investigators induced hyperthermia with microwaves to improve the radiation response of tumors (Liebesny 1921; Mueller 1910, 1920; Rhodenburg and Prime 1921; Teilhaber 1913).

Arons and Sokoloff (1937) reported on 30 cases of human tumors in the literature up to 1937. Diathermy hyperthermia was either applied immediately prior to radiation or between two daily radiation fractions. Measurements of tumor temperatures were not performed during the treatments with microwaves. The authors suggested that the favorable effects in these patients were obtained by nonthermal influences of microwaves.

More recently, Selawry et al. (1958) reviewed the investigations of a number of authors who combined ionizing radiation and hyperthermia. In conclusion the addition of heat to irradiation was considered to improve the treatment results. Heat had been induced by microwaves, water baths, or hot air. It was given simultaneously or in every possible sequence to irradiation. Thermometry was lacking or poorly performed. Control groups did not exist in these clinical trials.

During the past 15 years knowledge of the biological effects produced by hyperthermia alone or the combination of ionizing radiations plus heat has increased considerably. Experimental investigations either in vitro or in vivo yielded much biochemical, cellular, and physiological information (for reviews see Streffer and van Beuningen, this volume; Vaupel and Kalinowski, this volume). In particular the experiments proved the radiosensitizing effect of temperatures of around 42° C and which technically can be induced in cancer patients. This biological information together with the above-described clinical observations stimulated radiotherapists to study more systematically as to whether tumor patients can gain benefit from the combined treatment with ionizing radiation and hyperthermia.

Rationales for the Clinical Combination of Radiotherapy with Hyperthermia

Although modern megavoltage radiations have significantly improved the cure for cancer, fatal local failure still occurs in an unwelcomely high number of patients. About one-third of all cancer deaths may be due to lack of local control (Suit 1982). In addition, failure to control the primary site will obviously result in growth of micrometastases since residual tumor at the primary site is a continuous source of viable cells. As seen in biological experiments and historical investigations, radiation plus heat can exert a remarkable anticancer effect. Thus one expects that the combined treatment with both modalities increases the rate of local cure and positively influences the fate of the cancer patient.

With regard to the clinical benefit the combination of radiation and heat is used because of two different types of interaction (Dewey et al. 1980; Field 1983; Field and Bleehen 1979; Hahn 1982; Overgaard 1978, 1983; Streffer 1985; Streffer and van Beuningen, this volume; Suit and Gerweck 1979). Firstly, heat directly kills cells especially at higher temperatures (>43° C). It seems that in principle this effect does not differ quantitatively between a malignant and a normal cell. However, therapeutic benefit can be obtained since in a tumor certain special factors come into play.

During hyperthermia tumors tend to become hotter than surrounding normal tissues. Because of a more sluggish blood supply (Storm et al. 1980) one has a slow wash out of heat. The resulting selective heating of the tumor is a therapeutic advantage. Furthermore as parts of solid tumors are poorly vascularized they suffer from hypoxia, nutritional deprivation, and acidosis. Experimental investigations (for review see Streffer and van Beuningen, this volume) have proven a strong enhancement of the heat-induced cytotoxic effect under such environmental conditions. One can speculate that a moderate heat treatment, which is tolerated in well-vascularized normal tissues, kills a large number of hypoxic cells in many solid tumors.

In addition, as pointed out heat itself reduces the blood flow in tumors. This again leads to hypoxia and acidosis and thus increases the potential for hyperthermic cell killing (Storm 1983). A final aspect with regard to hypoxic cells is that they represent the most radioresistant fraction of cells in a solid tumor (Hall 1978). After irradiation alone there might be risk that a recurrence develops from this critical cell population. In a combined treatment the risk decreases because ionizing radiations and heat complement each other.

Secondly, hyperthermia acts a radiosensitizing agent. This means that radiation damage in the cells is considerably increased by the additional application of heat. Apparently the extent of the radiosensitizing action does not differ between normal and neoplastic cells. In contrast to direct cell killing the radiosensitizing effect is obtained at comparatively lower temperatures (<43° C) (Streffer 1985). The mechanisms of radiosensitization are complex and not completely understood. Under some conditions heat synergistically interacts with radiation by inhibiting the repair of radiation-induced sublethal or potentially lethal damage (Ben-Hur et al. 1974; Li et al. 1976). Furthermore, heat enhances killing of cells which are irradiated in relatively radioresistant phases of the cell cycle, e.g., in S-phase (Westra and Dewey 1971).

There are several practical reasons for which the radiotherapist preferably works with temperatures of the radiosensitizing range. At present the distribution of 42°-43° C in a tumor is technically less difficult than the homogeneous production of higher temperatures. Furthermore at comparatively low temperature levels the risk of side effects in normal tissues, which increases at about 45° or >45° C, remains low. Finally, due to modern physical treatment planing, the 100% radiation dose is fairly precisely deposited within the tumor volume. Therefore an additional moderate heat treatment preferably radiosensitizes the irradiated tumor. Summing up, most of the surrounding normal tissue is spared from radiosensitization and at temperatures of 42°-43° C also from the cell-killing effect of hyperthermia. However, as described above, it should be emphasized that application of 42°-43° C does not exclude the direct killing of the highly heat sensitive hypoxic cells in the tumor. Thus in clinical practice when working with temperatures at about 42°-43° C one deals with the radiosensitizing effect as well as with the direct cell-killing action of heat.

Technical Aspects

Up to now most of the clinical investigations concentrated on the local treatment of superficial lesions within about 3 cm of the body's surface. A smaller number of studies concern regional hyperthermia of deep-seated tumors, e.g., in the pelvis. The latter reflects the inherent limitations of existing equipment. There are still problems of safe and homogeneous power deposition at the depths necessary for the treatment of abdominal or thoracic tumors. The fact that the temperature measurement must be performed invasively is a further serious disadvantage.

Techniques of hyperthermia production and temperature measurement are described in this volume (Hand, this volume) and elsewhere (Bruggmoser et al. 1986). The clinician administers external microwaves with specially designed applicators. They usually contain dielectric materials. Thus heat distribution within the treated volume can be improved. Radiofrequency fields are induced either with plates or with coil electrodes, corresponding to capacitive or inductive radiofrequency. In some clinical trials heat has been generated by ultrasound.

To obtain an adequate temperature the appropriate modality and frequency must be selected. These parameters and the coupling between applicator and surface of the patient determine the penetration of heat. The design and size of the applicator, external electrodes, transducers, and antennas are of importance with regard to effective handling and application of heat.

Clinical Studies

Studies in the Federal Republic of Germany

In the Federal Republic of Germany recent clinical studies concerning the effect of hyperthermia plus radiation started during the 1970s. Dietzel has been one of the propagators of this combined cancer treatment modality (Dietzel 1975a, b; Dietzel et al. 1975). At the University Hospital of Heidelberg radiotherapists have been treating different malignancies with irradiation plus heat since 1972 (Hymmen and Wieland 1976; Weischedel and Wieland 1981). The authors observed a good response of the tumors, a low frequency of local recurrencies, and no increase in the rate of metastases. Impressed by the work of von Ardenne (1971), Pomp (1978) developed the Pomp-Siemens hyperthermia cabin for whole body hyperthermia. The latter has been used in several European institutions. An important challenge was the second International Symposium on Cancer Therapy by Hyperthermia and Radiation in 1977 at Essen (Streffer et al. 1977).

In a pilot study (Gerhard and Scherer 1983) in the Department of Radiotherapy at the University Hospital Essen we have treated patients with advanced cancer diseases (Karnofsky >0.7). We used generators with frequencies of 13, 27, 434, and 2540 MHz. Depending on site and depth of the tumors hyperthermia was produced in a number of treatments by application of two generators combining two fields of different frequencies. Thus penetration and distribution of heat could be improved. Hyperthermia followed 1-3 h after irradiation. In a number of cases the radiation dose was low as the patients had undergone previous radiotherapy. We aimed for temperatures of 40°-42.5° C over a period of 30 min. In a smaller number of patients in which hyperthermia was given alone or in combination with chemotherapy the heat treatment lasted for 1 h. The temperatures were measured invasively or intracavitarily only in some cases. Normally the heat supply was regulated by subjective tolerance of the patient.

Forty-five percent of the treated tumors showed regressions. In 26% the growth of the tumors stoped for at least 3 months (Table 1). Progress occurred especially when hyperthermia was the only treatment modality. Most favorable was the combination of radiation plus heat. Specifying the treatment results in terms of tumor entities, we obtained the best result for inoperable cancer of the breast (four cases: T4 N2 M0) and recurrences of the chest wall. Also melanomas showed a relatively good response, especially when the tumor volume was more than 1 ml (up to 200 ml). In head and neck tumors the remissions lasted for a relatively short time. The other entities, the majority being deep-seated tumors, did not satisfactorily respond probably due to insufficient supply and distribution of heat. Especially in the case of presacral recurrences of rectum carcinomas the latter might have come into play.

At the Department of Radiotherapy at the University of Erlangen and Nuremberg, 261 patients underwent combined radiotherapy and local radiofrequency (RF) hyper-

Table 1. Treatment results in 164 tumors of 126 patients. (Gerhard and Scherer 1983)

Complete regression in 25 tumors (15%)
21 tumors: radiation plus hyperthermia
2 tumors: chemotherapy plus hyperthermia
2 tumors: hyperthermia
Partial regression in 49 tumors (30%)
40 tumors: radiation plus hyperthermia
5 tumors: chemotherapy plus hyperthermia
4 tumors: hyperthermia
Cessation of growth (>3 months) in 42 tumors (26%)
38 tumors: radiation plus hyperthermia
2 tumors: chemotherapy plus hyperthermia
2 tumors: hyperthermia
Progress of growth in 48 tumors (29%)
5 tumors: radiation plus hyperthermia
7 tumors: chemotherapy plus hyperthermia
36 tumors: hyperthermia

thermia. A dose of 10 Gy was given in four to five fractions/week. Tumor doses varied between 30 and 60 Gy (Herbst and Sauer 1983; Sauer and Herbst 1984). Hyperthermia (13.56 MHz) was applied immediately before irradiation. It was given daily during a pilot phase, later two times or three times a week. If possible, temperature was measured by thermocouples pierced into superficial lesions. Temperatures higher than 42° C for at least 30 min were aimed for. The best response rate was found in head and neck tumors with N3 lymphnode metastases and in superficial tumors. Brain tumors and especially abdominal tumors yielded disappointing results. Initialy glioblastomas showed a good response but after completion of irradiation they recurred rather soon. In cases of rectal carcinomas a subjective relief occurred but with no essential reduction of the tumor size as identified by computed tomography. It was an important experience in Heidelberg (Weischedel and Wieland 1981), in Essen (Gerhard and Scherer 1982), and in Erlangen (Sauer and Herbst 1984) that the tolerance of the combined application of radiotherapy and hyperthermia is normally excellent. Severe complications were rare. The percentage of patients in which the treatment was terminated due to subjective intolerance was small.

Based on the pilot experiencies a multicentre prospective randomized study supported by the Federal Minister for Research and Technology has started in the Federal Republic of Germany. Several centers will test the effect of conventional radiotherapy versus radiotherapy plus external hyperthermia in advanced carcinomas of the cervix uteri, hypopharynx, oropharynx, and larynx. Hyperthermia is performed with a machine generating frequencies of 13.56 and 27.12 MHz (HMS 200, HMS Elektronik, Leverkusen, FRG). It follows immediately after irradiation. Temperature measurements are mandatory.

International Studies

Local Hyperthermia of Superficial Tumors

Treatment results published in 66 studies between 1977 and 1980 have been reviewed by Overgaard (1983). A total of 3024 patients, most of them with superficial lesions, had undergone hyperthermia. When heat was given alone (276 patients) complete response was obtained in only 13% of the patients. The values for partial response and no response were 39% and 48% respectively. After combined application of hyperthermia and radiation the frequency of complete response was about twice as high as after radiotherapy alone. The latter could be demonstrated very clearly in studies designed in the same way as that of Kim et al. (1982a). This investigation was carried out in patients with multiple recurrent malignant melanoma. In the same patient with several tumors of comparable size and site some nodules received the combination therapy whereas others were irradiated alone.

The clinical experience of the past years with localized hyperthermia and irradiation has been summarized by Perez and Meyer (1985) in a valuable review. Most of the recent studies describe hyperthermia performed with microwaves. At M.D. Anderson Hospital (Corry et al. 1982) and Stanford University School of Medicine (Marmor et al. 1980) heat was induced by ultrasound. Active cooling of skin depended on superficial tumors being involved or not. In nearly all the studies thermometry has been described in detail. Temperature was determined for the tumors as well as for the skin. The sensors were introduced directly or through catheters. Insertion into the anticipated sites followed after performing local anesthesia. Table 2 summarizes treatment techniques of selected reports with regard to hyperthermia and irradiation.

Usually the patients were irradiated with four to five fractions/week, 2–3 Gy per fraction (Table 3). When a scheme of two to three fractions/week was applied and the total doses were relatively high then single doses amounted to 3–6 Gy. Total doses varied from rather low (12 Gy) to rather high (80 Gy). In most studies megavoltage electron or photon beams were used (Table 2). A smaller number of patients received orthovoltage energy.

Considering the heating parameters tumor temperatures ranged in most cases between about 42° and 44° C (Table 3). These temperatures were maintained over a period of 30–60 min, generally two or three times/week. In most cases heat treatment followed 30–60 min after irradiation. Only a few institutions performed hyperthermia before radiotherapy (Table 3).

Table 2. Techniques in the combined treatment of superficial tumors

Authors	Hyperthermia type (MHz)	Thermometry in tumor and normal tissue	Skin cooling	Radiation type
Bicher et al. (1980)	MW (300, 915, 2450)	Yes	Yes	X-rays
Marmor and Hahn (1980)	US (2, 3)	Yes	Yes	250 kV (most)
U et al. (1980)	MW (915, 2450)	Yes	No	Co-60, 280 kV
Fazekas and Nerlinger (1981)	MW (2450)	Yes	No	Electron
Luk et al. (1981)	MW (915, 2450)	Some	No	Photon, electron
Overgaard (1981)	RF (27.12)	Yes	?	Electron, Co-60
Abe et al. (1982)	MW (2450)	Yes	Yes	?
Corry et al. (1982a)	US (1–3)	Yes	Yes	Co-60
Kim et al. (1982b)	RF inductive (27.12) RF conductive (13.56)	Yes	No	Electron (most)
Arcangeli et al. (1983)	MW (500)	Some	Some	5.7 MeV X-rays
Scott et al. (1983)	MW (434, 915, 2450)	Yes	Yes	Electron (most)
Li et al. (1984)	MW (915, 2450)	?	?	250 kV, Co-60 8 MeV X-rays
Lindholm et al. (1984)	MW (915, 2450)	Yes	Yes	Photon, electron
van der Zee et al. (1985)	RF, MW (27, 433, 2450)	Yes	Some	Electron (most)

MW, microwave; *US*, ultrasound; *RF*, radiofrequency.

Table 3. Treatment parameters in the combined treatment of superficial tumors

Authors	Irradiation			Hyperthermia				Sequence
	Daily fractions	Days/ week	Total dose	Temperature	Time (min)	Days/ week	Total No. of treatments	
Bicher et al. (1980)	4 Gy	2	16 Gy	42°-45° C	90	2	8	4× HT alone RT-HT
Marmor and Hahn (1980)	2-4 Gy	2-5	20-60 Gy	43° C	45	2-5	Not stated	HT-RT-HT
U et al. (1980)	2-6 Gy	2-5	18-42 Gy	42°-44° C	40-50	1-3	2-9	RT-HT
Fazekas and Nerlinger (1981)	1.8-3.5 Gy	3	20-35 Gy	42.5°-43.5° C	30-40	2	6-8	RT-HT
Luk et al. (1981)	3 Gy	?	15-42.5 Gy	42.5° C	60	3	6-9	RT-HT
Overgaard (1981)	5-10 Gy	1-2	15-40 Gy	43° C	30	After every X-ray		HT immediately after RT or 3-4 h after RT
Abe et al. (1982)	?	?	12-70 Gy	41°-43.5° C	20-30	1-3	?	RT-HT
Corry et al. (1982a)	4 Gy	3	24-40 Gy	43.5° C	60	3	8-10	HT-RT
Kim et al. (1982b)	3.3-6.6 Gy	1-2	38.5-42.4 Gy	42°-43.5° C	30	1-2	6-13	HT-RT
Arcangeli et al. (1983)	2/1.5/1.5 Gy	5	60 Gy	42.5° C	45	3	7	RT (2nd)-HT
	5 Gy	2	40 Gy	42.5° C	45	2	8	RT-HT
	6 Gy	2	30 Gy	45° C	30	2	5	RT-HT
Scott et al. (1983)	2 Gy	5	60-66 Gy	42°-44° C	30-45	2	10-14	RT-HT
	4 Gy	2	48-52 Gy	42°-44° C	30-45	2	10-13	RT-HT
Li et al. (1984)	2 Gy	5	40-80 Gy	41°-44° C	40	2	10-12	RT-HT
Lindholm et al. (1984)	3 Gy	5	30 Gy	41°-45° C	45	2	4	RT-HT
van der Zee et al. (1985)	2-5 Gy	2	14-25 Gy	39.5°-43.6° C	40	2	3-10	RT-HT

RT, radiotherapy; *HT*, heat therapy.

The response rates observed in the studies to which we refer in Tables 2 and 3 are shown in Table 4. Diseases of varying histology were evaluated for response: adenocarcinoma of the breast, squamous cell carcinoma of the head and neck, malignant melanoma, sarcoma, and in a smaller number of patients some other histologies. The rate of complete responses in matched comparable lesions has been determined after irradiation plus hyperthermia and after irradiation alone. As mentioned before in a certain number of cases different nodules of the same patient received either irradiation alone or irradiation combined with hyperthermia. After irradiation alone on average the complete response was around 30%. The corresponding value for the combined modality was twice as high and amounted to 63%.

The relatively low rate of complete responses in the investigations of van der Zee et al. (1985) might have been due to a large number of treatments of bulky or superficially extended tumors. However, considering the studies in Table 4 one can conclude that irradiation plus heat is clearly more effective than irradiation alone. Furthermore it seems that in principle the advantage of the combined treatment modality also exists when moderate total radiation doses are applied together with hyperthermia.

Table 4. The rate of complete responses *(CR)* after irradiation plus hyperthermia or after irradiation alone

Authors	Criteria of response	RT+HT	RT alone	Same patient comparisons	Evaluable patients (treatment trials)
Bicher et al. (1980)	Total response	62%	–	–	37
Marmor and Hahn (1980)	Superior response versus PR or >6 weeks difference in time to regrowth	47%	7%	15	15
U et al. (1980)	CR within 1 month of therapy	86%	14%	7	7
Fazekas and Nerlinger (1981)	CR	50%	–	–	32
Luk et al. (1981)	CR	41%	–	–	37
Overgaard (1981)	CR persistent (>2 months)	44%	38%	21	25
Abe et al. (1982)	CR	55%	–	–	22
Corry et al. (1982a)	CR	62%	0%	13	21
Kim et al. (1982b)	CR during follow-up	80%	33%	59	86
Arcangeli et al. (1983)	CR at completion of therapy or soon after	73%	42%	26	26
		64%, 78%[a]	35%	17	17
		67%, 77%[b]	38%	16	16
		87%	35%	15	15
Scott et al. (1983)	CR at 6 months	87%	39%	31	31
Li et al. (1984)	CR	68%	29%	31	30
Lindholm et al. (1984)	CR	56%	29%	17	17
van der Zee et al. (1985)	CR (PR)	14% (73%)	–	–	44

[a] HT once (64%) or twice (78%) per week.
[b] HT immediately (77%) or 4 h (67%) after RT.
RT, radiotherapy; *HT*, heat therapy; *PR*, partial response.

Table 5. Long-lasting tumor control in patients initially exhibiting a complete response after irradiation and hyperthermia. (Perez and Meyer 1985)

	No. of treated lesions	Tumor regression		Tumor control
		Complete	Partial (> 50%)	No evidence of recurrence for 6–42 months
Epidermoid cancer, Head and neck	53	26 (49%)	17 (32%)	22/26 (85%)
Epidermoid cancer head and neck (nonmeasurable)	9	Not available	Not available	7/9 (77%)
Epidermoid cancer in other locations	3[a]	0	3 (100%)	0
Adenocarcinoma	37	19 (51%)	11 (30%)	18/19 (95%)
Melanoma	23	16 (70%)	6 (26%)	16/16 (100%)
Sarcoma	5	4 (80%)	1 (20%)	3/4 (75%)

[a] Three treatments only.

Perez et al. (1983) gave a report on tumor controls lasting from several months to several years. Tumor doses ranged either from 10 to 30 Gy or from 30 to 40 Gy. Irradiation was given twice weekly in doses of 4 Gy (usually electrons of 9–16 MeV). Heat treatment started within 30 min after irradiation (in most cases 915 MHz, 41°–43° C for 60–90 min, twice a week). The results are summarized in Table 5 (Perez and Meyer 1985). They suggest enhanced tumor response and control after moderate doses of irradiation and hyperthermia.

The tolerance of normal tissues exposed to local hyperthermia and irradiation is acceptable. Table 6 lists a spectrum of side effects as reported by different groups. A number of authors found no significant enhancement of skin reactions. However, skin lesions induced by heat in combined treatments seemed to depend on the temperature level, on the single and total radiation dose, and on the period between irradiation and hyperthermic treatment (Luk et al. 1981; Arcangeli et al. 1983). Active skin cooling was an effective protection method when high radiation doses per fraction (6 Gy) and high temperatures (45° C) following immediately after irradiation were applied (Arcangeli et al. 1983). In cases of skin damage the blisters and burns healed within weeks or months. Regarding local pain and sensory reactions relief was normally obtained within minutes of cessation of therapy.

When compared with experimental animals (Fajardo 1984; Hume 1985) in humans the knowledge of the toxicity of heat is only limited. This holds true especially for late effects. Therefore the Radiation Therapy Oncology Group (RTOG, United States) started prospective clinical trials with the aim of evaluate besides therapeutic efficacy also the tolerance of normal tissues to a combination of radiotherapy and hyperthermia (Luk et al. 1984; Perez and Meyer 1985).

Disregarding the open questions concerning the toxicity of heat one has to be aware that the normal tissue is a most important factor which determines the amount of therapeutic gain. It is obvious that the treatment conditions (single and total radiation dose, temperature or heat dose, fractionation scheme, normal tissue protection by cooling, technical parameters, etc.) must be chosen in such a way that the thermal enhancement of radiation damage in normal tissues is comparatively low. Only when the thermal enhancement ratio (TER) for tumor control is higher than the TER for lesions of surrounding normal structures can an improved therapeutic gain be obtained.

Table 6. Side effects of localized hyperthermia treatment when combined with irradiation

Authors	Treatment courses	Incidence (%)	Effects
Bicher et al. (1980)	37 lesions	5	Skin burns
Fazekas and Nerlinger (1981)	59 lesions	–	No immediate or late enhancement of radiation-induced skin damage
Luk et al. (1981)	48 patients	9	Blisters (healing spontaneously); tumor temp. <42.5° C
		54	Blisters (healing spontaneously); tumor temp. 42.6°–43.9° C
		33	Blisters (healing spontaneously); tumor temp. >44° C
		56	Burns; tumor temp. >44° C
Overgaard (1981	14 lesions	–	No thermal enhancement of skin response, RT after 4 h HT
	10 lesions	–	Thermal enhancement ratio for skin response between 1.2 and 1.4, RT and HT simultaneously
Abe et al. (1982)	22 patients	–	No evidence for increase of skin damage after RT+HT
Corry et al. (1982b)	30 patients	13	Blisters (healing withing weeks)
		20	Pain (relief within minutes of cessation of therapy)
		10	Radicular pain
Kim et al. (1982a)	>100 lesions	<10	In general no disproportionally enhanced skin and subcutaneous reaction
Marmor et al. (1982)	52 patients	23	Pain
		19	Superficial burn
		10	Tumor ulceration
		2	Vomiting
		2	Laryngospasm
U et al. (1982) [cited in Perez and Meyer (1985)]	53	13	Superficial burn
Arcangeli et al. (1983)	123 lesions	33–64	Moist desquamation after RT+HT (27%–38% after RT alone); skin reaction especially when using high single RT doses and HT immediately after RT. Protection by cooling
Scott et al. (1983)	58 patients	–	Skin reaction rarely increased in the heated field
Lindholm et al. (1984)	51 lesions	Often	Moderate to severe local pain during treatment (2450 MHz)
		Occasionally	Necrosis of up to a few cm in involved skin (healing up to 5 months)
Scott et al. (1985)	200 patients	5	Neural anomalies; numbness and tingling in the ipsilateral arm during chest treatment was most common
van der Zee et al. (1985)	44 lesions	23	Blisters (healing within 2.5 months)

Regional Hyperthermia of Deep-Seated Tumors

In contrast to superficial malignant tumors the experience concerning regional heating of deep-seated malignancies is less. This is due to the fact that adequate treatment at depth represents a complex problem. Especially, one comes upon technical difficulties in determining temperatures of relevant tumor and normal tissue sites due to limited accessibility. Corry and Barlogie (cited in Perez et al. 1984) described the characteristics of the ideal regional hyperthermia system in the following way: (1) evaluable depth of penetration with half penetration distances of up to 12 cm; (2) evaluable field size from 2 to 30 cm; (3) field localization to the tumor to minimize normal tissue toxicity and optimize therapeutic range; (4) simple and safe patient machine interface which is preferably noninvasive, not requiring contact with the body surface and permitting operation with an awake and alert patient; (5) operation independent of anatomic location; and (6) allowance for simple and accurate thermometry and dosimetry.

Several devices have been designed for regional hyperthermia in deep sites. However, there is no system which fulfills all the criteria quoted above. Technically the developed apparatuses are based on the application of electromagnetic waves (Dewhirst et al. 1982; Dewhirst et al. 1984; Gibbs 1984; Gibbs et al. 1984a; Turner 1982; Turner 1984), magnetic-loop induction (Baker et al. 1982; Elliott et al. 1982; Hagmann and Levin 1983; Oleson 1984; Oleson et al. 1983a, b; Ruggera and Kantor 1984; Storm et al. 1979a, b, 1981, 1982a, b, 1984; Strohbehn 1982), focused ultrasound (Fessenden 1984; Lele 1985; Marmor 1983), and interstitial radiofrequency or microwave current fields (Dewhirst et al. 1982; Meoz et al. 1983; Strohben et al. 1984). The most widely used methods are inductive electromagnetic heating by circumferential coil (magnetic loop) and electromagnetic wave hyperthermia by annular phased array.

Storm et al. (1979a, b, 1982a, b, 1984) reported on their experiences with radiofrequency magnetic-loop applicators. These are self-resonant, noncontact circular structures with built-in impedance matching circuitry (Magnetrode, developed by Drs. F. K. Storm and D. Horten and the Henry Medical Electronics, Inc., Los Angeles, United States of America). The coil is a single turn of a rolled conducting sheet that overlaps in a nonconducting manner. The area of the overlap and the gap distance provide the proper amount of capacitance. The element parameters have been selected to produce very large circulating currents in the loop structure. The currents create a strong electromagnetic field into which the body or limp is immersed. Since the human body is nonmagnetic, interaction is solely with the electric field, which consists of concentric circular field lines. A matching network has been interposed between the applicator and the 13.56-MHz generator to correct the problems of impedance change caused by the introduction of the torso or limb to be heated. Storm and associates (1979a, b) stated that this device allows safe hyperthermia to be applied to any depth without preferential surface tissue heating.

Selective heating was observed (Storm et al. 1979b) in both primary and metastatic tumors located in deep sites. Thus in a primary intraabdominal sarcoma (10 × 15 × 30 cm) and in a metastatic colon carcinoma in liver (10 × 10 cm) temperatures of 50° C and 53° C respectively have been measured using commercial needle thermistors (Yellow Springs Instrument Co., Yellow Springs, Ohio, United States). In both cases the temperatures of the surrounding normal tissues and especially of the skin (about 39° C) were significantly lower. The response appeared to be related to tumor size as differential heating was possible more often in larger lesions. Selective heating > 45° C was obtained in 43% of tumors < 5 cm ($n = 21$) in greatest diameter and in 73% of tumors > 5 cm ($n = 15$). In tumors successfully heated, moderate to marked necrosis occurred.

Baker et al. (1982) performed regional hyperthermia using the Magnetrode system. The 107 patients suffered from a variety of advanced cancers. Heat treatment was combined with radiation therapy, chemotherapy, or both. Surgery followed in a number of cases. When a normal 5- to 6-week course of radiation therapy was planned, hyperthermia was employed twice weekly. In the case of previous radiation treatment when a reduced dose of only 25 Gy in 2 or 3 weeks was intended, hyperthermia was given 3 times/week. In both situations, application of heat (30 min) followed 1 h after irradiation. The anatomic sites which have been heated are the abdomen (liver, pancreas, other), the pelvis, the chest wall, the lung and mediastinum, the head and neck, the axilla, the groin, and the extremities.

At the first hyperthermia treatment temperature measurements of the skin, subcutaneous tissues, occasionally the abdominal wall, and all accessible tumors were performed. Rigid and flexible thermistor probes (Yellow Springs Instrument, Yellow Spring, Ohio) were passed through Teflon catheters placed in the tissues under local anesthesia. In subsequent treatments power output was regulated on the basis of temperatures previously observed and the tolerance of the patient. In some patients thermometry was repeated at intervals. A significant rise in temperatures of the skin occurred infrequently. Often they fell as the treatment proceeded. The temperature of the subcutaneous fat was elevated but seldom above 43° C. Temperatures were also measured in the stomach, the common bile duct, and the bladder, rectum, and colostomies. They were never significantly elevated and usually did not exceed the core temperature. Thermistors could be placed in one or two sites of 99 tumors. The maximum tumor temperatures ranged between about 37.5° C and about 50.0° C. In more than one-half of the sampled sites temperatures at or above 42° C were measured. When two sites in a tumor were compared differences of up to 2° C occurred.

Baker et al. (1982) observed an overall complete response of 16%, partial response of 52%, and no response of 32%. Although significant differences in the measured temperatures did not occur, in the group of deeper-seated tumors (abdomen, pelvis, lung, mediastinum) the rate of complete response was lower than in head and neck tumors, in tumors of the chest wall, axilla, groin, and extremities (Table 7). Striking pain relief was obtained after only a few treatments in both responders and nonresponders. Complications were not frequent and seldom serious. In conclusion Baker and associates (1982) stated con-

Table 7. Responses according to anatomic site. (Baker et al. 1982)

Site	Number treated	Complete response	Partial response	No response
Abdomen				
Liver	14	0	9	5
Pancreas	8	0	1	7
Other	9	0	4	5
Pelvis	28	1	19	8
Chest wall	8	3	4	1
Lung, mediastinum	10	0	7	3
Head and neck	17	9	5	3
Axilla, groin, extremities	13	4	7	2
Total	107	17 16%	56 52%	34 32%

firming the findings of Storm et al. (1979a, b) that the Magnetrode is capable of heating tumors at any depth with few adverse side effects. However, there are reports in the literature which suggest that this characterization of the circumferential coil device must be corrected. The latter will be discussed below.

A further device for heat treatment in deep-seated tumors is the annular phased array. Basically the system is a radiofrequency generator and a temperature-monitoring feedback control system. The annular phased array (APA) utilizes the principle of phased reinforcement to produce an enhanced penetration deep within the body.

Two octogonal, parallel arrays with 16 elements radiating radiofrequency form an open core through which the patient is placed. Three distilled water-filled vinyl bags couple the electromagnetic fields to the patient and provide surface cooling since the water can be circulated through a heat exchanger. The APA, which is available commercialy (BSD Medical Corporation, Salt Lake City, Utah, United States of America) is used in conjunction with the BSD-1000 radiofrequency generator. The signal of the generator (50–110 MHz) is fed into a 2-kW amplifier. The latter drives the APA, which is located in an adjoining radiofrequency isolation chamber. The thermometry system consists of high resistance lead thermistors essentially noninteractive in radiofrequency fields. The thermistors are inserted into 16-gauge Teflon catheters which under radiographic control must be placed invasively into the tumor and normal tissues. By a mechanized semiautomated "thermal mapping system" (Gibbs 1983) the thermistors can be moved in the catheters at fixed steps of 0.5 cm or greater every 4 s. Finally the values of recorded temperatures are plotted in a temperature-distance graph showing heat distribution along a catheter.

In a careful pilot study (Sapozink et al. 1984) 46 patients with advanced abdominal or pelvic malignancy were treated with the APA. A number of patients received concurrent radiation with photons from a linear accelerator. It was intended to maintain minimum tumor temperatures of 42°–43° C for a period of 30 min. However, priority was given to avoiding injury. Thus heavy sedation and analgesia were not performed and power was reduced or the treatment terminated when significant discomfort or major physiological stress occurred. The observations made during these treatments have been summarized (Gibbs 1984b; Sapozink et al. 1984; Stewart et al. 1984) as follows.

Due to adverse clinical circumstances extensive thermometry was performed in only 22 of 46 treated patients. Several reasons led to exclusion of patients from thermometry analysis. However, only in three cases did unfavorable tumor versus normal tissue heating cause discontinued treatment. On average, the time for reaching 41° C at some point of the tumor was 19 min in pelvic and 26 min in abdominal lesions (one-half of the patients with abdominal, one-half with pelvic disease). All tumors showed temperatures of at least 41° C. In 12 of 22 cases 43° C was either attained or readily attainable (average induction times to reach 43° C, 27 min in the pelvis and 32 min in the abdomen). During the entire course of treatments 41° C could be obtained or exceeded at 75% of the pelvic and 63% of the abdominal tumor points. The corresponding values for a temperature level of 43° C were 29% and 22% respectively. The mean period during which temperatures of 43° C could be maintained was only 14 min for the pelvis and 12 min for the abdomen. Temperature differences of 10° C and peaks of 51° C were occasionally observed in large necrotic tumors. It has been pointed out that such large blood flow-related temperature variation made assessment of uniformity of energy deposition impossible.

As a heating parameter the authors defined a "quality factor" (Q), being the product of the mean sampling fraction of temperatures exceeding a given index temperature and the length of time it was possible to maintain such temperature divided by the target time of 30 min. An optimal Q-value would be equal to one if all tumor temperatures equaled or exceeded the index temperature for 30 min. The following Q-values were obtained for 41° C and 43° C respectively: 0.7 and 0.13 in the pelvis, 0.55 and 0.12 in the abdomen.

Temperature could not be measured at all intratumor points. Therefore the measure of accuracy to which the temperature monitoring reflected the actual temperature distributions throughout the tumor remains unclear to a certain degree. However, when compared with temperature measurement in other techniques of deep heating that of the Salt Lake City Group (Gibbs 1983, 1984b; Sapozink et al. 1984) is the most reliable.

With regard to patient discomfort no significant persistent side effects occurred (Sapozink et al. 1984). Forty-six patients have been evaluated. Acute local side effects were defined as symptoms referable to normal tissue within or near the heat treatment volume. They subsided when the power was reduced. Acute systemic side effects corresponded to generalized signs or symptoms. Apparently in most cases they were due to systemic heating. Subacute side effects lasted for longer than 48 h after completion of a heat treatment. Table 8 summarizes the side effects observed for pelvic and abdominal treatments.

It was the primary goal of the investigations of Sapozink et al. (1984) to study the heating and thermometry technique with the APA system. However, in 23 patients therapy response has been quantified. The subjective response rates were 83% and 54% respectively

Table 8. Side effects due to pelvic (28 patients) or abdominal (18 patients) treatment. (From Sapozink et al. 1984)

	Local (acute)				Systemic (acute)		Subacute	
	Pain		Other					
Pelvis	Bladder spasm	8	Nausea	4	Tachycardia > 120	5	Minor superficial	2
	Pelvis	8	Urinary retention[c]	2	Systemic heating > 39° C	2	burns	4
	Abdomen	8	Dyspnea[d]	2	Dysphoria	2	Pain persistent > 48 h	1
	Back[a]	5	Vomiting	1	Anxiety	1	Radiation proctitis	
	Groin	5	Hypoesthesia	1			Related to tumor	
	Thigh	4	Diarrhea	1			invasion	
	Leg	4					and necrosis femoral	
	Foot	2					artery	
	Penis	2					leak requiring surgery	1
	Testis	1					Perianal abscess	1
	Perineum	1					Vaginal vault necrosis	1
	Ureteral[b]	1					Vaginal bleeding	1
	Flank	1						
Abdomen	Back[e]	6	Nausea	4	Tachycardia > 120	10	Pleural effusion	
	Abdomen	3	Vomiting	1	Systemic heating > 39° C	10	following	
	Flank	2	Dyspnea	1	Fatigue	3	probe placement	1
	Abdominal	2			Hypertension, diastolic		Pain persistent > 48 h	1
	wall				pressure > 100 mmHg	3		
	Rectum	1			Hypotension, systolic			
	Pelvis	1			pressure < 90 mmHg	1		
	Leg	1			Claustrophobia	1		
	Colicky	1						

[a] In two patients pain was positional.
[b] Pretreatment of ureteral obstruction.
[c] Both patients had received narcotic analgesics chronically.
[d] Both patients had marked chronic obstructive pulmonary disease.
[e] In three patients pain was positional.

for pelvic and abdominal treatments. Median survival reflected that the patients were in advanced disease at the beginning of the combined treatment. In the case of pelvic disease it was 15 months, in the case of abdominal disease 4 months.

The toxicity and the heating patterns of the Magnetrode and the APA have recently been compared (Gibbs 1984b; Sapozink et al. 1985). Heat treatment was successively performed with each device in 22 patients suffering from advanced pelvic or abdominal malignancies. The temperature distributions were recorded in identical sites. Thermometry catheters were placed invasively into tumor and normal tissues under computed tomography guidance or in accessible cavities of the body. The authors specified the placement of catheters by physical examination, plain radiography, and computerized tomography. Spatial thermometry was performed along individual catheters left in place between comparative treatments.

It has been observed that in comparison to the Magnetrode the APA produced broader regional heating, especially in deep sites. Temperature and power densities decreased with increasing depth in the Magnetrode treatments. The latter confirms the results of Oleson and associates (1983b). These authors treated visceral or extensive superficial malignant tumors of 31 patients with the Magnetrode and obtained heating throughout to >-42.5° C only in relatively superficial tumors. In the comparison between the APA and the Magnetrode Sapozink et al. (1985) noted that Magnetrode hyperthermia was limited particularly when performed in the pelvis. An ineffective central power deposition and severe sacrococcygeal pain were observed.

Considering the spatial thermal dose (TD) the investigations (Gibbs 1984b, Sapozink et al. 1985) revealed higher minimum tumor TDs and more favorable mean tumor/normal tissue TD ratios with the APA than with the Magnetrode. Thus it has been concluded that the APA is superior to the Magnetrode for pelvic and probably for abdominal treatment. However, there are also problems with abdominal heating using the APA, such as power output limitations, normal tissue heating usually of liver, and a high rate of systemic heating. With the Magnetrode in abdominal heating local pain was common, and the predominant power-limiting factor was normal tissue heating involving the abdominal wall, peripheral liver, or the flank. The Magnetrode has been recommended for effective heating near the surface of the body especially in the muscle annulus of the abdomen or in the shoulder girdle (Gibbs 1984b; Oleson et al. 1983b).

Considering these latter results, the statement by Storm et al. (1979a, b) that the Magnetrode allows safe hyperthermia to any depth must obviously be corrected. The investigations of Gibbs (1984b) and Sapozink et al. (1985) clearly demonstrate the importance of spatial temperature measurements. The aim should be to record temperature with time at different points throughout the tumor and in critical normal tissues. In a well-controlled treatment the physician should have a detailed idea of the three-dimensional heat distribution in the tumor, of the tumor/normal tissue thermal dose ratio, and of the time interval in which the desired temperature is reached and maintained. Relating tumor temperatures only to tumor size alone as in the investigation of Storm et al. (1979) neglects the fact that within the tumor considerable temperature variations occur with a lesser likelihood of heating in deeper sites (Oleson et al. 1983b).

A very important point in clinical hyperthermia of deep-seated tumors is the psychological support and encouragement for the patient. It has been pointed out the more knowledgeable the patient is about the procedures, the better he or she will be prepared to withstand the physiological changes and the confinement during treatment (Perez et al. 1984). Perez and associates (1984) recommend performing regional hyperthermia as inpatient procedures. To decrease anxiety diazepam may be given (10 mg i.m. or orally). Analgesics may be required. Before initiation of therapy the stomach and bladder must be empty.

During the treatment period general condition, systemic temperature, pulse, and blood pressure are monitored. Systemic temperature (esophageal thermometer) should not exceed 40.5° C. According to Perez et al. (1984) the following parameters are considered to be reasons to terminate a hyperthermic treatment: tachycardia >160–180/min (depending on age), cardiac arrhythmia, significant systolic or diastolic blood pressure change, chest pain, breathing difficulties, excessive anxiety, or hallucination. Patients with one of the following conditions are not introduced into a treatment with deep hyperthermia: major systemic infections, poorly insulin controlled diabetes mellitus, unstable cardiac situation, severe liver, kidney, brain, blood vessel diseases, psychotic or depressive history.

Prognostic Variables

The significance of prognostic variables as listed in Table 9 has been extensively investigated under experimental conditions (for reviews see Hand, this volume; Streffer and van Beuningen, this volume; Vaupel and Kalinowski, this volume). In addition there are important clinical observations constituting a base on which clinical studies have been initiated and will start in the future.

Techniques of hyperthermia and thermometry are summarized in this volume (Hand). Advantages and disadvantages of the currently available devices with regard to the clinical situation have recently been reviewed (Abe and Hiraoka 1985). For local and regional heating various techniques including electromagnetic waves (radiofrequencies and microwaves) and ultrasound can be used. It is the greatest disadvantage of ultrasound that tumors surrounded by aerated tissues or bone cannot be heated due to reflection of ultrasound at gas/tissue and bone/tissue interfaces. However, microwave devices and capacitively or inductively coupled generators of radiofrequency currents are also not completely satisfactory for effective heat deposition in deep-seated tumors. The latter has already been described.

Table 9. Prognostic variables in hyperthermia treatment

I.	Techniques	1. External: local or regional	
		2. Interstitial	
		3. Systemic	
II.	Physics	1. Heat	a) Temperature
			b) Thermal dose
			c) Sequence to RT
			d) Fractionation (thermotolerance)
		2. Radiation	a) Single dose
			b) Total dose
			c) Fractionation
III.	Tumor	1. Histology	
		2. Volume	
		3. Physiology (blood flow, hypoxia, etc.)	
IV.	TER normal tissue		

Recent investigations concerning invasive methods show possibilities by which the problem of heat treatment of deep-seated tumors might at least partly be solved. The following techniques of interstitial and intracavitary hyperthermia exist: interstitial and intracavitary insertion of small antennae which emit electromagnetic waves (Mendecki et al. 1982); interstitial implantation of ferromagnetic seeds heated by a magnetic induction device (Stauffer et al. 1982); and interstitial needle implants which heat by the passage of radiofrequent currents (Manning et al. 1979; Cosset et al. 1984). Considerable response rates (CR at about 70%) have been reported after applying these techniques in combination with radiotherapy (Manning et al. 1982; Cosset et al. 1984). Recently it has been shown that intraoperative interstitial microwave-induced hyperthermia and brachytherapy is feasible and produces controlled, localized hyperthermia with temperatures of 50° C or more in tumors (Coughlin et al. 1985).

Considering the influences of physical parameters many important clinical investigations have been carried out by Arcangeli et al. (1983). Thus these authors showed a tumor control of 75% after irradiation plus hyperthermia. In contrast irradiation alone led to a response rate of only about 40%. In the combined modality the enhancement of tumor control appeared to be related to the magnitude of given heat as well as to the size of single radiation fractions. An increase of both increased the tumor control (Agcangeli et al. 1983, 1984). The total radiation dose has also been found to be an important factor. Treating a variety of superficial metastatic and recurrent lesions (Bicher and Wolfstein 1984; Luk et al. 1984a; Perez et al. 1984b; van der Zee et al. 1984) the best tumor response was obtained after moderate doses of about 40 Gy. In most cases higher doses could not be applied due to previous radiotherapy. It is remarkable that in breast carcinoma an advantageous effect also resulted after rather low doses (14-25 Gy; van der Zee et al. 1985).

Adequate heating has been reported to be one of the main treatment parameters influencing the response after irradiation plus hyperthermia. Investigating temperature variables, Oleson et al. (1984) found a good correlation between CR (complete disappearance of tumor) and the minimum tumor temperature averaged over all treatments. In a series of 70 patients with breast carcinoma van der Zee et al. (1984) showed a positive correlation between the overall response and the mean tumor temperature. A thermal enhancement ratio of 2.6 was calculated for a mean temperature of 41.6° C. In principle the observations of Luk et al. (1984) and Sim et al. (1984) confirm these results. The latter group (Sim et al. 1984) demonstrated the importance of achieving a minimum temperature of 42.5° C through the whole tumor. Tumors in which several points were above the minimum selected temperature had twice the complete response rate when compared with tumors in which no point reached that temperature. Defining the therapeutic enhancement factor (TEF) as the ratio of thermal enhancement (TE) of tumor response to TE of skin damage, Arcangeli et al. (1983) obtained a TEF of 2.1 at selective tumor heating with a temperature of 45° C. In this case heat was applied immediately after irradiation. The radiation dose per fraction amounted to 6 Gy (total dose 30 Gy). A minor but also satisfactory TEF of 1.58 resulted when conventional single radiation doses of 1.5-2.0 Gy (total dose, 60 Gy) were given in combination with 42.5° C (Arcangeli et al. 1983).

The relationship between a certain temperature level and time of heat exposure has been studied in vitro and in vivo. Above 43° C in most systems an increase of 1° C allowed a twofold decrease in time for the same biological effect. Below 43° C for isoeffects even a three-to fourfold decrease in time was required (Field 1978; Henle and Dethlefsen 1978). Establishing a thermal dose Sapareto and Dewey (1984) developed a computer model integrating temperature profiles during the period of heat treatment. Recently, Gerner (1985) described that efforts to define a unit of thermal dose still fall into different categories. When quantifying in tumors as well as in normal tissues the effect of hyperthermia at different temperatures and heating times a generally accepted heat dose would be a most valuable reference parameter. The results of clinical hyperthermia trials in different centers could be compared in the same way as is done in radiation therapy on the base of radiation dose. It is obvious that with regard to a reliable heat dose careful multipoint thermometry has to be performed. Gibbs et al. (1985) discussed the great importance of temperature measurements in detail.

The sequence of irradiation and heat is a further factor which influences the efficacy of the combined treatment modality. van Beuningen and Streffer (1985) have reviewed experimental observations concerning this point. The highest thermal enhancement ratios (TER) were obtained for tumors as well as for normal tissues when heat and radiation were applied simultaneously. This has been confirmed clinically in treatments of cutaneous or lymph node metastases of malignant melanoma (Overgaard 1981). Simultaneous radiation and heat (42.5° C - 60 min) gave thermal enhancements of approximatively 2.5 in both tumor and surrounding skin. However, in such a situation a therapeutic gain cannot be obtained due to damage of the normal tissue. Thus Overgaard (1985) proposed giving heat treatment immediately after radiation only if normal tissue heating can be avoided. Otherwise an interval of about 3-4 h should pass between radiation and hyperthermia. A biological reason for this latter procedure is the observation that normal tissues recover from radiation damage within about 4 h. Thereafter in contrast to tumors the radiosensitzing effect of heat no longer occurs in normal tissues (for review see van Beuningen and Streffer 1985). The latter is confirmed by clinical findings. In the study by Overgaard (1981) the TER for control of melanoma metastases decreased to about 1.4 with heating 4 h after irradiation and with 72-h intervals between the fractions. Although under these conditions the tumor TER is comparatively low, a therapeutic gain is obtained since using this schedule no thermal enhancement of skin damage was found.

However, since maximal radiosensitization of the tumor is achieved by a simultaneous application of radiation and heat, one should aim for a clinical situation in which tumor temperature increases immediately after irradiation. Only when the risk for the normal tissue is high (see above) should a period of about 4 h elapse. This problem may be circumvented in special cases by using different portals for the radiation and heat treatments.

Hyperthermia is performed within a fractionated treatment schedule. Therefore thermotolerance might become a problem. Depending on the temperature of the priming treatment heat sensitivity of tumors as well as of normal tissues can decrease as observed under experimental conditions. In general, the larger the priming heat dose the larger is the induced thermal resistence (for review see Jung 1983).

Although thermotolerance is a phenomenon occurring very generally it seems to be highly variable. There is a wide variation in both the extent of thermotolerance and the time course after prior heating. In different experiments the decay of thermotolerance varied between a few hours and several days (for review see Field 1985).

Studies on fractionated hyperthermia in murine tumors led to the conclusion that thermal resistance was probably one of the most critical factors for the success of clinical hyperthermia (Urano et al. 1982). On the other hand it has been pointed out that in the application of moderate radiosensitizing temperatures thermotolerance might be only of minor significance (Streffer and van Beuningen 1983). The definitive answer to the question of thermotolerance in heat treatments must be obtained from clinical investigations. As shown in Table 3 most clinicians performed hyperthermic treatment only twice a week. However, Arcangeli et al. (1984b) concluded from a complex clinical study that for the combination of radiation and hyperthermia there is no firm evidence of a reduction in the thermal enhancement ratio.

Finally, the importance of tumor factors for the efficacy of heat treatment should be briefly stressed. In Chap. 2 the significance of physiological parameters such as blood flow, hypoxia, and nutritional state has been described. The relevance of these variables has been discussed in detail by Vaupel and Kalinowski, this volume. There is a general agreement that tumor circulation is decreased as a result of hyperthermia treatment. However, considerable differences in the times and temperatures required to induce this effect exist (Reinhold et al. 1985). After an increase in oxygenation up to a "breaking point" at 41° C, in the higher temperature range a decrease in oxygenation occurs as observed in a clinical study (Bicher et al. 1984). Thus by this mechanism highly heat sensitive hypoxic cells are generated.

Considering the histology of a tumor it is well known that radiosensitivity significantly depends on this parameter. However, in contrast to their relative radioresistance, sarcomas and melanomas responded as well to irradiation and hyperthermia as epethilial tumors (for review see Perez and Meyer 1985).

More detailed information exists especially on breast carcinoma and neck nodes (Squamous cell carcinoma). The breast is rather easy to heat. A relatively homogeneous heating can be achieved (Hofman et al. 1984). Heating of neck nodes is also not problematic due to their superficial location (Arcangeli et al. 1985; Hofman et al. 1984; Perez et al. 1983; Valdagni et al. 1985). Overgaard (to be published) has established dose response relationships for breast and neck nodes (N2-N3) on the basic of data collected from the literature and partly personal experience. Comparing radiation alone vs radiation and heat at a level of 50% complete response, the TER values were 1.5 and 1.4, respectively. For malignant melanoma a thermal enhancement of 2.0 was obtained. This indicates that the combination of radiotherapy and heat is well suited for this tumor disease.

With regard to heat sensitivity, the tumor size seems to be most important. There are reports which described a correlation between this variable and the temperatures reached in hyperthermic treatments. Perez et al. (1984) observed that lesions smaller than 2 cm in diameter were heated to a temperature of 42.5° C or higher in 85% of the instances. However, for tumors measuring 2.1-4.0 cm or for those of greater diameter the corresponding values amounted to only 71% and 50%, respectively. Similar observations were made by Oleson et al. (1984) and Kim et al. (1982a). It has been pointed out that the poorer results in larger tumors are most likely due to physical factors, namely inadequate power deposition rather than negation of the biological advantages of hyperthermia (Perez and Meyer 1985).

Conclusions and Prospects

With regard to the regression of superficial tumors the combination of radiotherapy plus hyperthermia is superior to radiation alone and especially to exclusive application of heat. On the other hand, heating of deep-seated tumors still represents a complex technical problem. It is a challenge for physicists and engineers to develop systems which allow safe heat deposition in the depth, leading to temperatures of 42°-43° C. The required temperatures should be maintained for 30 min within the complete tumor volume regardless of whether it is spherically or irregularly shaped. Furthermore the option should also be fulfilled when the blood flow of the tumor is inhomogeneous and that of the surrounding normal tissue is highly variable (Bleehen 1985).

In principle, sufficient heat deposition can be obtained in the depth of the pelvis with the annular phased array. However, deep heating of the abdomen, and particularly of the chest, by external hyperthermic treatment meets with considerable difficulties. In the future for special deep sites intracavitary or interstitial hyperthermia will also be a promising way.

It is a great disadvantage in clinical hyperthermia that thermometry must be performed invasively. However, extensive temperature measurements should be the base on which heat supply is regulated during a treatment course. Furthermore, the definition of a heat dose obviously depends on thermometry. For reliable comparison of treatment results a generally accepted thermal dose concept might be helpful. As pointed out by Dewey (1985) the latter could comprise the equivalent time at a reference temperature. However, at present definitions of a unit of thermal dose still fall into different categories (Gerner 1985). In conclusion with regard to technical aspects methods of deep heating have to be improved, possibilities of noninvasive thermometry have to be found, and a practical concept for thermal dose should be developed for clinical hyperthermia. We are aware that the realization of this latter point presents considerable problems as heat distribution in the tumor is rather inhomogeneous.

Considering clinical investigations during future years randomized multicenter studies will test hyperthermia as an adjuvant to radiotherapy in the treatment of different tumor entities. A German protocol is briefly described on p. 114. J. Overgaard (Section of Experimental Radiotherapy, Institute of Cancer Research, Radiumstationen, DK-8000 Aaarhus C, Denmark) coordinates protocols initiated by the European Society for Hyperthermic Oncology (ESHO). The protocols concern locally advanced breast carcinoma, advanced neck nodes and malignant melanoma. All these national and international protocols besides tumor control also assess early and late tolerance of normal tissues exposed to radiation alone or radiation combined with hyperthermia. The insight into late normal tissue effects is necessary before starting with phase III trials in patients who have good prognosis for long survival. When hyperthermia is established as an efficient adjuvant to radiation therapy, in a further step heat-sensitizing methods such as the application of certain drugs (Issels et al. 1984; Wiegant et al. 1985) or the alteration of metabolic pathways (von Ardenne 1970/1971[19]); Kim et al. 1978; Tähde and Rajewsky 1982 can be tested systematically within the combined modality of radiation plus heat.

References

Abe M, Hirooka M (1985) Review, localized hyperthermia and radiation in cancer therapy. Int J Radiat Biol 47: 347-359

Abe M, Hirooka M, Takahashi M (1982) Combination of hyperthermia and radiation in cancer therapy. Strahlentherapy 158: 275-280

Arcangeli G, Nervi C (1984) The lack of clinical evidence of tumor thermotolerance after some schedules of combined heat and radiation. In: Overgaard I(es) Proceedings of the 4th international symposium on hyperthermic oncology, vol. 1, Summary papers. Taylor and Francis, London, pp 231-234

Arcangeli G, Civadilli A, Nervi C, Creton G (1983) Tumor control and therapeutic gain with different schedules of combined radiotherapy and local external hyperthermia in human cancer. Int J Radiat Oncol Biol Phys 9: 1125-1134

Arcangeli G, Nervi C, Cividalli A, Lovisolo GA (1984) Problem of sequence and fractionation in the clinical application of combined heat and radiation. Cancer Res (Suppl) 44: 4857-4863

Arcangeli G, Guersa A, Lovisolo G, Civadelli A, Marino C, Mauro F (1985) Tumour response to heat and radiation: prognostic variables in the treatment of nech node metastases from head and neck cancer. Int J Hyperthermia 1: 207-217

Arons J, Sokoloff B (1937) Combined roentgenotherapy and ultrashort wave. Am J Surg 36: 533-543
Baker HW, Snedecor PA, Goss IC (1982) Regional hyperthermia for cancer. Am J Surg 143: 586-590
Ben-Hur E, Elkind MM, Bronk BV (1974) Thermally enhanced radioresponse of cultured Chinese hamster cells: inhibition of repair of sublethal damage and enhancement of lethal damage. Rad Res 58: 38-51
Bicher HI, Mitagvaria PN (1984) Changes in tumor tissue oxygenation during microwave hyperthermia clinical relevance. In: Overgaard I (ed) Hyperthermic oncology 1984, vol 1. Summary papers. Taylor and Francis, London, pp 169-173
Bicher HI, Sandhu TS, Hetzel FW (1980) Hyperthermia and radiation in combination: a clinical fractionation regime. Int J Radiat Oncol Biol Phys 6: 867-870
Bicher HI, Wolfstein RS (1984) Microwave hyperthermia as an adjuvant to radiation therapy. Summary experience of 208 multifraction treatment cases. In: Overgaard I (ed) Proceedings of the fourth international symposium on hyperthermic oncology, vol 1, summary papers. Taylor and Francis, London pp 363-366
Bleehen NM (1985) Hyperthermia in the next stage in the oncologist's odyssey. In: Overgaard I (ed) Proceedings of the 4th international symposium on Hyperthermic oncology, vol 2. Review lectures, symposium summaries and workshop summaries. Taylor and Francis, London, pp 371-383
Bruggmoser G, Hinkelbein W, Engelhardt R, Wannenmacher M (eds) (1986) Locoregional high-frequency hyperthermia and temperature measurement Springer, Berlin Heidelberg New York Tokyo (Recent results in cancer research, vol 101)
Bruns P (1888) Die Heilwirkung des Erysipels auf Geschwülste. Beitr Klin Chir 3, 443
Busch W (1866) Über den Einfluß, welchen heftigere Erysipele zuweilen auf organisierte Neubildungen ausüben. Verhandl. des naturhistorischen Vereines der preußischen Rheinlande und Westphalens. 23, 28-30
Coley WB (1893) The treatment of malignant tumors by repeated inoculations of erysipelas, with a report of ten original cases. Am J Med Sci 105: 487-511
Coley WB (1894) Treatment of inoperable malignant tumors with the toxins of erysipelas and the *Bacillus prodigiosus*. Am J Med Sci 108: 50-66
Coley WB (1911) A report of recent cases of inoperable sarcoma successfully treated with mixed toxins of erysipelas and *Bacillus prodigiosus*. Surg Gynecol Obstet 13: 174-190
Corry PM, Spanos W, Tilchen EW, Barlogie B, Barkley HT, Armour EP (1982a) Combined ultrasound-radiation therapy treatment of human superficial tumors. Radiology 145: 165-169
Corry PM, Barlogie P, Tilchen EJ, Armour EP (1982b) Ultrasound-induced hyperthermia for the treatment of human superficial tumors. Int J Radiat Oncol Biol Phys 8: 1225-1229
Cosset JM, Dutreix J, Dufour J, Janora P, Damia E, Haie C, Clarke D (1894) Combined interstitial hyperthermia and brachytherapie institute Gustave Roussy technique and preliminary results. Int J Radiat Oncol Biol Phys 10: 307-312
Coughlin CT, Wong TZ, Strohbehn JW, Collacchio TA, Sutton JE, Belch RZ, Douple EB (1985) Intraoperative interstitial microwave-induced hyperthermia and brachytherapy. Int J Radiat Oncol Biol Phys 11: 1673-1678
Dewey WC (1985) Summary of biological studies. In: Overgaard I (ed) Proceedings of the 4th international symposium on hyperthermic oncology, vol 2: review lectures, Symposium sumaries and Workshop summaries. Taylor and Francis, London, pp 341-351
Dewey WC, Freeman MC, Raaphorst GP, Clark EP, Wong RSL, Highfield DP, Spiro IJ, Tomasovic SP, Denman DL, Coss RA (1980) Cell biology of hyperthermia and radiation. In: Meyn RE, Withers HR (eds) Radiation biology in cancer research. Raven, New York, pp 589-621
Dewhirst MW, Connor WG, Sim DA (1982) Preliminary results of a phase III trial of spontaneous animal tumors to heat and/or radiation: early normal tissue response and tumor volume influence on initial response. Int J Radiat Oncol Biol Phys 8: 1951-1961
Dewhirst MW, Sim DA, Sapareto S (1984) Importance of minimum tumor temperature in determining early and long-term responses of spontaneous canine and feline tumors to heat and radiation. Cancer Res 44: 43-50
Dietzel F (1975a) Tumor und Temperatur. Aktuelle Probleme bei der Anwendung thermischer Verfahren in Onkologie und Strahlentherapie. Urban and Schwarzenberg, Munich
Dietzel F (1975b) A summary and overview of cancer therapy with hyperthermia in Germany. Proceedings of the international symposium on cancer therapy by hyperthermia and radiation. Washington, D.C., April 28-30. The American College of Radiology, pp 75-84

Dietzel F, Ringleb D, Schneider U, Wricke H (1975) Microwave in radiotherapy of tumors - alternative to heavy particles? Strahlentherapie 146: 438-441
Elliott RS, Harrison WH, Storm FK (1982) Hyperthermia: electromagnetic heating of deep-seated tumors. IEEE Trans Biomed Eng 29: 61-64
Fajardo L-G (1984) Pathological effects of hyperthermia in normal tissues. Cancer Res. (Suppl) 44: 4826-4835
Fazekas JT, Nerlinger RE (1981) Localized hyperthermia to irradiation in superficial recurrent carcinomas: a preliminary report on 46 patients. Int J Radiat Oncol Biol Phys 7: 1457-1461
Fessenden P (1984) Ultrasound methods for inducing hyperthermia. Frontiers Radiat Ther Oncol 18: 62-69
Field SB (1978) The response of normal tissues to hyperthermia alone or in combination with x-rays. In: Streffer C et al. (ed) Cancer therapy by hyperthermia and radiation. Urban and Schwarzenberg, pp 37-48
Field SB, Bleehen NM (1979) Hyperthermia in the treatment of cancer. Cancer Treat Rep 6: 63-94
Field SB (1983) Cellular and tissue effect of hyperthermia and radiation. In: Steel GG et al. (eds) The biological basis of radiotherapy. Elsevier, Amsterdam, pp 287-303
Field SB (1985) Clinical implications of thermotolerance. In: Overgaard I (ed) Proceedings of the 4th international symposium on hyperthermic oncology, vol 2. Review lectures, symposium summaries and workshop summaries. Taylor and Francis, London, PP 235-244
Gerner EW (1985) Definition of thermal dose. Biological isoeffect relationships and dose for temperature-induced cytotoxicity. In: Overgaard I (ed) Proceedings of the 4th international symposium on hyperthermic oncology, vol 2. Review lectures, symposium summaries and workshop summaries. Taylor and Francis, London, pp 245-251
Gerhard H, Scherer E (1983) Lokalhyperthermie in der Behandlung fortgeschrittener Tumoren: kurative und palliative Wirkungen. Strahlentherapie 159: 521-531
Gibbs FA (1983) "Thermal mapping" in experimental cancer treatment with hyperthermia: description and use of a semiautomatic system. Int J Radiat Oncol Biol Phys 9: 1057-1063
Gibbs FA (1984a) Noninvasive electromagnetic heating techniques and the operational characteristics of the annular phased array (APA). Front Radiat Ther Oncol 18: 56-61
Gibbs FA (1984b) Regional hyperthermia: a clinical appraisal of noninvasive deep-heating methods. Cancer Res (Suppl) 44: 4765-4770
Gibbs FA, Sapozink MD, Gates KS, Stewart JR (1984) Regional hyperthermia with an annular phased array in the experimental treatment of cancer. Report of work in progress with a technical emphasis. IEEE Trans Biomed Eng 31: 115-119 (1984)
Gibbs FA, Sapozink MD, Steward JR (1985) Clinical thermal dosimetry: why and how. In: Overgaard J (ed) Proceedings of the 4th international symposium on hyperthermic oncology, vol 2. Review lectures, symposium summaries and workshop summaries. Taylor and Francis, London, pp 155-167
Giles U (1938) The historic development and modern applications of artificial fewer. New Orleans Med Soc J 91: 655-670
Hagmann MS, Levin RL (1983) Numerical evaluation of the helical-coil applicator for hyperthermia. Abstr Radiat Res 94: 534
Hahn GM (1982) Hyperthermia and cancer. Plenum, New York
Hall EJ (1978) Radiobiology for the radiologist. Harper and Row, Hagerstown
Henle KJ, Dethlefsen LA (1978) heat fractionation and thermotolerance: a review. Cancer Res 38: 1843-1851
Herbst M, Sauer R (1983) Zur Tumorbehandlung mit Hyperthermie und Radiotherapie. Strahlentherapie 159: 93-98
Hofman P, Langendijk JJW, Schipper J (1984) The combination of radiotherapy with hyperthermia in protocolzed clinical studies. in: Overgaard I (ed) Proceedings of the fourth international symposium on hyperthermic oncology, vol 1, summary papers. Taylor and Francis, London, p 379-382
Hume SP (1985) Experimental studies of normal tissue response to hyperthermia given alone or combined with radiation. In: Overgaard I (ed) Proceedings of the 4th international symposium on hyperthermic oncology, vol 2. Review lectures, symposium summaries and workshop summaries. Taylor and Francis, London, pp 53-70

Hymmen U, Wieland C (1976) Leistung und Wirkungsmechanismus einer lokalen kombinierten Strahlentherapie - Wärmeanwendung. Med Klin 71: 1183-1187

Issels RD, Biaglow JE, Epstein L, Gerweck LE (1984) Enhancement of cystamine cytotoxicity by hyperthermia and its modification by catalase and superoxide dismutase in Chinese hamster ovary cells. Cancer Res 44: 3911-3915

Jähde E, Rajewski MF (1982) Sensitization of clonogenic malignant cells to hyperthermia by glucose-mediated, tumor-selective pH reduction. J Cancer Res Clin Oncol 104: 23-30

Johnson HJ (1940) The action of short radiowaves on tissues: III A comparison of the thermal sensitivities of transplantable tumors in vivo and in vitro. Am J Cancer 38: 533-550

Jung H (1983) Modification of thermal response by fractionation of hyperthermia. Strahlentherapie 159: 67-72

Kim SH, Kim JH, Hahn EW (1978) Selective potentiation of hyperthermia killing of hypoxic cells by 5-thio D-glucose. Cancer Res 38: 2935-2938

Kim JH, Hahn EW, Ahmed SA (1982a) Combination hyperthermia and radiation therapy for malignant melanoma. Cancer 50: 478-482

Kim JH, Hahn EW, Antich PP (1982b) Radiofrequency hyperthermia for clinical cancer therapy. Nat Cancer Inst Monogr 61: 339-342

Lambert RA (1912) Demonstration of the greater susceptibility to heat of sarcoma cells as compared with actively proliferating connective-tissue cells. JAMA 59: 2147

Lele PP (1985) Ultrasound: is it the modality of choice for controlled localized heating of deep tumors? In: Overgaard J (ed) Hyperthermic oncology, vol 2. Review lectures, symposium summaries and workshop summaries. Proceedings of the 4th international symposium on hyperthermic oncology. Taylor and Francis, London, pp 129-154

Li GC, Evans RG, Hahn GM (1976) Modification and inhibition of repair of potentially lethal x-ray damage by hyperthermia. Radiat Res 67: 491-501

Li RY, Zhang TZ, Lin SY, Wang HP (1984) Effect of hyperthermia combined with radiation in the treatment of superficial malignant lesion in 90 patients. In: Overgaard J (ed) Proceedings of the 4th international symposium on hyperthermic oncology, vol 1, summary papers. Taylor and Francis, London, 395-397

Liebesny P (1921) Experimentelle Untersuchungen über Diathermie. Wien Klin Wschr 34: 117-118

Lindholm CE, Kjellen E, Landberg T, Nilsson P, Hertzman S, Persson B (1984) Microwave-induced hyperthermia and radiotherapy. Clinical results. In: Overgaard J (ed) Proceedings of the 4th international symposium on hyperthermic oncology, vol 1, summary papers. Taylor and Francis, London, pp 341-344

Luk KH, Purser PR, Castro JR, Meyler TS, Philips TL (1981) Clinical experiences with local microwave hyperthermia. Int J Radiat Oncol Biol Phys 7: 615-619

Luk KH, Pajak TF, Perez CA, Johnson RA, Conner N, Dobbins T (1984a) Prognostic factors for tumor response after hyperthermia and radiation. In: Overgaard J (ed) Proceedings of the 4th international symposium on hyperthermic oncology, vol 1, summary papers. Taylor and Francis, London, pp 353-356

Luk KH, Francis ME, Perez CA, Johnson RJ (1984b) Combined radiation and hyperthermia: comparison of two treatment schedules based on data from a registry established by the radiation therapy oncology group (RT06). Int J Radiat Oncol Biol Phys 10: 801-809

Manning MR, Cetas TC, Miller RC, Oleson JR, Connor WG, Gerner EW (1982) Clinical hyperthermia: results of a phase I trial employing hyperthermia alone or in combination with external beam or interstitial radiotherapy. Cancer 49: 205-216

Marmor JB (1983) Cancer therapy by ultrasound. In: Advances in radiation biology, vol 10. Academic, New York, pp 105-133

Marmor JB, Hahn GM (1980) Combined radiation and hyperthermia in superficial human tumors. Cancer 46: 1986-1991

Marmor JB, Pounds D, Hahn GM (1982) Clinical studies with ultrasound-induced hyperthermia. Nat Cancer Inst Monogr 61: 333-337

Mendecki J, Friedenthal E, Botstein C (1980) Microwave applicators for localized hyperthermia treatment of cancer of the prostate. Int J Radiat Oncol Biol Phys 6: 1583-1588

Meoz RT, Corry P, Frazier OH, Barlogie B, Armour E (1983) Hyperthermia and radiation therapy for the treatment of intrathoracic human malignant tumors: initial clinical and technical observations (abstract). Radiat Res 94: 541

Müller C (1910) Eine neue Behandlungsmethode bösartiger Geschwülste. Münch Med Wschr 57: 1490-1493
Müller C (1920) Über Stand und Ziele der Röntgentiefentherapie der Karzinome. Strahlentherapie 10: 749-757
Nauts HC (1985) Hyperthermic oncology: historic aspects and future trends. In: Overgaard J (ed) Hyperthermic oncology, vol 2. Review lectures, symposium summaries and workshop summaries. Proceedings of the 4th international symposium on hyperthermic oncology. Taylor and Francis, London, pp 199-209
Oleson JR (1984) A review of magnetic induction methods for hyperthermia treatment of cancer. IEEE Trans Biomed Eng 31: 91-97
Oleson JR, Cetas TC, Corry PM (1983a) Hyperthermia by magnetic induction: experimental and theoretical results for coaxial coil pairs. Radiat Res 95: 175-186
Oleson JR, Heusinkveld RS, Manning MR (1983b) Hyperthermia by magnetic induction: II. Clinical experience with concentric electrodes. Int J Radiat Oncol Biol Phys 9: 549-556
Oleson JR, Sim DA, Manning MR (1984) Analysis of prognostic variables in hyperthermia treatment of 161 patients. Int J Radiat Oncol Biol Phys 10: 2231-2239
Overgaard K (1934) Über Wärmetherapie bösartiger Tumoren. Acta Radiol 15: 89-100
Overgaard K, Okkels H (1940) The action of heat on Wood's sarcomas. Acta Radiol 21: 577-582
Overgaard J (1978) The effect of local hyperthermia alone and in combination with radiation on solid tumors. In: Streffer C et al (eds) Cancer therapy by hyperthermia and radiation. Urban and Schwarzenberg, Baltimore, pp 49-61
Overgaard J (1981) Fractionated radiation and hyperthermia: experimental and clinical studies. Cancer 48, 1116-1123
Overgaard J (1983) Hyperthermic modification of the radiation response in solid tumors. In: Fletcher GH et al. (eds) Biological bases and clinical implications of tumor radioresistance. Masson, New York, 337-352
Overgaard J (1986) Hyperthermia as an adjuvant to radiotherapy. Review of the randomized multicenter studies of the European Society for Hyperthermic Oncology. Presented at the Hyperthermia Symposium, Strahlenklinik Universitätsklinikum Essen, April 1986. Strahlenther Onkol (to be published)
Perez CA, Meyer JC (1985) Clinical experience with localized hyperthermia and irradiation. In: Overgaard J (ed) Proceedings of the 4th international symposium on hyperthermic oncology, vol 2, review lectures, symposium summaries and workshop summaries. Taylor and Francis, London, pp 181-198
Perez CA, Nussbaum G, Emami B, Vongerichten D (1983) Clinical results of irradiation combined with local hyperthermia. Cancer 52: 1597-1603
Perez CA, Emami B, Nussbaum GH (1984a) Regional (deep) heating, clinical studies in progress. Front Radiat Ther Oncol 18: 108-125
Perez CA, Emami B, Vongerichten D (1984b) Clinical results with irradiation and local microwave hyperthermia in cancer therapy. In: Overgaard J (ed) Proceedings of the 4th international symposium on hyperthermic oncology, vol 1, summary papers. Taylor and Francis, London, pp 398-402
Pomp H (1978) Clinical application of hyperthermia in gynecological malignant tumors. In: Streffer C et al (eds) Cancer therapy by hyperthermia and radiation. Urban and Schwarzenberg, Baltimore, pp 326-327
Reinhold HS, Wike-Hooley JL, van den Berg AP, van den Berg-Blok A (1984) Environmental factors, blood flow and microcirculation. In: Overgaard J (ed) Proceedings of the 4th international symposium on hyperthermic oncology, vol 2. Review lectures, symposium summaries and workshop summaries. Taylor and Francis, London, pp 41-52 (1984)
Rhodenburg GL, Prime F (1921) The effect of combined radiation and heat on neoplasms. Arch Surg 2: 116-129
Ruggera PS, Kantor G (1984) Development of a family of RF helical coil applicators which produce transversely uniform axially distributed heating in cylindrical fat-muscle phantoms. IEEE Trans Biomed Eng 31: 98-105
Sapareto SA, Dewey WC (1984) Thermal dose determination in cancer therapy. Int J Radiat Oncol Biol Phys 10: 787-796

Sapozink MD, Gibbs FA, Gates KS, Stewart JR (1984) Regional hyperthermia in the treatment of clinically advanced deep seated malignancy: results of a pilot study employing an annular array applicator. Int J Radiat Oncol Biol Phys 10: 775-786
Sapozink MD, Gibbs FA, Thomson JW, Eltringham JR, Steward JR (1985) A comparison of deep regional hyperthermia from an annular array and a concentric coil in the same patients. Int J Radiat Oncol Biol Phys 11: 179-190
Sauer R, Herbst M (1984) Hyperthermia for overcoming radioresistant-clinical experiences. Tumor Diag Therap 5: 116-121
Schmidt HE (1909) Zur Röntgenbehandlung tiefliegender Tumoren. Fortschr Roentgenstr 14: 134-136
Scott RS, Johnson RJR, Kowal H, Krishnamesetty RM, Story K, Clay L (1983) Hyperthermia in combination with radiotherapy: a review of five years experience in the treatment of superficial tumors. Int J Radiat Oncol Biol Phys 9: 1327-1333
Scott RS, Clay C, Storey KV, Johnson JR (1985) Transient microwave induced neurosensory reactions during superficial hyperthermia treatment. Int J Radiat Oncol Biol Phys 11: 561-566
Selawry OS, Carlson JC, Moore GE (1958) Tumor response to ionizing rays at elevated temperatures. Am J Roentgenol 833-839
Sim DA, Dewhirst MW, Oleson JR, Grochowski KJ (1984) Estimating the therapeutic advantage of adequate heat. In: Overgaard J (ed) Proceedings of the 4th International Symposium of Hyperthermic Oncology, vol 1. Summary papers. Taylor and Francis, London, pp 359-362
Stauffer PR, Cetas TC, Jones RC (1982) System for producing localized hyperthermia in tumors through magnetic induction heating of ferromagnetic implants. Nat Cancer Inst Monogr 61: 483-487
Stewart JR, Gibbs FA, Sapozink MD, Gates KS (1984) Regional hyperthermia using an annular phased array system. Front Radiat Ther Onc 18: 103-107
Storm FK (1983) Hyperthermia in cancer therapy. Hall Medical, Boston
Storm FK, Silberman AW (1984) Clinical thermochemotherapy: a controlled trial in advanced cancer patients. Cancer 53: 863-868
Storm FK, Harrison WH, Elliot RS, Halzitheofilou C, Morton DL (1979a) Human hyperthermic therapy: relation between tumor type and capacity to induce hyperthermia by radiofrequency. Am J Surg 138: 170-174
Storm FK, Harrison WH, Elliott RS, Morton DL (1979b) Normal tissue and solid tumor effects of hyperthermia in animal models and clinical trials. Cancer Res 39: 2245-2251
Storm FK, Harrison WH, Elliot RS, Morton DL (1980) Hyperthermic therapy for human neoplasmas: thermal death time. Cancer 46: 1849-1854
Storm FK, Elliott RS, Harrison WH, Kaiser LR, Morton DL (1981) Radiofrequency hyperthermia of advanced human sarcomas. J Surg Oncol 17: 91-98
Storm FK, Harrison WH, Elliott RS, Silberman AW, Morton DL (1982a) Thermal distribution of magnetic loop induction in phantoms and animals: effect of the living state and velocity of heating. Int J Radiat Oncol Biol Phys 8: 865-871
Storm FK, Kaiser LR, Goodnight JE, Harrison WH, Elliott RS, Gomes AS, Morton DL (1982b) Thermo-chemotherapy for melanoma metastases in liver. Cancer 49: 1243-1248
Streffer C (1985) Mechanism of heat injury. In: Overgaard J (ed) Hyperthermic oncology. vol 2, review Lectures, Symposium summaries and workshop summaries. Proceedings of the 4th international symposium on hyperthermic oncology. Taylor and Francis, London, pp 213-222
Streffer C, van Beuningen D (1983) Hyperthermia as a radiosensitizing and cytotoxic agent: cell biological and biochemical considerations. Verh Dtsch Krebsges 4: 87-97
Streffer C, van Beuningen D, Dietzel F, Roettinger E, Robinson JE, Scherer E, Seeber S, Trott KR (eds) Cancer therapy by hyperthermia and radiation. Proceedings of the 2nd international symposium, Essen, June 2-4, 1977. Urban and Schwarzenberg, Baltimore
Strohbehn JW (1982) Theoretical temperature distributions for soleoidal-type hyperthermia systems. Med Phys 9: 673-682
Strohbehn JW, Douple EB, Coughlin CI (1984) Interstitial microwave antenna array system for hyperthermia. Front Radiat Ther Oncol 18: 70-74
Suit HD (1982) Potential for improving survival rates for the cancer patient by increasing the efficacy of treatment of the primary lesion. Cancer 50: 1227-1234
Suit H, Gerweck LE (1979) Potential for hyperthermia and radiation therapy. Cancer Res 39: 2290-2298

Teilhaber A (1913) Zur Frage von der operationslosen Behandlung des Carcinoms. Berl klin Wschr 50: 348–349

Turner PF (1982) Deep heating of cylindrical or elliptical tissue masses. In: Third int. symp: cancer therapy by hyperthermia, drugs and radiation. Natt Cancer Inst Monogr 61: 493–495

Turner PF (1984) Regional hyperthermia with an annular phased array. IEEE Trans Biomed Eng 31: 106–114

Noell KT RU, Woodward KT, Worde BT, Fishburn RI, Miller LS (1980) Microwave-induced local hyperthermia in combination with radiotherapy of human malignant tumors. Cancer 45: 638–646

Urano M, Rice LC, Montoya V (1982) Studies on fractionated hyperthermia in experimental animal systems II. Response of murine tumors to two or more doses. Int J Radiat Oncol Biol Phys 8: 227–233

Valdagni R, Gosetti G, Amicheti M, Pani G, Knapp DS (1985) Irradiation and hyperthermia in N3 metastatic meck modes: clinical results with an analysis of treatment parameters. Strahlentherapie 161: 551–552

van Beuningen D, Streffer C (1985) Die Hyperthermie. In: Heuck F, Scherer E (eds) Handbuch der medizinischen Radiologie, vol XX. Springer Berlin Heidelberg New York Tokyo, pp 641–681 (1985)

von Ardenne M (1970/71) Theoretische und experimentelle Grundlagen der Krebs-Mehrschritt-Therapie, 2nd ed. VEB Verlag Volk und Gesundheit, Berlin

van der Zee J, van Rhoon GC, Wike-Hooley JL, van den Berg AP, Reinhold HS (1984) Thermal enhancement of radiotherapy in breast carcinoma. In: Overgaard J (ed) Proceedings of the 4th international symposium on hyperthermic oncology, vol 1. Summary papers. Taylor and Francis, London, pp 345–348

van der Zee, van Rhoon GC, Wike-Hooley JL (1985) Clinical derived dose effect relationship for hyperthermia given in combination with low dose radiotherapy. Br J Radiology 58: 243–250

Weischedel U, Wieland C (1981) Kombinierte Wärme-Strahlentherapie maligner Tumoren. Onkologie 4: 6–9

Westermark F (1898) Über die Behandlung des ulcerierenden Cervixcarcinoms mittels konstanter Wärme. Zentralbl Gynaek 22: 1335–1339

Westermark N (1927) The effect of heat upon rat tumors. Scand Arch Physiol 52: 257–322

Westra A, Dewey WC (1971) Heat shock during the cell cycle of Chinese hamster cells in vitro. Int J Radiat Biol 19: 467–477

Wiegant FAC, Tuyl M, Linnemans WAM (1985) Calmodulin inhibitors potentiate hyperthermic cell killing. Int J Hyperthermia 1: 157–169

Hyperthermia and Drugs

R. Engelhardt

Medizinische Universitätsklinik, Hugstetter Strasse 55, 7800 Freiburg, FRG

Introduction

Hyperthermia has the potential to modify the cytotoxicity of anticancer drugs and to convert some other drugs to cytotoxic agents having no considerable cytotoxicity at normal temperature. This has been demonstrated by in vitro and in vivo experiments, which are compiled in the first part of this review. The pattern of modification of the dose response curves of the drugs and the degree of change of their effectiveness by heat depends on a great variety of factors; not all these are known at present. In this review an attempt is made to compile as many data as possible on these modifying factors. They will be discussed in the context of each specific drug in order to avoid misleading generalizations, but rather to give a "hyperthermic portrayal" of these drugs.

The topic of thermochemotherapy has been reviewed by outstanding scientists throughout the past decade: Har-Kedar and Bleehen (1976), Wallach (1977), Hahn (1978), Field and Bleehen (1979), Hahn (1979), Marmor (1979), Hahn and Li (1982), Hahn (1982), and Li (1984). The aim of this review is not only to present a fairly complete and current compilation of the published phenomena in heat-drug interaction, but in addition the data are ordered according to each drug in order to provide an easier insight from the point of view of the clinical oncologist.

This is the reason for listing key data of the publications in drug-specific tables. In this way it is hoped to facilitate comparison of the data, thereby making it easier to understand the source of the results and their importance in respect to possible clinical application. In these tables the tumor system or model used as well as the assay applied are given. Information on the drug dose and on the heating criteria (temperature and duration of heating as well as heating technique) are also listed. The type of heat-drug interaction is qualified by showing no enhancement (-) or enhancement (+), which may be additive or synergistic. In one column of these tables key words are given to indicate the "special topic" of this paper, particularly if one of those factors is investigated which characterizes the heat-drug interaction in more detail. The comments on each drug roughly follow the guidelines given below:

First the type of heat-drug interaction is mentioned as noninterfering, with no enhancement, or showing enhancement, which may be additive or synergistic. Then the question is discussed whether one of the two agents, i.e., heat and drug, induces tolerance to the other, which is marked in the table by a downward arrow.

For in vivo experiments sometimes the heating technique seems to be of relevance and is then also discussed. The same is true for the route of drug application, which is also mentioned in the table.

Recent Results in Cancer Research. Vol 104

Thereafter, data of the influence of the duration of the exposure either to heat or to the drug and the results of the timing and sequencing as well as the heating rate are discussed. It will be seen that quite often these factors are the reason for conflicting results.

With some drugs it has been shown that their thermal enhancement is influenced by modifying factors such as pH, O_2-tension, and nutrient supply. This is of importance, as larger tumors tend to reveal a microenvironment within the tumor, characterized by these "milieu factors." So the size of the tumors was also found to influence the degree of thermal enhancement.

Heterogeneity in terms of the microenvironment as well as in terms of tumor-cell subpopulations logically will influence the heat-drug interaction as demonstrated for some individual drugs.

Some of the sections address themselves especially to the question of the mechanism of heat-drug interference. In this respect it sometimes seems important to keep in mind peculiarities of the mechanism of the drug action under normothermic conditions; these are therefore briefly summarized at the beginning of these sections on the specific drugs. The sparse data on changes in pharmacokinetics of the drugs are also discussed in these sections.

With regard to the clinical application of thermochemotherapy, data on changes in the toxicity of the drugs are of course of special interest. Depending on the type of normal tissue tested, the results will be of relevance either to local and regional application of heat and drugs or to their systemic application.

A special section is dedicated to those drugs which are not cytotoxic at normal temperature but act as thermosensitizers or mimick heat effects in combination with cytotoxic drugs. These drugs may help in improving the selectivity of heat-drug interaction.

The second part of this review considers clinical reports on thermochemotherapy in human cancer patients. As in the first part, only full publications and so-called extended abstracts have been taken into account. But in addition the clinical reports had to be selected on being evaluable at least to some degree according to the standard criteria for reporting cancer clinical trials (Miller at al. 1981).

Alkylating Agents

The cytotoxic effects of alkylating agents are most likely the result of their interactions with DNA (Ludlum 1977). The consequences of base alkylation include misreading of the DNA code, single-strand breakage, cross-linking of DNA (especially by the so-called bifunctional alkylating agents, and inhibition of DNA, RNA, and protein synthesis preferentially in rapidly dividing tissues. Alkylating agents share a common molecular mechanism of action, but they differ greatly in their pharmacokinetic features, lipid solubility, chemical reactivity, and membrane transport properties.

Classic Alkylating Agents

Cyclophosphamide

Cyclophosphamide (CTX) has to be activated by hepatic microsomal enzymes to produce the two powerful intracellular alkylating metabolites acrolein and phosphoramide mustard. Prevention of cell division is caused primarily by cross-linking DNA strands. Although activated derivatives are available, in vitro experiments have not been reported.

Table 1. Results of preclinical investigations on thermal enhancement of cyclophosphamide and ifosfamide

System/ Model	Assay	Dose	Temperature duration	Enhancement	Special topic	Reference
1. Cyclopshamide						
TCT-4909 bladder tumor in rats	Tumor volume doubling time	50 mg/kg (i.p.)	44.2 °C 20 min (ultrasound)	+	Timing, drug uptake	Longo et al. 1983
Marrow stem cells and RIF-1 tumor in C3H/He mice	Spleen colony formation	50–200 mg/kg (i.p.)	41° C 45 min (whole body hypertherm-ia)	+	Therapeutic ratio, selectivitiy, toxicity	Honess and Bleehen 1982
BT4A neurogenic tumor in BD IX rats	Tumor growth, tumor response	50–200 mg/kg (i.p.)	44° C 1 h (waterbath)	+	Local toxicity, timing, sequencing	Dahl and Mella 1983
Perfused rat liver	Activation of CTX	125–500 μg/ml	42 °C	+	Metabolism, toxicity	Collins and Skibba 1983
Rat liver microsomes and slices	Production of alkylating activity		40.5° C–41.8° C up to 1 h	↓	Metabolism	Clawson et al. 1981
C-1300 neuroblasto-ma in mice (s.c.)	Tumor regression, survival, relapse-free rate	50 mg/kg (i.p. 2 cycles 1 × /week	41.5° C 4 min (microwave)	+	Papaverine	West et al. 1980
Fibrosarcoma FSa-II C3H/Sed in mice	Median tumor growth time	200 mg/kg (i.p.)	41.5° C 90 min (waterbath)	+	Glucose, pH, timing, sequence	Urano and Kim 1983
FSa + NFSa fibrosarcoma in C3H-mice	DNA damage	200 mg/kg (i.p.)	(41.5° C) 42° C 1 h (waterbath)	+	Glucose, pH	Murray et al. 1984
Murine osteosarcoma in C3H/HeN mice	Alkaline phosphatase, cell counts	200 mg/kg (i.p.) (day 1)	42.5° C ± 0.1° C 30 min (day 1, 3, 5)	+	Timing, tumor size	Hiramoto et al. 1984
Lewis lung carcinoma in BDF mice	Tumor growth delay	12.5/25.0 mg/kg (i.p.)	42.5° C 30 min (waterbath)	+	Multiple drug	Hazan et al. 1984
MX-1 human breast carcinoma in nude mice	Tumor volume	50 mg/kg	43° ± 0.5° C 30 min (ultrasound)	+	Timing	Senapati et al. 1982
Lymphosar-coma (Gardner) in C3H mice	Response rate, survival	50 mg/kg (i.p.)	41.5° C 4 min (microwave)	+		Malangoni et al. 1978
MBT-2 transitional cell carcinoma in C3H mice	Tumor volume, survival time	50 mg/kg q 3 weeks	43.5° C 60–90 min (water bath)	+	Timing, “heatdose”	Haas et al. 1984

Table 1 *(continued)*

System/ Model	Assay	Dose	Temperature duration	Enhancement	Special topic	Reference
Lewis lung carcinoma in C3H mice	Cure rate, survival	150 mg/kg (i.v.)	42° C (hot air)	+	Timing	Yerushalmi and Hazan 1979
Lewis lung carcinoma in BDF mice	Tumor growth delay, cure rate	17–140 mg/kg (1–3 times) (i.p.)	42.5° C 30 min (1–3 times) (waterbath)	+	Timing, fractionnation	Hazan et al. 1981
Yoshida sarcoma in Sprague-Dawley rats (colon)	Survival time	50 mg/kg	43° ±0.5° C 30 min (waterbath)	--	Timing	Lorenz et al. 1983
Tumor-bearing mice	LD_{10}	200–350 mg/kg	41° C 60 min	+	Toxicity	Rose et al. 1979
Lewis lung carcinoma in DDF mice	Survival rate	136–208 mg/kg (i.p.)	38.8°–38.9 °C 45 min	--	Tumor control without heating	Rose et al. 1979
2. Ifosfamide						
OG cells (uterine cervical carcinoma)	Cell survival	Up to 1 µg/ml	43° C 1 h	+		Fujiwara et al. 1984

From in vitro results of other alkylating agents one would expect CTX to show gradual enhancement with a gradually raised temperature.

Thermal enhancement of CTX was established in several animal tumor systems (see Table 1). The temperature level and the heating time necessary to obtain synergism were reported within a wide range between 41° C for 45 min (Honess and Bleehen 1982) and 44.2° C for 20 min (Longo et al. 1983). The shortest heating period which was effective in the sense of enhancement was 4 min at 41.5° C when ultrasound was used. These observations supported speculations on a nonthermal effect of ultrasound (Malangoni et al. 1978) (see also mechlorethamine).

There are only two reports on negative results concerning the thermal enhancement of CTX. Rose et al. (1979) were not able to enhance the control rate of Lewis lung carcinoma in mice by 38.8°–38.9° C hyperthermia for 45 min. The high tumor control rate without hyperthermia may explain this negative result and additionally the temperature might have been too low. The latter seems to be the more likely, as in the same tumor model both Hazan et al. (1981) and Yerushalmi and Hazan (1979) were able to demonstrate an enhanced antitumor effect at 42.5° C.

The other negative report came from Lorenz et al. (1983). In their experiments it seems that timing of the heat treatment was not likely to produce enhancement: the heating was applied twice, 24 h and 8 days after drug application.

All the other investigators dealing with the problem of timing (see Table 1) found the best enhancement when both modalities were applied simultaneously or within a ½–1 h, provided heat follows drug application. Heating may enhance the drug effect to a lesser degree when applied up to 20 h after the drug. This is due to the long serum half-life of the drug. But the enhancement may then be undetectably small (Maeta et al. 1984) and there are no differences between heating 4 h and 18 h after drug administration (Haas et al. 1984). Further enhancement can be obtained by fractionation of both heat and drug as reported by Hazan et al. (1981), who found this modality three to four times superior to drug effect alone, using cure rate and tumor growth delay as parameters. Haas et al. (1984) demonstrated the relevance of "heat dose": whereas 60 min heating enhanced the drug effect in a less significant manner, 90 min heating gave statistically significantly better results (Haas et al. 1984). The thermal enhancement of drug combinations was reported by Hazan et al. (1984). Dosages of cytoxane and *cis*-platinum, not being effective without heat and only being slightly enhanced by heat when given each drug alone, became highly effective, even producing cures, when applied together and followed immediately by heating.

Tumor size seems to exert an influence on the effectiveness of the combined application of CTX and heat. Hiramoto et al. (1984) reported a greater enhancement in tumors 21 days after implantation compared with tumors 16 days after implantation. This may be due to the fact that bigger tumors are more likely to have areas of poor nutritional and O_2 support and low pH. Acidity of sarcomas in mice could be boosted by glucose infusions (Murray et al. 1984), shifting the spontaneous pH value in the tumor tissues from 7.1 to 6.6 (after glucose). The thermal enhancement ratio (TER) was thereby increased from 1.31 (without glucose) to 2.86 (with glucose infusion 60 min before heating). For methodological reasons (enhanced DNA degradation by heat plus glucose) the authors (Murray et al.) were not able to demonstrate an enhancement of CTX-induced DNA cross-linking, because the glucose-induced acidity may have enhanced the heat toxicity rather than the thermally enhanced drug effect. Papaverine was reported to improve slightly the combined effect of CTX (plus vincristine) and heat, perhaps by its vasoactivity and thereby increasing the blood and drug supply of the tumor (West et al. 1980).

As to the mechanism of the thermal enhancement, studies were performed to measure the uptake of (ring -4-^{14}C)-labeled drug into the tumor. In these experiments the drug was administered 10 min after a 20-min heating period and no increase but rather a decrease in uptake was found (Longo 1983). Metabolism of CTX at 40.5°–41.8° C was studied by Clawson and coworkers (1981). Microsomal activation was found to be depressed at 41.8° C in microsome preparations and at 43° C in liver slices.

Collins and Skibba (1983) confirmed the reduction in CTX activation at 42° C using the perfused rat liver model. The reduction was more pronounced at higher drug levels. Additionally the biosynthetic function of the liver was reduced by CTX plus heat perhaps by competition for biochemical intermediates. These findings may explain enhanced urinary excretion of unchanged CTX in patients undergoing combined treatment of CTX and hyperthermia (Ostrow et al. 1982).

Local toxicity was increased by combined CTX-heat treatment at 44° C/60 min as reported by Dahl and Mella in 1983. Thermal enhancement of CTX toxicity was even more pronounced in bone marrow stem cells than it was in radiation-induced fibrosarcoma (RIF-1) tumors when 41° C heating for 45 min (whole body hyperthermia) was applied as reported by Honess and Bleehen in 1982, showing a therapeutic ratio (TR) of 0.91, representing a therapeutic loss rather than a therapeutic gain.

Ifosphamide

Isfosphamide chemically is a structural analog of cyclophosphamide. Thermal enhancement of this drug has been demonstrated in vitro using OG cells (Fujiwara et al. 1984).

Mechlorethamine (HN_2)

One of the first reports on a thermal enhancement of an alkylating agent is that by Suzuki (1967) (Table 2). Whereas 5 mg/kg mustargen (HN_2) given to rats bearing Yoshida solid sarcoma did not reveal tumor regression, this happened when combined with 42° C local heating. Using double the amount of drug, tumor regression occurred without heat and the addition of heat then caused no appreciable difference in the amount of tumor regression. The author reports a marked increase in local toxicity by heating. Enhancement of both local and systemic toxicity was diminished when local hyperthermia (41° C/1 h) was combined with systemic cooling (20°-24° C). Tumor weight (Ehrlich-tumor in mice) after 4 weeks was only half as great in animals treated with HN_2 plus heat than in normothermic controls.

Drug metabolism of HN_2 was studied by Collins and Skibba (1979). Degradation and reactivity of the drug was increased, as can be seen in the 11% reduction in the *t*/2 in the perfusate. Toxicity in terms of several different parameters of hepatic function was only minimally enhanced. Drug uptake studies were reported by Shingleton and coworkers (1962). In dogs ^{14}C-labeled HN_2 was used to demonstrate a significantly greater uptake in heated muscle tissue (41°-43° C) than in cooled muscle tissue (30°-32° C). Preferential uptake occurred in VX-2-carcinoma compared with surrounding normal tissue, this difference even being enhanced by heating. Enhancement of the HN_2 effect by ultrasound treatment without an obvious rise in temperature was found by Kremkau and coworkers (1976). Although a thermal mechanism of action had to be taken into account, the authors claim a mainly nonthermal ultrasound effect, increasing the drug uptake 1.8 times compared with untreated controls.

Table 2. Results of preclinical investigations on thermal enhancement of mechlorethamine

System/ model	Assay	Dose	Temperature duration	Enhancement	Special topic	Reference
Mouse leukemia L-1210	Survival time of animals	0.15-1.0 μg/ml (in vitro)	26°-42.5° C 1 h (ultrasound)	+	Non-thermal enhancement	Kremkau et al. 1976
VX2 carcinoma in rabbits	^{14}C label counts	1.25-2.5 mg/kg	30°-32° C vs. 41°-43° C (radiofrequency)	+	Drug uptake, selectivity	Shingleton et al. 1962
Perfused rat liver	t1/2	10-50 μg/ml	42° C 1 h	+	Drug metabolism	Collins and Skibba 1979
Ehrlich tumor in DDD mice	Tumor weight	15 mg/kg (i.v.)	41° C 1 h (waterbath)	+	Differential heating	Kamura et al. 1979
Yoshida solid sarcoma in rats	Tumor growth	5 mg/kg (i.v.)	42° C (waterbath)	+	Toxicity	Suzuki 1967

Melphalan

Melphalan was used for about 2 decades as the drug of choice in hyperthermic limb perfusion treatment for malignant melanoma patients. In vitro and in vivo experiments have not been reported on before 1977. At that time Goss and Parsons reported on the more pronounced sensitivity of human malignant melanoma cell lines compared with human fibroblast lines, when incubated with melphalan. The cell lines used revealed marked differences in sensitivity to drug and heat, given each modality alone as well as given in combination. Interestingly, drug sensitivity was always accompanied by heat sensitivity and synergistic enhancement was found in sensitive lines only. But even then an increase in the difference between the melanoma and the fibroblast cell lines by heating occurred. Heat treatment at 42° C gave the greatest differential effect, even more than 44° C. Heterogeneity of enhancement was reflected by a range of 1–50 times less survival of the melanoma lines compared with the survival of the fibroblast lines. Similar results were reported by Neumann et al. (1985). They found thermal enhancement of melphalan in two human melanomas. But only in one of these tumors was the thermal enhancement more pronounced than in the human bone marrow progenitor cells (Neumann et al. 1985), thus providing the possibility of a therapeutic gain.

In vivo experiments by Senapati et al. (1982) confirmed the phenomenon of thermal enhancement of melphalan (Table 3) especially at suboptimal dosis.

Honess and Bleehen (1985) have demonstrated tumor selectivity of thermally enhanced toxicity of melphalan in an animal model using whole body hyperthermia of 41° C for 50 min. Bone marrow stem cells were used as the reference tissue. When "treating" KHT and RIF leg tumors therapeutic ratios (TR values) were 2.0 and 1.6, respectively. Treating the same tumors as lung microtumors, TR values were 1.6 for KHT and 1.4 for RIF tumors. These results not only underline the possibility of a therapeutic gain by systemic application of heat and drug (melphalan) but also strengthen the observation of others (Hiramoto et al. 1984; Li and Hahn 1984) that bigger lesions are more sensitive to this combined modality treatment.

Studies in pharmacokinetics of melphalan in mice (Honess 1985) could not explain the therapeutic gain cited above. Peak levels of the drug in plasma and tumor tissue 20 min after injection were elevated by heating but even more in the plasma than in the tumor. The increased area under the curve (AUC) values may be due to the loss of weight observed during heating.

The metabolism of the drug, studied in isolated perfused rat liver, revealed no changes in the *t*/2 under hyperthermic conditions and no adverse effects on liver function were found (Collins and Skibba 1983).

The effect of a steroid anesthetic (Saffan; alphaxalone and alphadolone 3:1) was pointed out by Joiner et al. (1982). Although in both the tumors investigated (B16 melanoma and Lewis-lung carcinoma) TER values of 1.55 and 2.43, respectively, occurred by heating, dose-modifiying factors (DMF values) of 1.78 and 2.69, respectively, were measured in unheated but anesthesized animals.

Chlorambucil

This drug, chemically characterized as a lyophilic weak acid, enters the cells by diffusion. The intra-/extracellular ratio of distribution of this drug was shown to be increased by lowering the pH from 7 to 6 (Mikkelsen and Wallach 1982; Table 4). As low pH values are

Table 3. Results of preclinical investigations on thermal enhancement of melphalan

System/ model	Assay	Dose	Temperature duration	Enhancement	Special topic	Reference
Human fibroblastoma + MM cell lines in vitro	% survival (colony formation)	1-10 µg/ml	42°-44° C 4 h	+	Human cell lines, selectivity	Goss and Parsons 1977
Human tumors in vitro	Clonogeneity	0.01-1.0 µg/ml	40.5° C 2 h	+	Human tumors	Neumann et al. 1985
Human tumors + human CFU-C	Clonogeneity	0.01-1.0 µg/ml	40.5° C 2 h	+	Human tumors, selectivity	Neumann et al. 1985
B16-malignant melanoma + Lewis lung carcinoma in mice	Surviving fraction (per tumor)	1.5-7.5 mg/kg (i.p.)	41.4°-42° C 1 h (waterbath)	+	Effect of anesthetics	Jointer et al. 1982
MX-1 human breast carcinoma in nude mice	Tumor volume	4-6 mg/kg	43° ± 0.5° C 30 min (ultrasound)	+	Dose dependence	Senapati et al. 1982
Perforated rat liver	Metabolism, biosynthetic function of liver	5 µg/ml	42° C	-	Metabolism, toxicity	Collins and Skibba 1983
RIF-1 tumor in C3H mice	Drug concentration in plasma and tumor	7.5 mg/kg (i.p.)	41° C 45 min (whole body hyperthermia)	+	Pharmacokinetics	Honess et al. 1985
KHT + RIF lung microtumor in mice + marrow stem carcinoma	Surviving fraction of lung colonies	Up to 20 mg/kg (i.p.)	41° C ca. 50 min (whole body hyperthermia)	+	Tumor size, selectivity	Honess and Bleehen 1985

Table 4. Results of preclinical investigations on thermal enhancement of chlorambucil

System/ Model	Assay	Dose	Temperature, duration	Enhancement	Special topic	Reference
SV40 transformed hamster lymphoid cells + V79 cells	Drug uptake	10 µm	37° C	(+)	Drug uptake (cells)	Mikkelsen and Wallach 1982

often found in human tumors and can be further lowered by hyperthermia, drug uptake might be enhanced by this mechanism in vivo.

Thio-tepa

Thio-tepa is an ethylenimine-type alkylating compound. Mahaley and Woodhall in 1961 found thio-tepa to be of enhanced toxicity by hyperthermia (Table 5). So were some other alkylating compounds, but logically cytoxane was not.

The most intensive investigations on the mechanism of the thermal enhancement of this drug, a prototype of an alkylating agent, were reported by Johnson and Pavelec in 1973. Cell inactivation rate was increased by a 3° C temperature increase the same amount as by doubling the drug concentration. Thermal enhancement was found to be compatible with either release of ethylenimine radicals or alkylation as well as with both of these classes of reactions. Similar findings were reported more recently be DeSilva et al. (1985), referring to the mechanism of the thermal enhancement of BCNU. At 45° C thermal denaturation of proteins and nucleic acids dominates the inactivation by acting additively to the thermally enhanced drug effect.

Timing and sequence for heat and drug application has been studied by Longo et al. (1983). They found synergism when heat and drug were given within 1 h of each other, with a maximal synergism when drug administration followed heating within 30 min similar to their results with cytoxane. It is obvious again that in this situation a direct thermal acceleration of the chemical reaction cannot be the reason for the enhancement. Other mechanisms have to be taken into account. At the temperature used (44.2° C) thermal damage to DNA synthesis and denaturation of proteins will take place.

Alkyl Sulfonates

In 1973 Bronk et al. reported on thermal enhancement of the cytotoxicity of MMS (methyl-methane-sulfonate) at 42° C (Table 6). They found an increase in single-strand breaks which paralleled the enhanced lethality at elevated temperatures. At 41.5° C repair seemed to be inhibited.

Table 5. Results of preclinical investigations on thermal enhancement of thio-tepa

System/ model	Assay	Dose	Temperature, duration	Enhancement	Special topic	Reference
VX2 carcinoma cells in rabbits	Tumor weight	0.0015 *M* (in vitro)	20°/37°/ 42° C 1-4 h	+	Several alkylating agents	Mahaly and Woodhall 1961
CH cells, V strain	Survival (%)	5 + 10 μg/ml	35°-42° C 30-240 min	+	Mechanism	Johnson and Pavelec 1973
TCT-4909 bladder tumor in rats	Tumor growth	2 mg/kg (i.p.)	44.2° C 20 min	+	Timing, sequence	Longo et al. 1983
Perfused dog limb	^{32}P-labeled drug	-	36.7°/ 38.9° C	+	Drug uptake, toxicity	Rochlin et al. 1961

Methylenedimethanesulfonate (MDMS) was used by Dickson and Suzangar (1974) to demonstrate lack of synergism against Yoshida sarcoma in vivo (Table 7). But as a curative drug dosis was used, and as heat was applied 20 h after drug administration, enhancement logically would not have been detectable. In Yoshida tumor slices and tumor cells the authors measured respiration, glycolysis, and uptake of ^{14}C-labeled thymidine, uridine, and leucine. They found all the parameters depressed by MDMS as well as by heat and by the combined application. The results, however, do not seem conclusive with regard to the question of synergism.

Nitrosoureas (BCNU, CCNU, Me-CCNU, ACNU)

The chloroethylnitrosoureas are characterized by being highly lipid soluble, and by chemical decomposition in aqueous solution, yielding two reactive intermediates, chloroethyldiazohydroxide and an isocyanite group. The former by further decomposition yields the reactive chloroethyl carbonium ion to alkylate DNA; the latter is believed to inhibit DNA repair by reaction with amine groups in a carbamoylation reaction.

The first attempt by Hahn (1974) to demonstrate thermal enhancement of BCNU failed. He later found the reason, this being the enhanced hydrolysation rate of the nitrosoureas at elevated temperatures and the drug's inactivation by binding to serum components. He, therefore, proposed to use the "relative dose" as being the time integral of dose concentration over the exposure period (Hahn 1978). In both exponentially growing and plateau-phase cells (HA-1) a shift in shoulder width and a shift in slope occurred in the

Table 6. Results of preclinical investigations on thermal enhancement of methylmethansulfonate

System/ model	Assay	Dose	Temperature, duration	Enhancement	Special topic	Reference
Human fibroblasts + CH fibroblasts in culture	Single-strand breaks, viability of cells	0.1–0.5 m*m*	42° C 6 h	+	Mechanism	Bronk et al. 1973

Table 7. Results of preclinical investigations on thermal enhancement of methylenedimethanesulfonate

System/ model	Assay	Dose	Temperature, duration	Enhancement	Special topic	Reference
Yoshida tumor in rats (foot)	Tumor volume, regression rate	2.5 mg (i.p.)	42° C 1 h	–	Timing	Dickson and Suzangar 1974
Yoshida tumor slices and cells	Respiration, glycolysis, DNA-precursor uptake	40–100 μg/ml	38°–42° C Up to 6 h	+/–	Mechanism	

dose-response curves when the temperatures were raised. The Arrhenius plots for cell inactivation by the four nitrosoureas (BCNU, CCNU, Me-CCNU, ACNU) were all consistent with alkylation reactions, as the slope of the lines yielded the activation energy of about 20 kcal/mol. The enhancement pattern showed a gradual increase in the cytotoxicity of the drug with gradually increasing temperatures.

Twentyman at al. in 1978 confirmed the enhancement in vitro and in vivo, using the EMT-6 tumor model in mice with survival fraction (SF) and tumor regrowth as parameters. Further confirmation of and more detailed information on the mode of enhancement were reported by Hahn's and several other groups (Tables 8-10). There are only three reports on negative results, which will be discussed later (Rose et al. 1979;Lorenz et al. 1983; Osieka et al. 1978).

Although the thermal enhancement is found to be similar and of the same pattern (no threshold) in all the three nitrosoureas, the amount of thermal enhancement is different for each drug. Thermal dose-modifying factors (TDMFs) were calculated by Hahn (1979) in exponentially growing cells, taking 50%/10%/1% survival as the end points; the TDMF values of 42° C were 5.1/3.9/3.3 for BCNU, 6.9/4.5/3.9 for CCNU, and 2.3/2.7/2.8 for Me-CCNU. The largest differences between exponentially growing cells and plateau-phase cells were found in CCNU and the smallest differences in Me-CCNU. The effect of rate of heating was investigated by Herman (1982). There was no loss of enhancement when heating the cells from 37° C to 42.4° C was performed within 30 min instead of immediately. But prolongation of the time to get the cells heated up over the limit of 60 min or even 180 min reduced the synergistic effect. This is consistent with Hahn's findings cited below, of absence of a thermally induced drug resistance at 37° C (Hahn 1979), but of lack of enhancement of BCNU cytotoxicity at 43° C after a preheating period of several hours (Hahn 1983).

Lowering the temperature to 30° C gave little reduction in BCNU cytotoxicity, and precooling the cells to 30° C before heating them up to 42.4° C within 30 min was no different from starting the heating at 37° C (Herman 1983).

Thermal enhancement of BCNU cytotoxicity was strikingly enhanced at low pH values. Lowering the pH value of the medium from pH 8.5 to 6.5 increased the cell kill in combined exposure experiments at both 43° C and 41° C, whereas the cytotoxicity of either modality alone was nearly unaffected (Hahn and Shiu 1982).

The pH dependence of the thermal enhancement of BCNU may in part explain the findings of Li and Hahn (1984), who described synergism as being more pronounced in larger than in smaller tumors.

Table 8. Results of preclinical investigations on thermal enhancement of BCNU

System/ model	Assay	Dose	Temperature, duration	Enhancement	Special topic	Reference
HA-1 (exponential growth)	Surviving fraction	5-15 µg/ml	41° C 1 h	(+)/-	Mechanism	Hahn 1974
EMT-6 tumors in mice	Tumor regrowth and surviving fraction	20 mg/kg	41.6° C 1 h	+	In vivo/in vitro	Twentyman et al. 1978

Table 8 *(continued)*

System/ model	Assay	Dose	Temperature, duration	Enhancement	Special topic	Reference
HA-1 (exponential growth) + plateau phase)	Surviving fraction ratio	Up to 12 units ("relative dose")	37°-43° C 1 h	+	Definition of "relative dose", mechanism	Hahn 1978)
HA-1	Surviving fraction	–	43° C 1 h	+	Timing, preheating	Hahn 1979
EMT-6 + KHT tumor in C_3H mice	Tumor doubling time, cure rate	5–10 mg/kg	(41° C) 42° C 43° C 1 h	+	Timing, fractionation	Marmor 1979
Mouse CFU-C	Colony formation	5×10^{-5} -5×10^{-7} *M*	42° C 15 min–2 h	+	Timing, dose dependence, toxicity	O'Donnell et al. 1979
EMT-6 in vitro	Surviving fraction	2–10 µg/ml	43° C 1 h	+	Thermally induced drug tolerance	Morgan et al. 1979
CHO cells (exponential growth)	% survival	1–16 µg/ml	42.4° C	+	Heating rate	Herman 1982
BT4A neurogenic tumor in BDIX rats	Growth delay, cure rate	20 mg/kg (i.p.)	44° C 1 h	+	Toxicity	Dahl and Mella 1982
RIF-1 tumor in C3H/He mice + mouse marrow stem cells	Surviving fraction Spleen colony technique	15–60 mg/kg (i.p.)	41° C 45 min (whole body hyperthermia)	+ +	Therapeutic ratio	Honess and Bleehen 1982
HA-1 CH cells	Surviving fraction	3.3 µg/ml	43° C 1 h (41° C)	+	pH	Hahn and Shiu 1983
CHO cells (exponential growth)	% survival	1–15 µg/ml	42.4° C (30° C)	+	Precooling	Herman 1983
HA-1 cells	Surviving fraction	Up to 12 (relative dose)	43° C 1 h	+	Timing, fractionation, thermally induced drug-heat resistance	Hahn 1983
Yoshida sarcoma in colon of rats	% survival of the rats	25 mg/kg (i.p.)	43° C 30 min (×2/day 2+8)	–	Fractionation, timing	Lorenz et al. 1983
Yoshida sarcoma in colon of rats	% survival of the rats	25 mg/kg (i.p.)	43° C 1 h	(+)/–	Timing	Lorenz et al. 1984
9L brain tumor cells	Surviving fraction	5–60 µ*M*	42°-44° C 1 h	+	Mechanism	DeSilva et al. 1984 (cited in Dewey 1984)
RIF tumor in C3H mice	growth delay, cure note	10.0 mg/kg	44° C 30 min	+	Tumor size	Li and Hahn 1984
Sprague-Dawley rats	Radiation myelitis	5 mg/kg (i.p.)	41.2°-41.8 °C 90 min	–	Toxicity	Neville et al. 1984

Table 9. Results of preclinical investigations on thermal enhancement of CCNU + MeCCNU

System/ model	Assay	Dose	Temperature, duration	Enhancement	Special topic	Reference
B16-malignant melanoma and Lewis lung carcinoma in mice	Surviving fraction (per tumor)	2.5–15 mg-/kg (i.p.)	41.4°–42° C 1 h (waterbath)	+	Timing, heterogeneity	Joiner et al. 1982
Human colon carcinoma (three lines) in nude mice	Tumor volume	18 mg/kg (i.p.)	41° C 45 min (whole body hyperthermia)	–	Drug uptake, fractionation, intrinsic sensitivity	Osieka et al. 1978
Lewis lung carcinoma in BDF mice	Survival, tumor response	27–39 mg/ kg (i.p.)	41.5°–42.5 °C ca. 20 min (Whole body hyperthermia)	–	whb, heating time	Rose et al. 1979
Ependymoblastoma in mice	Survival time, survivors day 60	2–16 mg/ kg	40° C 2 h (×3) (whole body hyperthermia	+	Dose dependence	Thuning et al. 1980
HA-1 CH cells	Surviving fraction	Up to 15 (relative dose)	37°–43° C 1 h	+	Thermal dose modifying factors of nitrosoureas	Hahn 1979

Table 10. Results of preclinical investigations on thermal enhancement of ACNU

System/ model	Assay	Dose	Temperature, duration	Enhancement	Special topic	Reference
L5178Y cells	Surviving fraction	–	42° C 1 h	+		Mizuno et al. 1980
B 16 malignant melanoma (mice) + C 24 malignant melanoma (human) in mice	Tumor volume	10 mg/kg	43° C 30 min (waterbath)	+	Fractionation	Yamada et al. 1984

Time sequence is important in heat-drug combination of the nitrosoureas. The most pronounced enhancement is found in vitro (Hahn 1979; O'Donnell et al. 1979) and in vivo (Marmor 1979; Joiner et al. 1982) when heat and drug are applied simultaneously or when the drug is given within the first 30 min of heating.

Preheating and particularly postheating are less effective or even ineffective.

Additionally, the duration of heating has to exceed 30 min to obtain a more pronounced enhancement of the cytoxicity, demonstrated by Joiner et al. (1982) in their in vivo experiments. There is only one report on enhancement of BCNU in an in vivo model (Lorenz et al. 1984) when heating was performed 3 h after drug administration. The other results of the same author are less surprising insofar as there was no thermal enhancement when heating was performed 24 h after drug administration (Lorenz et al. 1983).

No thermal enhancement of CCNU toxicity was found by Rose et al. (1979). This may be explained by the fact that the heating period was too short (20 min at treatment temperature). The results of Osieka et al. (1978) are also negative in terms of thermal enhancement of CCNU. Fractionation of the heat was ineffective, perhaps because of having too long intervals between the fractions (24 h), and the drug was administered only once.

The fractionated application of both drug (BCNU) and heat was found to be more effective not only with regard to the overall effect but particularly to the amount of the enhancement in vivo (Marmor 1979).

Regarding fractionated administration of both heat and drug, thermally induced thermotolerance as well as thermally induced drug tolerance has to be taken into account. Hahn (1983) has summarized the data important in this respect: In his experiments CHO cells were sensitized to BCNU toxicity at 37° C by preheating at 43° C. The potentation was lost after 24 h. The same preheating induced sensitization to 41° C BCNU toxicity for 6 h, and 43° C BCNU toxicity was enhanced only when the drug was administered immediately after preheating.

Morgan et al. (1979) using EMT-6 cells also found a 12-h-lasting sensitization to 37° C BCNU toxicity by 43° C preheating. But preheating the cells for 3 h at 40° C reduced cell killing by BCNU at 43° C.

So far it has not been possible to give a conclusive explanation for the different reaction patterns to the different types of preheating. Additionally, differences were demonstrated between the different drugs (see Adriamycin, bleomycin, actinomycin-D, and MMS).

The amount of drug, given as a single dose, was demonstrated to be important by Thuning et al. (1980). In an ependymoblastoma in mice they found an increasing enhancement at higher dosages. This parallels findings by Marmor (1979) in respect to Adriamycin.

The estimation of a therapeutic gain of the thermally enhanced BCNU cytotoxicity was investigated by Honess and Bleehen (1982). In their RIF-1 tumor model in mice they found a dose-modifying factor (DMF) of 2.0–1.6 which was dose dependent in the same manner as previously mentioned (Thuning et al. 1980). As the DMF for mouse marrow stem cells under the same conditions was 2.1, the therapeutic ratio was negative at 0.69.

Mouse marrow stem cell toxicity was found to be thermally enhanced by only 15 min heating (42° C) (O'Donnell et al. 1979). This effect was also found to be dose dependent.

Normal tissue toxicity was reported to be not enhanced in a tumor model in rats (Dahl and Mella 1982). Using local heating up to 44° C the authors found striking enhancement of the therapeutic effect.

Douglas et al. (1981) reported sudden myelopathy in patients treated with whole body hyperthermia, -irradiation, and BCNU. Neville et al. (1984) therefore investigated the effect of BCNU plus heating on radiation myelitis in rats. No enhancement of the toxicity by addition of BCNU was found.

Concerning the mechanism of thermally enhanced cell killing, the linear slope of the Arrhenius plot and the activation energy being consistent with an alkylating process support the idea of a thermodynamically increased rate of the main critical process in the action of the nitrosoureas, i.e., alkylation. This assumption is, however, inconsistent with the finding that MMS, a monofunctional alkylating agent, demonstrates a nonlinear Arrhenius plot for temperature-dependent cell-killing curves. Therefore, other mechanisms are thought to be involved. The lack of increased uptake of (chloroethyl-u-14C)-labeled MeCCNU in the tumors, reported by Osieka et al. (1978), may be seen in the context of the lack of enhancement of the cytotoxicity the authors found in these experiments. A more detailed insight was given by DeSilva et al. (1984; cited in Dewey 1984). Comparing a BCNU-resistant and a BCNU-sensitive subline of 9L brain tumor cells they found that heat enhanced the drug toxicity more in the resistant subline than in the sensitive one.

Cell survival was a function of the amount of reactive intermediate or product formed, and the rate of formation of this reactive intermediate was found to be enhanced by hyperthermia. In addition, in the sensitive cell line the subsequent alkylation reaction was also thermally enhanced. The enhancement of the alkylation reaction was even more pronounced in the resistant subline, what in part may be due to a directly heat-induced DNA cross-linking.

Another nitrosourea drug, ACNU, has also been described to be thermally enhanced in respect to its cytotoxicity. In vitro, a ratio of a 10% survival dose of ACNU at 37° C to the same survival dose at 42° C of 1.5 was reported. As no corrections were made to the increased rate of hydrolysis at 42° C (see Hahn 1978), the enhancement may be even more pronounced. In vivo experiments, using both human and mice melanoma cell lines, revealed a marked synergism at 43° C particularly by fractionated (two fractions) application (Yamada 1984).

Antitumor Antibiotics

Anthracyclines

Anthracylines are a subset of anthraquinones which are known to chelate divalent cations, intercalate DNA, and engage in oxidation-reduction reactions (Friedmann 1980; Pigram et al. 1972). In addition it has been shown that they react directly with cell membranes, resulting in alterations in membrane function (Murphree et al. 1976; Mikkelsen et al. 1977).

Adriamycin: In Vitro Experiments. Thermal enhancement of Adriamycin in vitro was shown first by Hahn et al. in 1975. Exposing exponentially growing EMT-6 cells to 0.5 µg/ml Adriamycin at 37°, 41° and 42° C for up to 6 h, a striking synergism was found at 42° C in terms of cell survival. At 41° C the enhancement was only small. In plateau-phase cultures sensitization occurred at 43° C.

The basic phenomenon of thermal enhancement was confirmed by further investigations in Hahn's group and by others (Tables 11–13) using different cell lines, temperatures, heating and preheating intervals, and different parameters and thereby characterizing the enhancement in more detail.

Table 11. Results of preclinical investigations on thermal enhancement of Adriamycin - in vitro

System/model	Assay	Dose	Temperature, duration	Enhancement	Special topic	Reference
EMT-6 cells mammary carcoma in mice	Surviving fraction surviving fraction Ratio (in vitro)	Up to 30 mg/kg	41° C 42° C (up to 6 h)	(+) ++	Drug uptake	Hahn et al. 1975
HA-1 cells	Surviving fraction	0.2–2.4 μg/ml	43° C 30 min >30 min	+	Drug uptake, preheating	Hahn and Strande 1976
L1210 mouse leukemia cells	Fluorimetry	–	40.5° C 41.5° C	+ +	Drug uptake	Klein et al. 1977
HA-1 CH cells	Cell survival	Up to 0.8 μg/ml	43° C		Drug tolerance induced by heat and ethanol	Li and Hahn 1978
HA-1 CH cells	Cell survival	0.2–1.0 μg/ml	43° C 1 h	++	Time dependence	Hahn 1979
EMT-6 tumors in BALB-C mice	Cell survival (in vitro)	2.5–5 mg/kg	43° C 30 min	(+)	Dose dependence	Marmor et al. 1979
EMT-6 cells	Surviving fraction	0.1–10 μg/ml	43° C 1 h	+	Drug tolerance induced by heat	Morgan et al. 1979
L5178Y cells (exponential growth) FM3A cells (exponential growth)	Surviving fraction	0.1–0.3 μg/ml Up to 2 μg/ml	42° C <30 min 43° C <30 min >30 min	+ +	Time dependence, drug tolerance induced by heat	Mizuno et al. 1980
EMT-6 spheroids	Surviving fraction spheroid growth	1 μg/ml	43° C 1 h 42° C 6 h	– +	Timing and duration of heat	Morgan and Bleehen 1981
BT4C cells	Surviving fraction	Up to 3.0 μg/ml	42° C 41° C 1 h	+ (+)	Drug tolerance induced by heat	Dahl 1982
FM3A cells	Surviving fraction	0.4 μg/ml	41° C 43° C	+ –	Uptake, lidocaine, drug resistance	Mizuno and Ishida 1982
V-79 carcinoma cells (CH)	Surviving fraction	Up to 4 μg/ml	42.5° C	(+)	Low temperature (25° C)	Roizin-Towle and Hall 1982
Perfused rat liver	Fluorimetrically	2.4 μg/ml	42° C 1 h 41° C 3 h		Metabolism, toxicity	Skibba et al. 1982
SHG human melanoma cell line	Cell counts/ml	10–28 μg/ml	40° C up to 168 h	+	Enhancement by procaine + lidocaine	Chlebowski et al. 1982
CHO cells	Cell kill	0.2–3.0 μg/ml	42.4° C 30 min	+	Heating rate	Herman et al. 1982
CHO cells	Clonogeneity	Up to 3.0 μg/ml	42.4° C 30 min	+	Precooling	Herman 1983

Table 11 *(continued)*

System/ model	Assay	Dose	Temperature, duration	Enhancement	Special topic	Reference
P388 in BDF mice	Surviving fraction (in vitro), survival time	10 μg/ml	42° C 1 h (in vitro)	+ +	In vivo/in vitro	Adwankar and Chitnis 1984
HA-AD CH fibroblasts	Surviving fraction	5–30 μg/ ml	42° C 43° C 1 h	–	Resistance, intrinsically, thermotolerant by heat and drug	Li 1984
HA-1 CH cells	Surviving fraction	3–8 μg/ml	45° C 15 min		Thermotolerant by heat or drugs, ethanol	Li 1984
P388 mouse leukemia	Cell survival	Up to 10 μg/ml	43° C		Drug uptake	Van der Linden et al. 1984
a) Sensitive				a) +	Intrinsic	
b) Resistant				b) –	Resistance	
Ehrlich ascites cells	[^{3}H]-Adriamycin uptake	1 μCi per cell	37°–43° C 30–240 min	+	Drug uptake	Yamane et al. 1984
Human tumors + bone marrow	Clonogeneity	Up to 10^{-2} μg/ml	40.5° C	+/–	Selectivity	Neumann et al. 1985

Table 12. Results of preclinical investigations on thermal enhancement of Adriamycin in vivo

System/ model	Assay	Dose	Temperature, duration	Enhancement	Special topic	Reference
New Zealand rabbits	Fluorometric Adriamycin Adriamycin-ol a-glycones	5 mg/kg (i.v.)	42.3° (chamber)		Drug distribution	Mimnaugh et al. 1978
Ehrlich ascites cells in DDD mice	Tumor growth	2.0 mg/kg (i.v.)	41° C 1 h	(+)	Differential heat	Kamura et al. 1979
KHT tumor in C3H mice	Tumor growth delay, cure rate	Up to 10 mg/kg	43° C 30 min	–	Toxicity	Marmor 1979
C3D2F1/ BOM mammary carcinoma	Survival	25 mg/kg	40.5° C 2 h 42.5° C 1 h	+	Toxicity	Overgaard 1976
16C mammary carcinoma in C3H/He mice	Tumor growth delay	10 mg/kg (i.v.)	43° C 1 h (microwave)	+	Timing, uptake, metabolism	Magin et al. 1980
MAM16C adenocarcinoma in mice	Tumor growth inhibition	10 mg/kg (i.v.)	43° C 1 h (microwave)	+	Drug distribution, metabolism timing	Magin et al.

Table 12 *(continued)*

System/ model	Assay	Dose	Temperature, duration	Enhancement	Special topic	Reference
Rat liver	Metabolites	Up to 5 µg/ml	41.5° C	+	Livertoxicity, drug metabolism	Skibba et al. 1982
BT4A in BDIX rats	Tumor response, growth delay	7 mg/kg	44° C 1 h	+	Toxicity	Dahl 1983
In Fisher rats: a) methyl-cholantren-induced sarcoma b) transitional cell carcinoma	Tumor growth, survival	2 mg/kg (i.p.)	41.5° C 30 min (whole body hyperthermia) Weekly application	a) – b) – a) + b) –	Fractionation, intrinsic resistance	Rotstein et al. 1983
MBT-2 (transitional cell carcinoma) in C3H mice	Tumor growth rate, survival	1 mg/kg	43.5° C 1 h 90 min	+ ++	Time dependence	Haas et al. 1984
Rats	Blood cell counts, bone marrow histology	6 plus 8 mg/kg	41° ±0.5° C 10 min 20 min (whole body hyperthermia)	+ ++	Toxicity	Hinkelbein et al. 1984
MAM16C in mice a) sensitive b) resistant	Tumor growth inhibition	10 mg/kg (i.v.)	43° C	a) – b) –	Intrinsic resistance	Van der Linden et al. 1984

Table 13. Results of preclinical investigations on thermal enhancement of 4-epi-Adriamycin

System/ model	Assay	Dose	Temperature, duration	Enhancement	Special topic	Reference
BT4A in BD IX rats	Tumor growth delay	7 mg/kg	44° C 1 h	+	Toxicity	Dahl 1983
K-562 blasts	[^{3}H]uridine incorporation	Up to 0.05 µg/ml	40.5° + 42° C 1 h	+	Temperature dependence	Engelhardt et al. 1985

Thermally induced tolerance not only to heat but also to Adriamycin was described in HA-1 (Chinese hamster) cells when the heating period at 43° C exceeded 30 min (Hahn and Strande 1976). This was confirmed by Hahn in 1979 when they described synergism occurring only during a heating period up to a maximum of about 60 min, followed by an increasing resistance to Adriamycin. This resistance could be demonstrated also by preheating the cells at 42° C before they were exposed to the drug at 37° C. In EMT-6 cells Morgan et al. (1979) also described protection to Adriamycin toxicity at 37° C after preheating at 43° C but no heat-induced drug tolerance occurred after 3 h preheating at 40° C.

The time dependence of the enhancement was also reported by Dahl (1982) using a BT4-C cell line. When heating at 41° C a sensitization occurred only within the 1st h, revealing a biphasic slope of the survival curve (Dahl 1982). Finally, FM3A cells, exponentially growing, heated at 43° C also developed Adriamycin resistance after a sensitive phase of 30 min (Mizuno et al. 1980).

No enhancement occurred in an Adriamycin-resistant cell line (HA-AD CH fibroblast) used by Li (1984 although heat and -ray sensitivity existed. On the other hand Adriamycin resistance was found in a heat-resistant variant. In HA-1 cells chemically or heat-induced thermotolerance conferred different degrees of Adriamycin tolerance. Chemicals to induce thermotolerance were sodium arsenite, cadmium chloride, and ethanol (6%). Induction of both thermotolerance and tolerance to Adriamycin by ethanol (6%) had already been reported in 1978 by Li and Hahn.

Enhancement of Adriamycin toxicity by membrane-acting local anesthetics (procaine and lidocaine) was described by Chlebowsky et al. (1982) and was enhanced further by 40° C hyperthermia. Perhaps because of the rather low temperature used throughout the incubation period of up to 168 h no thermally induced tolerance occurred to either the drug or the heat.

The temperature dependence of Adriamycin cytotoxicity was also investigated at hypothermic temperatures: Kamura et al. (1979) tried to use the reduced activity of Adriamycin at 20°-24° C for reducing the side effects of the drug by systemic cooling, while heating the tumor (Ehrlich ascites tumor cells in DDD mice) to 41° C for 60 min.

Herman reported on the importance of heating rate: When CHO cells were heated from 37° to 42.4° C within 3 min, 30 min, or 3 h, respectively, the thermal enhancement decreased with increasing length of heating. Cooling the cells (30° C) lowered the effect of Adriamycin. Herman reported that precooling at 30° C in one series of experiments (Herman et al. 1982) was effective in sensitizing, the cells against the hyperthermic enhancement (41° C, 1 h) but failed to do so in another series of experiments (Herman 1983). Also Roizin-Towle and Hall (1982) reported on markedly depressed activity of Adriamycin in V-79 carcinoma cells (Chinese hamster) at 25° C. Enhancement occurred very pronouncedly between 25° and 37° C and was still visible between 37° and 42.5° C.

A rather sophisticated model system of multicellular tumor spheroids was used by Morgan and Bleehen (1981) In EMT-6 spheroids they demonstrated enhancement at 42° C after 6 h of heating but not at 43° C after 1 h. These findings are in contrast to all of the findings cited above, showing enhancement at 43° C. The lack of thermally induced resistance to Adriamycin, however, seems to be consistent with the findings of Morgan et al. (1979) and Hahn et al. (1975), who used the same cell type (EMT-6 cells) also in monolayer systems instead of spheroids. Another approach was reported by Neumann et al. (1985): They developed a variation of the Salmon-Hamburger assay for testing specimens from spontaneous human tumors augmented in the nude mice system. Adriamycin turned out to be not enhanced at 40.5° C in most of the specimens tested with the exception of a chondrosarcoma, which was tested immediately after resection. There was also no enhancement in human bone marrow progenitor cells (CFU-C) Neumann et al. 1985).

The biphasic mode of thermal enhancement described above and the possibility of inducing Adriamycin tolerance not only thermally but by ethanol incubation (Li and Hahn 1978) induced speculations on the cell membrane being the site of action in terms of an increased uptake of the drug within the first 30-60 min of thermal exposure. Hahn and Strande (1976) reported an increased uptake of Adriamycin in heated HA-1 cells. And Mizuno and Ishida (1982) reported on time dependence of this increased uptake within the first 30 min of incubation (FM3A tumor cells) In Ehrlich ascites cells Yamane et al.

(1984) also found an increase in [^{3}H]Adriamycin uptake when temperature was increased from 37° to 40°, 42°, and 43° C. This enhancement of uptake was most expressed at the highest temperature (43° C) and after 30 min of incubation. The effect was less pronounced after 60 min and completely abolished after 120 min. Increased cell membrane permeability was thought to be the reason for the increased intracellular [^{3}H]Adriamycin contents. Increased Adriamycin uptake was also found by Klein et al. (1977) using L1210 mouse leukemia cells and taking [^{3}H]thymidine incorporation as a parameter for cytotoxicity. Interestingly, they found increasing uptake and decreasing [^{3}H]thymidine incorporation at 40.5° and 41.5° C. The uptake of Adriamycin in P388 mouse leukemia cells was studied by van der Linden et al. (1984) at 43° C. Using Adriamycin-resistant and -sensitive sublines, the sensitive one showed an enhanced effect in the colony formation test which was consistent with a two- to threefold increase in Adriamycin uptake by the cells. In contrast the resistant subline was not affected with respect to the colony formation although the cells accumulated one to two times more Adriamycin at 43° C than at 37° C. Under in vivo conditions (MAM16C) tumors failed in both their sensitive and their resistant subline to enhance accumulation of drug and there was no effect in tumor growth inhibition when heated to 42° C. Although it seems likely from the investigations mentioned above that the Adriamycin uptake is enhanced under hyperthermic conditions, the mechanism of thermal enhancement in this drug does not yet seem to be elucidated completely.

Adriamycin: In Vivo Studies. Hahn et al. reported in 1975 not only on the in vitro but also on the in vivo enhancement of Adriamycin. Using the EMT-6 sarcoma in BALB-D mice, heat was applied at 43° C for 30 min and Adriamycin was given from 10 to 25 mg/kg. When measuring the effect in vitro (cloning assay), they reported superiority of heat plus Adriamycin to either modality alone.

The enhancement was confirmed by Overgaard in 1976, treating mammary carcinoma bearing mice. Enhancement was reported on heating levels of 40.5° C for 2 h and 42.5° C for 1 h, using cure rate as the end point. Because of the 90% kill rate of the animals treated with Adriamycin alone (25 mg/kg), the statistical evaluations, however, were crucial, and the reduced toxicity the author mentioned in the combined treatment group was somewhat speculative.

In 1979, Marmor et al. reported further data on the in vivo treatment of EMT-6 tumors in BALB-C mice: When RF heating of the tumors was performed at 43° C for 30 min the Adriamycin effect on cell survival was enhanced only at a dose of 10 mg/kg and not at 2.5 or 5 mg/kg. When they used KHT tumors in C3H mice, 43° C for 30 min gave no enhancement at 2.5 and 5.0 mg/kg Adriamycin given either simultaneously or at intervals of up to 24 h. The explanation for this negative result is most probably the insufficient dosage which has to be chosen for tolerance reasons. The toxicity of Adriamycin (10 mg/kg) and heat was not clinically tolerable as 14 out of 25 animals died after 4–7 days. No tumor cures occurred. As the KHT tumor gives nearly 100% metastases within 10–13 days after implantation, the influence of heat and Adriamycin on the degree of lung metastases was also examined: Heat alone did not increase the average severity of metastases and the effect of Adriamycin was not potentiated by heat.

Kamura et al. (1979) reported on a small enhancement by 41° C heating for 1 h (Ehrlich ascites tumor cells in DDD mice) using tumor growth as the end point. In their experiments the animals were cooled to 20°–24° C to reduce the toxicity of the drug.

Bone marrow toxicity of Adriamycin and heat was studied by Hinkelbein et al. (1984). Two doses of Adriamycin, administered within 2 weeks (6 mg/kg and 8 mg/kg, respectively), were combined with heat at 41° ±0.5° C. Toxicity was markedly increased after

10 min of heating and even more increased after 20 min of heating. They demonstrated "heat-dose" dependence of the thermal enhancement of Adriamycin in terms of bone marrow toxicity in rats. In MAM16C adenocarcinoma growing in mice, 43° C revealed a marked tumor growth inhibition as reported by Magin et al. (1980). But the same tumor (MAM16C) was used by van der Linden et al. (1984) without any enhancement of the tumor growth inhibition at 43° C regardless of the Adriamycin resistance or Adriamycin sensitivity of the sublines investigated.

Negative results were also reported by Mizuno et al. (1981) investigating the FM3A tumor in C3H mice (2 mg/kg Adriamycin) at 43° C. This negative result was not consistent with the enhanced Adriamycin uptake at 43° C during the first 30 min of heating (in vitro).

Regarding changes of drug uptake by heat, Magin et al. (1980) found no increase in ^{14}C uptake in the heated tumors, although tumor growth delay was enhanced as mentioned previously. But the authors reported a shift to an increase of the metabolites, confirming the results of Skibba et al. (1982). These authors, using the perfused rat liver to study the drug metabolism, additionally found the bile flow decreased, thus increasing the AUC values for total drug equivalents. Adriamycin and heat were minimally detrimental to hepatic biosynthetic functions.

The studies by Mimnaugh et al. (1978), concerning Adriamycin distribution in New Zealand rabbits also revealed no change in plasma clearance by 42.3 °C whole body hyperthermia. But α-phase was shortened from $t/2\alpha$ 1.24 to 0.92 min. Tissue concentration of Adriamycin was slightly increased in skeletal muscle as were the agylcones in the duodenum.

Rotstein et al. (1983) reported on the possibility of turning negative enhancement results into positive ones by repeating the treatment three times at weekly intervals. They used methylcholanthrene-induced sarcoma implanted in Fisher rats. Adriamycin dosage was 2 mg/kg i.p.. Heating was performed as whole body hyperthermia at 41.5° C for 30 min. After a single treatment there was no effect on tumor growth or survival. But repeating the procedure as indicated previously there was a marked enhancement. In another tumor, a transitional cell carcinoma, treated in the same manner, enhancement was not seen even after the fractionated application. Positive results were reported with another transitional cell carcinoma (MBT-2) in C3H mice when 1 mg/kg Adriamycin was combined with 43.5° C hyperthermia. Sixty minutes heating gave a good enhancement which was even increased by 90 min of heating. End points were tumor-doubling time, growth rate, and survival (Haas et al. 1984).

Dahl confirmed his in vitro findings (Dahl 1982) by studying the BT4A tumor in BD IX rats (Dahl 1983). Adriamycin dose was 7 mg/kg and heating was performed at 44° C for 1 h. Tumor response, doubling time, and growth delay showed consistent results.

The new analog of Adriamycin, *4-epi-Adriamycin,* showed principally the same effects (Dahl 1983). Thermal enhancement of 4-epi-Adriamycin was also demonstrated in K-562 leukemic cells using [^{3}H]uridine incorporation as the assay (Engelhardt et al. 1985). Cytotoxicity of *daunorubicin* was found to be not thermally enhanced by Mizuno et al. (1982), heating L5178Y cells at 42° C (Table 14). Consistent with this result, Klein et al. (1977) found only additive effects in L1210 cells with regard to [^{3}H]methyl-thymidine incorporation.

Another anthracycline antitumor antibiotic, *aclacinomycin-A,* (Acla-A) was investigated by Mizuno et al. (1980) (Table 15). In contrast to the time-response curve for Adriamycin, suggesting drug tolerance after 30 min of heating, the time-response curve for Acla-A was almost linear at 43° C up to 2 h. In both cell lines investigated, the shoulder of the dose-response curve at 37° C was markedly reduced at 42° and 43° C respectively, indicating that the cellular ability to resist Acla-Atoxicity at low concentrations was reduced at elevated temperatures.

Table 14. Results of preclinical investigations on thermal enhancement of daunorubicin

System/ model	Assay	Dose	Temperature, duration	Enhancement	Special topic	Reference
L5178Y cells	Surviving fraction	0.05–0.2 μg/ml	42° C	–		Mizuno et al. (1980)
L1210 leukemia cells	[³H]-thymidine incorporation	10^{-4} $-10^{-5}M$	39°–41.5° C	–	Drug uptake	Klein et al. (1977)

Table 15. Results of preclinical investigations on thermal enhancement of aclacinomycin-A

System/ model	Assay	Dose	Temperature duration	Enhancement	Special topic	Reference
L5178Y cells	Surviving fraction ratio	1.0–7.0 μg/ml	42° C 1 h	+	Heating time	Mizuno et al. 1980
FM3A cells		1.0–4.0 μg/ml	43° C up to 2 h	+		
DND-1A human malignant melanoma and	Surviving fraction	Up to 50 × $10^{-7}M$	0° C 42° C	 +	Freeze temperature	Ohnoshi et al. 1985
DND-39A Burkitt's lymphoma cells		Up to 25 × $10^{-6}M$	0° C 41° C	 +		

Thermal enhancement of Acla-A was also reported by Ohnoshi et al. (1985). These authors interestingly found Acla-A cytotoxicity still measurable at 0° C in a human malignant melanoma cell line, whereas no Acla-A cytotoxicity was found a Burkitt's lymphoma cell line, thus demonstrating not only differences among anthracyclines but also between different cell lines.

Mitoxantrone, a synthetic amino anthraquinone, is a cell-cycle-phase nonspecific antitumor agent which binds to DNA by intercalation and inhibits nucleic acid synthesis. Herman (1983) reported markedly synergistic lethality by exposure of CHO cells to this drug at 42.4° C as compared with drug exposure at 37° C. The synergism of drug and hyperthermia on cell survival was concentration dependent, being less pronounced at a lower drug concentration. Thermal enhancement (42° C) of mitoxantrone was also found by Ohnoshi et al. (1985) in both a rather heat-resistant human malignant melanoma cell line and a heat-sensitive Burkitt's lymphoma line (Table 16).

Yang and Rafla (1985) reported the thermal enhancement of the cytotoxicity of mitoxantrone by long-term/low-temperature heating (16 h, 40° C) at least as pronounced as by short-term/high-temperature heating (1 h/43° C). The enhancement was markedly less pronounced in plateau-phase cells. Preheating induced thermotolerance but no drug tolerance occurred. These data may be of clinical interest, as whole body hyperthermia usually has to be performed at temperatures below 42° C and takes about 1 h to obtain the target temperature.

Table 16. Results of preclinical investigations on thermal enhancement of mitoxantrone

System/ model	Assay	Dose	Temperature duration	Enhance-ment	Special topic	Reference
CHO cells (exponential growth)	Survival (%)	Up to 1.5 μg/ml	42.4° C 1 h	+	Concentration dependence	Herman 1983
DND-1A human malignant melanoma + DND-39A Burkitt's lymphoma cells	Surviving fraction	Up to 10 × $10^{-8}M$	0°–42° C 1 h	+	Intrinsic heat resistance, heterogeneity	Ohnoshi et al. 1985
V-79 CH cells a) exponential growth b) plateau phase	Survival (%)	a) Up to 0.8 μg/ml b) Up to 14 μg/ml	40° C overnight + 43° C 1 h	+	Chromosome aberrations, preheating	Yang and Rafla 1985

Other Antitumor Antibiotics

Actinomycin-D. Interaction of heat and actinomycin-D (act-D) was investigated by Donaldson et al. (1978) (Table 17). The simultaneous administration of 43° C heat and act-D produced more than additive cytotoxicity if the duration of heat exposure was less than 30–60 min. With longer exposures the cells developed resistance to act-D. By preheating (43° C) the cells were also rendered insensitive to act-D at 37° C, and longer periods of preheating induced increased duration of protection, lasting up to 18 h. The application of heat following act-D exposure also protected against the 37° C cytotoxicity of act-D. [^{3}H]Act-D uptake, however, was increased by preheating, although this was combined with an inability of the cells to retain the drug.

Thermal enhancement of cytotoxicity of act-D was also found by Mizuno et al. (1980), using L5178Y cells heated for 1 h at 42° C. [^{3}H]Uridine incorporation was found to be reduced by act-D and heat (42.5° C/2 h) in 19 solid tumors in children (Willnow 1981). The pattern of thermal enhancement was strictly individual for each tumor, probably reflecting primarily the proliferative activity of the tumor probes. Reaction patterns, being an individual property of the tumors, were also found by Neumann et al. (1985). In two out of six human tumors they found the cytotoxicity of act-D to be enhanced by a 2-h exposure to 40.5° C heating, when using clonogenicity as the assay. Enhancement occurred, however, at dosages of act-D which were lethal for the CFU-C cultures, so that no therapeutic gain could be expected from these data (Neumann et al. 1985).

In vivo investigations were performed by Rose et al. (1979). Using P388-leukemia and ROS-osteogenic sarcoma in mice, cytotoxicity of act-D was enhanced by whole-body hyperthermia (38.8°–38.9° C) to the tumors as well as to the side effects, thereby revealing no therapeutic gain. In contrast to this, Yerushalmi reported twice in 1978 a prolongation of survival of mice bearing fibrosarcomas, after combined act-D and local heat treatment.

Mitomycin-C. Mitomycin-C (mito-C) cytotoxicity was found to be enhanced by 41°–42° C heating for 1 h, using a human colon cancer cell line (Barlogie et al. 1980)

Table 17. Results of preclinical investigations on thermal enhancement of actinomycin-D

System/ model	Assay	Dose	Temperature duration	Enhancement	Special topic	Reference
CH cells (plateau phase)	Surviving fraction	0.5 μg/ml	43° C up to 2 h	+	Timing, sequencing, drug uptake	Donaldson et al. 1978
L5178Y	Cell survival	-	42° C 1 h	+		Mizuno et al. 1980
19 solid human tumors	[^{3}H]-uridine incorporation	0.5 μg/ml	42.5° C 2 h	+	Tumor-individual pattern	Willnow 1981
Human tumors	Clonogeneity	Up to 10^{-2} μg/ml	40.5° C 2 h	+	Heterogeneity	Neumann et al. 1985
Human tumors + human CFU-C	Clonogeneity	Up to 10^{-2} μg/ml	40.5° C 2 h	+	Toxicity, therapeutic gain	Neumann et al. 1985
P388 leukemia in BDF1 mice	Animal survival	0.5-1.0 mg/kg i.p.	38.8°-38.9 °C 45 min whole body hyperthermia	+	No therapeutic gain	Rose et al. 1979
Osteogenic sarcoma in AKD2F1 mice	Animal survival	0.5-0.7 mg/kg i.p.	38.8°-38.9 °C 45 min whole body hyperthermia	+	No therapeutic gain	
SV-10+SV-40 fibrosarcoma in mice	Survival time	-	42° C	+	Therapeutic gain	Yerushalmi 1978 Yerushalmi 1978

Table 18. Results of preclinical investigations on thermal enhancement of *mitomycin-C*

System/ model	Assay	Dose	Temperature duration	Enhancement	Special topic	Reference
L5178Y cells	Surviving fraction	-	42° C 1 h	+		Mizuno et al. 1980
LoVo human colon carcinoma cells in culture	Survival %, Do value, DMF/10%	Up to 4 μg/ml	41° + 42° C 1 h	+	Temperature level, reaction pattern	Barlogie et al. 1980
EMT-6 mouse mammary tumor cells (exponential growth)	Surviving fraction	0.01-10 μ*M*	41°-43° C 1-6 h	+	Time temperature, dose dependence, hypoxia	Teicher et al. 1981
FM 3A cells	Surviving fraction	1.0 μg/ml	41° C 1 h	+	Lidocaine	Mizuno and Ishida 1982
Ascites hepatoma 100 B-cells in rats	Surviving fraction	1 mg/kg (i.p.)	41.5° C 1 h	+		Koga et al. 1984
Ehrlich ascites cells in DDD mice	Tumor weight	1 mg/kg (i.v.)	41° C 1 h (waterbath)	(+)	Differential heating	Kamura et al. 1979

Table 18 *(continued)*

System/ model	Assay	Dose	Temperature duration	En-hance-ment	Special topic	Reference
MBT-2 transitional cell carcinoma in mice	Tumor volume, survival time	2 mg/kg (i.v.)	43.5° C 90 min (waterbath)	–		Haas et al. 1984
Human lymphocytoma CH cells (M-3, K-1)	Chromosome aberration	1–2 μg/ml	43° C 1 h	+	Cell cycle	Vig et al. 1982

(Table 18). This effect was seen in exponentially growing cells as well as in plateau-phase cells. The enhancement pattern suggested a reduction in the ability of the cells to absorb sublethal damage and an increase in the sensitivity to the drug. This enhancement was measurable at 41° C but was more pronounced at 42° C, showing DMF values at the 10% survival level of 1.5 and 2.6 respectively.

A more than additive enhancement of the mito-C toxicity was also seen in L5178Y cells used by Mizuno et al. (1980) when the cells were heated for 1 h at 42° C.

No substantial enhancement at 41° C, but significant effects at 42° C, and very marked enhancement of the cytotoxicity of mito-C at 43° C were described by Teicher et al. (1981) in EMT-6 cells. The effects were more pronounced at higher doses of the drug and especially under hypoxic conditions, and longer times of exposure (up to 6 h) were also found to be more effective. An increased activation of the drug occurring under conditions of heat and hypoxia is thought to be the reason for the enhancement of the cytotoxicity of mito-C. A slightly enhanced cytotoxicity of mito-C was reported by Mizuno and Ishida (1982) at 41° C and this could not be further enhanced in the presence of lidocaine.

By combining mito-C with differential heating technique (local heating to 41° C for 1 h and systemic cooling to 20°–24° C), Kamura et al. (1979) were able to reduce Ehrlich ascites tumor weights in mice by about 50%, compared with controls. From the design of the experiments it is not quite clear whether this reflects additional effects or synergism.

Marked synergism of mito-C plus heat (41.5° C/1 h) was reported by Koga et al. (1984). Whereas either modality alone was of little effect in ascites hepatoma cells in rats, the combined application produced 30% survivors at day 60 as compared with zero survivors in the control groups. In contrast to the data cited above are the findings of Haas et al. (1984), who could not find any effect against MBT-2 transitional cell carcinoma in mice in terms of tumor volume or survival time, when treating the mice with mito-C normothermically or hyperthermically (43° C/90 min). Consistent with the majority of the findings, Vig et al. (1982) reported on an increased potential of mito-C to induce chromosome aberrations when applied at temperatures higher than 37° C. This is not restricted to any phase of the cell cycle and insofar as it is consistent with the findings of Barlogie, showing enhancement also in plateau-phase cells.

Bleomycin. Bleomycin is a mixture of small molecular weight (about 1500 daltons) peptides, predominantly the A2 peptide. The primary action of bleomycin is to produce single- and double-strand brakes in DNA (Takeshita et al. 1978). The sequence of events leading to DNA breakage includes oxidation of Fe^{2+} to Fe^{3+} and formation of superoxide or hydroxyl radicals. Repair processes seem to play an important role in determining the lethality of the lesions. A bleomycin-inactivating enzyme has been detected in both normal and malignant cells (Barrenco et al. 1975; Umezawa et al. 1974).

The first report on thermal enhancement of bleomycin came from Hahn (1974). In HA-1 (Chinese hamster) cells, exponentially growing, he found reduction in the survival fraction (SF) occurring at temperatures as low as 41° C and he supposed this effect to result from interference of heat with the repair of sublethal bleomycin damage. In subsequent studies (Hahn et al. 1975) it became obvious that the enhancement was markedly increased at 43° C; that is 500 times more than predicted by addition of the effects of heat and bleomycin alone. The threshold-like increase in thermally enhanced bleomycin cytotoxicity occurred between 41° and 43° C. The synergism was also seen in density-inhibited cultures, although to a lesser degree. The phenomenon of thermal enhancement has been confirmed thereafter under a wide variety of conditions in vitro and in vivo (Tables 19, 20). There is only the report of Wüst et al. (1973), who did not find even an ad-

Table 19. Results of preclinical investigations on thermal enhancement of bleomycin in vitro

System/ model	Assay	Dose	Temperature duration	Enhancement	Special topic	Reference
Jensen's sarcoma + embryonic tissues	[^{3}H]-thymidine [^{3}H]-uridine	-	39°-41° C 1-6 h	-		Wüst et al. 1973
HA-1 CH cells (exponential growth)	Surviving fraction	5-50 μg/ml	41° C 1 h	+	Mechanism	Hahn 1974
HA-1 CH cells	Surviving fraction	40 μg/ml	43° C	+	Sequence, drug uptake	Braun and Hahn 1975
HA-1 cells (exponential growth) EMT-6 mammary sarcoma a) in vitro b) in mice	Surviving fraction ratio	5-50 μg/ml	43° C (41° C)	+	Enhancement pattern, mechanism	Hahn et al. 1975
HeLa cells (exponential growth)	Surviving fraction	10 μg/ml	42° C (40° C) 1-20 h	+	Sequence, mechanism	Rabbani et al. 1978
HeLa cells	DNA single-strand breaks	-	43°-45° C 1 h	+	Mechanism	Kubota et al. 1979
EMT-6 cells	Surviving fraction	2-20 μg/ml	43° C 1 h	+	Preheating	Morgan et al. 1979
L5178Y mouse leukemia	Surviving fraction ratio	Up to 40 μg/ml	42° C 1 h	+	Sequence, timing	Mizuno et al. 1980
and FM3A cells	Surviving fraction ratio	20-30 μg/ml	37°-43° C 20-80 min	+		
FM3A cells and	Surviving fraction	30 μg/ml	39°-43° C 40 min	+	Ethanol	Ishida and Mizuno 1981

Table 19 *(continued)*

System/ model	Assay	Dose	Temperature duration	Enhancement	Special topic	Reference
HeLaS3 cells		40 μg/ml	1 h			
FM3A cells mouse mammary carcinoma	Surviving fraction	20 μg/ml	43° C 40 min	+	Sequence, timing, ethanol	Mizuno and Ishida 1981
EMT-6 spheroids	Surviving fraction, spheroids growth delay	5–20 μg/ml	43° C 1 h (40° C 6 h)	+	Sequence, drug tolerance	Morgan and Bleehen 1981
CHO cells (exponential growth)	Survival %	1–20 μg/ml	43° C 1 h	+	Heating rate	Herman et al. 1982
HeLaS3 cells (exponential growth)	Surviving fraction	10–40 μg/ml	43° C 1 h	+	Ethanol	Ishida and Mizuno 1982
V-79 cells CH (exponential growth)	Surviving fraction	Up to 300 μg/ml (10–50 μg/ml plus heat)	42.5° C 1 h	+	Comparison with other drugs	Roizin-Towle and Hall 1982
Human lymphocytoma M-3 + K-1 CH cells	Chromosome breakage, sister chromotid exchange	1–4 μg/ml	43° C 30–45 min	+	Mechanism	Vig et al. 1982
HA-1 CH cells (exponential growth)	Surviving fraction	15 μg/ml	43° C 1 h	+	pH	Hahn and Shiu 1983
CHO cells (exponential growth)	Colony formation, survival (%)	1–20 μg/ml	43° C 1 h (30° C)	+	Precooling	Herman 1983
V-79 CH cells	Surviving fraction	Up to 60 μg/ml	43° C	+	Mechanism, sequence	Lin et al. 1983
V-79 CH cells, BA1112 rhabdomyosarcoma in rats	Surviving fraction	20–60 μg/ml	43° C 1 h (10–60 min) 43° C 90 min (RF)		Sequence, drug uptake	Lin et al. 1983
KK-47 cells (human bladder carcinoma)	Colony formation, survival (%)	3 μg/ml	43° C 30 min	+	Sequence, sandwich-heating	Nakajima and Hisazumi 1983
OG cells cervix carcinoma	Cell proliferation	Up to 300 μg/ml	43° C 60 min	+		Fujiwara et al. 1984
V-79 CH cells, mouse L cells	Surviving fraction	0.1 μg/ml	42° C 44° C 40° C	+	Sequence	Kano et al. 1974
Human tumors	Clonogeneity	0.1–1 μg/ml	40.5° C	(+) 2+/3–	Low temperature	Neumann et al. 1985

Table 20. Results of preclinical investigations on thermal enhancement of bleomycin in vivo

System/ model	Assay	Dose	Temperature duration	Enhancement	Special topic	Reference
Lewis lung carcinoma in mice	Growth delay	15 mg/kg	43° C 1 h (days 4, 7, 10)	+	Fractionation, timing	Magin et al. 1979
KHT tumor in C3H mice EMT-6 tumors in BALB-C mice	Growth delay, cure rate, lung metastases, clonogeneity	7.0–15.0 mg/kg (i.v.)	41°–43° C 30 min (radiofrequency)	+	Fractionation, metastases	Marmor et al. 1979
Isogeneic adenocarcinoma in C3H mice	Growth delay	25–100 mg/kg (i.p.)	43.5° C 44.5° C 20 min (waterbath)	+	Sequence	Szczepanski and Trott 1981
BT4A tumor in BDIX rats	Growth delay	20 mg/kg (i.p.)	44° C 1 h (waterbath)	+	Resistance, toxicity	Dahl and Mella 1982
Isogenous squamous cell carcinoma CBA/HT mice	Tumor growth delay	25 and 50 mg/kg	43° C 15 min (ultrasound)	+	Sequence, fractionation	Hassanzadeh and Chapman 1983
RIF tumors in C3H mice	Growth delay, cure rate	7.5 mg/kg (i.p.)	44° C 30 min × 2 (7-day interval)	(+)	Tumor size	Li and Hahn 1984
FM3A tumors in female C3H/He mice	Tumor growth	15 mg/kg (s.c.)	43° C 1 h 2–3 times 41° C 1 h	+	Ethanol	Takiyama 1984

ditional effect in two tumor cell lines and several normal embryonic tissues, using [^{3}H]thymidine or [^{3}H]uridine incorporation as the parameter. Not known at that time was the fact that the temperature used was just below the threshold temperature. More recently, thermal enhancement of bleomycin cytotoxicity was reported at temperature levels below the usually accepted threshold temperature of 42° C. Neumann et al. (1985) found enhancement at 40.5° C using human tumor cells in long-term drug exposure.

Concerning the timing and sequencing of the two modalities (bleomycin and heat) extensive studies have been performed in order to find the most effective time sequence and to obtain information on the mechanisms underlying the phenomenon of synergism. Simultaneous application seems unequivocally to be the most effective route. But preheating at 41° C as well as at 43° C (with administration of the drug at 37° C) as well as postheating at 43° C were also reported to be effective (Braun and Hahn 1975). Similar findings were reported by Rabani et al. (1978): preheating was effective, even if an interval of up to 3 h was allowed, whereas postheating was ineffective even if it was performed immediately after drug administration. The importance of the interval between drug administration and heating was confirmed by several investigators (Magin et al. 1979; Marmor et al. 1979) and it seems that an interval of more than 4 h abolishes the enhancement (Lin et al. 1983). Morgan et al. (1979) found 3 h of preheating at 43° C to induce drug tolerance at 43° C but 1 h preheating at 43° C revealed marked sensitization to subsequent 37° C toxicity of bleomycin for up to 12 h.

There is only one report on postheating being more effective than preheating (Szczepanski and Trott 1981). But, as discussed by the authors, the postheating might have been rather a simultaneous application, because of the i. p. administration route of the drug and the interval of only 30 min between drug injection and heating; preheating was less effective. Yet there are some other reports of contradictory results on the effects of timing and sequencing: Hassanzadeh and Chapman (1983) found synergism only when the application of the two modes was simultaneous. When consecutively applied, even within 1 h, only additive effects were observed.

Putting together pre- and postheating ("sandwich" heating), Nakajima and Hisazumi (1983) obtained a reduction in survival percentage from the control (37° C) of 40.4% to 5.8%, which is more than simultaneous treatment (19.5%). Preheating produced an effect of 21.6% and postheating of 25.0% survival. This result is interesting in the context of the results of Kano et al. (1984), who reported the possibility of preventing thermally induced thermotolerance by pretreating the cells with bleomycin (or peplomycin). This would explain the fact that in Nakajima and Hisazumi's experiments the postheating effect was not abolished by thermally induced thermotolerance induced by the preheating phase. In V-79 cells and in mouse L-cells Kano further reported on enhanced cell kill by postheating at 40° C over 3 h. Preheating in this experiment was ineffective. The authors speculated on the possibility that bleomycin might have sensitized the cells against relatively low-level hyperthermia, which is ineffective when given alone.

The importance of the heating rate on the thermal enhancement of bleomycin was demonstrated by Herman et al. (1982). Heating the cells (CHO cells) from 37° to 43° C within 30 min is as effective as immediate heating. Heating them from 37° to 43° C within 3 h gave an effect one magnitude less. Precooling to 30° C and heating of the cells to 43° C within 30 min did not increase the thermal enhancement (Herman 1983).

Hahn and Shiu (1983) reported on the pH dependence of thermal enhancement of bleomycin. Whereas the cytotoxicity of the drug at 37° C was not changed between a pH of 6.5 and a pH of 8.5, a dramatic increase in the thermal enhancement (43° C) was seen when the pH shifted from 7.5 to 6.5.

Increased effectiveness was found by fractionated application, which might be of interest, bearing in mind clinical application schedules. Marmor et al. (1979) was the first to report tumor cures (EMT-6 tumors in BALB-C mice) after repeated (twice) treatments. Magin et al. (1979) also used fractionated schedules (days 4, 7, 10) when treating a Lewis lung carcinoma. Hassanzadeh and Chapman (1983) found DMFs of 1.95 when a three-fraction application was used. This value is practically the same (DMF, 2.0) as for one-time application of the same final drug dose. Advantages in reducing the toxicity should be expected by the fractionated application.

Mizuno and Ishida (1981) reported on ethanol mimicking the thermal enhancement of bleomycin in vitro. Concentrations below 4% applied together with hyperthermia in the range of 39°-42° C sensitized cells against bleomycin toxicity at 37° C nearly as much as 6% ethanol and even more than 43° C pretreatment alone. Ethanol had to be given before bleomycin treatment; simultaneously applied there is only little sensitization. As in preheating, pretreatment with ethanol revealed decreasing sensitization with increasing interval between either pretreatment and bleomycin exposure. The effect is abolished after a 6-h interval at 37° C. The recovery from the sensitization is inhibited by storage at 0° C, an effect described for sensitization also by heat. The recovery is not inhibited by prevention of protein synthesis or storage in phosphate-buffered saline. Takiyama (1984) applied ethanol in an in vivo model intra tumorally and found enhanced synergism in 3LL tumors at 41° C (bleomycin 15 mg/kg, s. c.).

Experiments in an intrinsically rather bleomycin resistant tumor were performed by Dahl and Mella (1982), who described impressive enhancement in their in vivo model: whereas bleomycin alone had no effect on tumor growth and heat alone revealed only a transient growth delay, the combination induced eight partial remissions out of nine tumors. These results are consistent with those of Marmor et al. (1979), Magin et al. (1979), and Szczepanski and Trott (1981). But Morgan and Bleehen (1981) reported negative results, using the clonogenic assay after treatment in vivo. The same tumor, however, investigated in monolayers or spheroids, had shown enhanced bleomycin toxicity at 43° C so that the negative results of Morgan and Bleehen in their in vivo model might have technical reasons or different sublines of the same tumor might have been investigated.

Thermal induction (40° C, 6 h) of drug tolerance to bleomycin was reported only by Morgan and Bleehen (1981) using their EMT-6 model. The cells did not develop tolerance to bleomycin at 37° C, only at 43° C.

The effect of bleomycin and hyperthermia has been found to be dependent on the size of the tumors treated (Li and Hahn (1984). Small-size tumors (0.1 cm^3) were less sensitive to the combined modality treatment (heat and drug) than medium- and large-size tumors (0.3 cm^3 and 0.7 cm^3, respectively). The medium- and large-size tumors may have been large enough to provide an intratumoral milieu, which is characterized by low pH, hypoxia, and nutritional deficiency, and what is known to promote the toxicity of the drug as well as of the heat. The same results have been obtained by the authors for BCNU as well as for the combination of the two drugs.

The mechanism of thermal enhancement of bleomycin is far from being fully understood. The thermally increased reaction rate seems unlikely to explain the reaction type (threshold). Vig et al. (1982) demonstrated that the site of action may be the chromosomes, because the frequency of chromosome aberration induced by bleomycin was higher at 43° C. No cell-cycle-phase specificity was seen and sister chromatid exchanges were not more frequent at 43° C than at 33° C.

These findings confirm those of Kubota et al. (1979), who described a markedly increased production of single-strand breaks by bleomycin at 43°-45° C than at 37° C whereas heat alone produces only a small number of strand breaks. Interestingly, the repair is nearly complete if cells are allowed to stay at 37° C after the treatment, but repair is markedly reduced at 43° C. This inhibition of repair is irreversible. As these effects can be induced in part by the membrane-interacting agent ethanol as well, changes in membrane permeability and consecutively increased drug uptake were throught to be a major mechanism underlying the enhancement. Braun and Hahn (1975), however, found slightly reduced [^{14}C]bleomycin uptake when cells were heated, which is consistent with the findings of Lin et al. (1983), who did not find [^{57}Co]bleomycin uptake enhanced in the AF_{12}rhabdomyosarcoma tissue in rats. Besides the main mechanism of thermal enhancement, which seems to be prevention of repair of sublethal bleomycin damage (Hahn et al. 1975; Meyn et al. 1980; Kubota et al. 1979), the thermal enhancement of the cytotoxicity of bleomycin additionally may be due to a thermal reduction in the activity of the bleomycin degradation-enzyme (Lin et al. 1983).

Peplomycin. Peplomycin is a new analog of bleomycin. Mizuno and Ishida (1981, 1982) reported on thermal enhancement of this drug (Table 21). Details of the reaction pattern according to different modes of timing and sequencing seem to be very similar to what was reported about bleomycin. The authors also described enhancement by ethanol pretreatment (5.7%/1 h), which they found to be even more effective than a 1 h heating at 43° C. The pattern of reaction to different concentrations and schedules of application of etha-

Table 21. Results of preclinical investigations on thermal enhancement of peplomycin

System/ model	Assay	Dose	Temperature duration	Enhancement	Special topic	Reference
FM3A cells	Survival fraction	Up to 20 μg/ml	43° C 1 h	+	Ethanol, drug uptake	Mizuno and Ishida 1981
FM3A cells and HeLa cells	Survival fraction	Up to 20 μg/ml 10 μg/ml	40° -42° C 1 h	+ +	Lidocaine, mechanism	Mizuno and Ishida 1982
V-79 CH cells	Survival fraction	0.05 μg/ ml	40° C	+	Prevention of thermotolerance	Kano et al. 1984

nol are also very similar to their findings on bleomycin. Ethanol did not change the uptake of peplomycin. Among some other membrane-active substances, lidocaine was also investigated and was found to enhance the cytotoxicity of peplomycin. The enhancement by lidocaine showed pH dependency and was not seen in Adriamycin, mito-C, and cis-DDP.

Miscellaneous Alkylator-Like Agents

Cis-Platinum

Cis (II)-platinum diaminedichloride(*cis*-DDP) is the only heavy metal compound in common use as a cancer chemotherapeutic agent. There is evidence that the formation of intrastrand cross-links may be the crucial feature of *cis*-DDP action, with the consequences of changes in DNA conformation and inhibition of DNA synthesis (Cohen et al. 1979).

Thermal enhancement of *cis*-DDP was first reported by Klein et al. (1977), using [^{3}H]methylthymidine incorporation as parameter in L1210 mouse leukemia cells. Hahn (1979) described an enhancement pattern following Arrhenius kinetics. Temperature dependence of the drug cytotoxicity was confirmed by many authors in quite a number of rather different in vitro and in vivo systems (Tables 22, 23) including hypothermic temperatures as low as 22° C (Herman 1983) and 22.2° C respectively (Roizin-Towle and Hall 1982).

Maximum enhancement seems to occur around 42° C with only a small further enhancement above that temperature (Fisher and Hahn 1982).

Enhancement was seen not only in exponentially growing cells but, although to a lesser degree, also in plateau-phase cells, the latter starting at a smaller Do value at 37° C (Fisher and Hahn 1982; Barlogie et al. 1980).

There are only a few reports of negative results: Wile et al. (1983), as stated by the authors themselves, obviously had chosen an unsuitable tumor insofar as the animals died from early occurring metastases, so that regional hyperthermia (isolation perfusion) could not reveal any efficacy. In terms of toxicity, however, an enhancement was observed to a remarkable degree, additionally hampering the development of a therapeutic gain. Hazan et al. (1984) also did not find a significant enhancement combining *cis*-platinum with heat but found a marked enhancement when combining the simultaneous administration of two drugs (*cis*-DDP + cyclophosphamide) with heat. Finally, Neuman et al. in 1985 using human tumors in an in vitro colony formation assay found only a small enhancement with

Table 22. Results of preclinical investigations on thermal enhancement of *cis*-DDP in vitro

System/ model	Assay	Dose	Temperature duration	Enhancement	Special topic	Reference
L1210 mouse leukemia cells	DNA synthesis, [^{3}H]thymidine incorporation	50+100 μg/ml	38° C 1 h	+	Lowest hyperthermia	Klein et al. 1977
HA-1 CH cells	Surviving fraction	0.8–6.0 μg/ml	39°–43° C 1 h	+	Mechanism	Hahn 1979
LoVo adenocarcinoma exponential growth + plateau-phase and CHO cells	% survival + (Do values) + (DMF)	Up to 10 μg/ml	41° + 42° C 1 h	+	Pattern of enhancement, thermoresistance (relative)	Barlogie et al. 1980
CHO cells (exponential growth)	Colony formation	0.5–5.0 μg/ml	43° C 1 h	+	Drug uptake, mechanism	Meyn et al. 1980
HA-1 cells (exponential growth + plateau phase)	Plating efficiency, surviving fraction	0.2–6.0 μg/ml	39°–44° C 1 h	+	Enhancement pattern, mechanism	Fisher and Hahn 1982
CHO cells	Clonogeneity	0.5–4.0 μg/ml	42.4° C 1 h	+	Heating rate	Herman et al. 1982
V-79 Chinese hamster	Cell survival clonogeneity	Up to 300 μ*M*/ml	22.2°–42.5° C 1 h	+	Timedependence	Roizin-Towle and Hall 1982
Seven different human melanoma lines	Clonogeneity	1, 2, 5 μg/ml	42° C 1 h	+	Heterogeneity	Zupi et al. 1982
L1210 mouse cells	Survival (%)	7.5 μ*M*	43° C 15 min	+	Sequence, mechanism	Bowden et al. 1983
CHO cells	Colony formation	0.5–4.0 μg/ml	22°–42.4° C 1 h	+	Precooling, heating rate	Herman 1983
TA3Ha mouse mammary adenocarcinoma in strain-A and athymic mice and	Tumor growth	0.5–4.0 mg/kg	43° C 30 min 1 h after drug × 3/5-day interval	+	Targeting, drug resistance, fractionation	Murthy et al. 1984
in vitro	Colony formation	10 μg/ml	43° C 30 min	+		
HEP-2 human epidermoid carcinom in nude mice and	Tumor growth	0.5–4.0 mg/kg	43° C 30 min 1 h after drug × 3/5-day interval	+		
in vitro	Colony formation	10 μg/ml	43° C 30 min	+		
Human tumors	Clonogeneity	Up to 10 μg/ml	40.5° C 2 h	(+)	Human tumors	Neumann et al. 1985
Human tumors and CFU-C	Clonogeneity	Up to 10 μg/ml	40.5° C 2 h	(+) –	Therapeutic ratio	Neumann et al. 1985

Table 23. Results of preclinical investigations on thermal enhancement of *cis*-DDP in vivo

System/ model	Assay	Dose	Temperature duration	Enhance-ment	Special topic	Reference
P388 leukemia + bone marrow *in* DBA/2 mice	Spleen colony assay	1–5 mg/kg	41° C 42° C 30 min	+	Sequence, therapeutic ratio	Alberts et al. 1980
Mouse sarcoma 180 in NHRI mice s.c.	Tumor volume, drug uptake	1.2 μg/g (+0.13 mg lipid /g) (i.v.)	42° C 1 h (waterbath)		Liposomes, drug uptake	Yatvin et al. 1981
MTG-B mouse mammary tumor in C3H/HeJ mice	Tumor growth, histology	5.0 mg/kg i.p.	44° +45° C 30–45 min (invasive heating)	+	Sequence	Douple et al. 1982
RIF tumor + bone marrow in mice	Clonogeneity	Up to 7.5 mg/kg	41° C 45 min (whole body hyper-thermia)	+	Therapeutic ratio	Honess 1983
VX-2 carcinoma in rabbits	Local tumor control	2 mg/kg	42° C 45 min	(+)	Toxicity	Wile et al. 1983
BT4A neurogenic rat tumor in BDIX rats	Tumor growth delay CR, PR	3 mg/kg	44° C 1 h	+	Timing, sequence, toxicity	Dahl and Mella 1984
MBT-2 transitional cell carcinoma in C3H mice	Animal survival, tumor volume, doubling time	8 mg/kg × 4 weekly	43.5° C 1 h 90 min	+	Timedepen-dence	Haas et al. 1984
Lewis lung carcinoma in BDF mice	Tumor growth delay	1.5–5.0 mg/kg	42.5° C 30 min (waterbath)	(+)	Multiple drug	Hazan et al. 1984
BT4A tumors in BDIX rats	Tumor growth time	2–4 mg/kg i.p.	44° C 1 h 41° C ca. 30 min (waterbath)	+ +	Drug dose dependence, whole body hyperthermia vs. local heat	Mella 1985

40.5° C hyperthermia in one out of six tumors. The negative results may indicate heterogeneity in the sensitivity for thermal enhancement of *cis*-DDP between different tumors, which was reported on different sublines of the same tumor as well (Zupi et al. 1984).

Synergism is markedly influenced by the heating rate, which was demonstrated by Herman et al. (1982). Heating CHO cells from 37° to 42.4° C within 3 h instead of 30 min revealed a lesser degree of enhancement. This was not due to drug tolerance, as the cytotoxicity was still *cis*-DDP dose dependent.

There is good accordance in all reports concerning the timing of drug and heat application: Maximum enhancement is found when heat and *cis*-DDP are applied simultaneously. Also good synergism is found when heating is applied before drug administration especially within a 1-h interval. No enhancement but even less than additive effects are seen when heating is performed after exposure to the drug (Bowden et al. 1983; Murthy et al. 1984; Dahl and Mella 1984; Zupi et al. 1984; Douple et al. 1982; Alberts et al. 1980).

Drug-resistant cell lines became sensitive when treated in combination with heat (Murthy et al. 1984). A relatively thermal-resistant cell line proved to be sensitive to the combination of *cis*-DDP and heat, and thermal sensitivity increased abruptly as the temperature was raised from 42° to 43° C (Barlogie et al. 1980).

Whereas local toxicity was found markedly increased by Wile et al. (1983), Dahl and Mella (1984) did not find skin lesions significantly increased. Differential increase in cytotoxicity as well as in drug uptake was reported by Alberts et al. (1980), when comparing P388 leukemia cells and normal bone marrow colony-forming units, suggesting the possibility of a therapeutic gain in systemic application of heat and drug.

This is in contrast to two other reports, comparing thermally enhanced cytotoxicity of *cis*-DDP against tumors or tumor cells with the cytotoxicity to bone marrow (Honess 1983; Neumann et al. 1985). These authors did not see any therapeutic gain in their experimental settings. An increase in systemic and especially in renal toxicity of *cis*-DDP parallel with the increasing antitumor effect was also reported by Mella (1985) when applying *cis*-DDP intraperitoneally with increasing dosis in combination with whole body hyperthermia up to 41° C. Bone marrow toxicity, however, was not mentioned in this paper.

Concerning the mechanism of thermal enhancement, final conclusions are not yet possible. Uptake of the drug seems to be increased but alone cannot account for the enhancement effects, as an enhancement is measurable even if heat is applied after the drug is washed out from the cellular medium (Murthy et al. 1984).

"Heat dose dependency" (Haas et al. 1984) of the enhancement and the pattern of reaction (steepening of the exponential portion of the survival curve) and gradual increase in the enhancement with increasing temperature favor the assumption of a thermodynamic effect by increasing the reaction rate of *cis*-DDP with the target molecule. But this seems not the only explanation because Meyn et al. (1980) reported a 10 times increased cytotoxicity and only a 6.5 times increased number of cross-links of DNA.

Although the drug-dose-dependent survival is thermally modified by abrogation of the shoulder of the survival function, indicating a reduction in the ability of the cells to absorb sublethal damage, DNA repair inhibition seems not to be involved to an essential degree, as Meyn has demonstrated that repair kinetics are not changed after hyperthermia. Heat-induced changes in the cell cycle are probably also not an important mechanism for the enhancement since exponentially growing and plateau-phase cells both show enhancement succeptibility and *cis*-DDP is not known to be a cell-cycle-specific-acting drug. So the mechanism of enhancement remains to be elucidated completely.

Interestingly, Murthy et al. (1984) reported on the possibility of converting noncytotoxic *trans*-DDP to the cytotoxic *cis*-DDP by hyperthermia, thus opening a possibility for targeting the drug effect to the locally heated tumor volume. Another attempt at drug-targeting by means of heating was made by Yatvin et al. (1981), using *cis*-DDP encapsulated in liposomes having a phase-transition temperature between 41° and 43° C. Drug uptake in the locally heated tumor (42° C/1 h) as well as tumor growth delay were greater when *cis*-DDP was administered in liposomes.

Table 24. Results of preclinical investigations on thermal enhancement of DTIC

System/ model	Assay	Dose	Temperature duration	En-hance-ment	Special topic	Reference
Sarcoma in mice	Survival time	100 mg/kg (i.p.)	40° C 1 h	–	Sex differences	Orth et al. 1977
Melanomas (human)	Colony inhibition (%)	0.1–10.0 µg/ml	42° C 1 h	+	Heterogene-ity, predictive assay	Mann et al. 1983

DTIC

Only two reports are available on thermal enhancement of the cytotoxicity of DTIC [5-(3,3-dimethyl-1-triazino)imidazole-4-carboxamide(dacarbazine)]. In vitro experiments may have been hampered by the assumption that DTIC had to be activated by liver microsomes. But activation seems to occur by tumor cells in vitro as well (Meyskens 1980). Even light activation may be sufficient as reported by Metelmann and von Hoff (1979) (Table 24).

Mann et al. (1983) reported on synergism in 15 out of 53 malignant melanoma specimens, taken from 53 patients, and treated in vitro by DTIC plus 42° C/1 h heating. The results predicted correctly the effect in four out of five patients treated by the same combination.

In an animal model Orth et al. (1977) saw no enhancement. But the design of the study was somewhat confusing, as female tumor-bearing mice revealed a marked enhancement by glucose and heat treatment (26.5 vs. 37 days median survival time) whereas male animals did not (30.8 vs. 33.2 days median survival time).

Carboquone

Carboquone (Esquione, Sankyo Co., Tokio) has been studied mainly by Japanese investigators: Kidera and Baba (1978) and Kamura et al. (1979) (Table 25). When this drug was combined with tourniquet administration and 42° C heating (30 min), the survival time was prolonged and cures were obtained. But toxicity increased as well as treatment-related mortality from 11.1% to 29.7% of the animals. No striking increase in toxicity occurred when carboquone administration and local heating were combined with systemic cooling (20°–24° C). In this experimental setting the thermal enhancement of the antitumor toxicity of carboquone was greater than that of any other drug tested (see Adriamycin, 5-FU, NH_2, mito-C). Mizuno et al. (1980) also found an enhancement ratio of 1.6 in vitro.

Peptichemio

This alkylating agent was used by Kostic et al. (1978) (Table 26). Synergism occurred depending on duration of heating. The effect was characterized by an increase in the steepness of the dose-reponse curve as well as a reduction in the shoulder, the latter being thought to indicate inhibition of repair of sublethal damage.

Table 25. Results of preclinical investigations on thermal enhancement of carboquone = carbazilquione

System/ model	Assay	Dose	Temperature duration	En-hance-ment	Special topic	Reference
Autochthonous sarcoma in mice (hind limb)	Survival time	12.5% of LD_{50} (1 mg/kg)	41° C 30 min	+	Tourniquet, toxicity	Kidera and Baba 1978
L5178Y cells	Surviving fraction		42° C 1 h	+		Mizuno et al. 1980
Ehrlich ascites tumor cells in DDD mice (hind limb)	Tumor weight	1 mg/kg (i.v.)	41° C 1 h (waterbath)	+	Systemic cooling	Kamura et al. 1979

Table 26. Results of preclinical investigations on thermal enhancement of peptichemio

System/ model	Assay	Dose	Temperature duration	En-hance-ment	Special topic	Reference
L cells (log-phase)	Surviving fraction	0.1–1.5 µg/ml	42° C 1–2 h	+		Kostic et al. 1978

Antimetabolites

Folate Antagonists

Methotrexate

Methotrexate (MTX) exerts its cytotoxic effect by binding tightly to the enzyme dihydrofolate reductase and thereby blocking the reduction of folic acid. MTX normally enters the cell through an active carrier-mediated cell membrane transport system. In some tumors having reduced transport capabilities MTX enters the cells passively when high extracellular levels are provided.

In HA-1 CH cells Hahn and Shiu (1983) did not find thermal enhancement of the cytotoxicity of MTX (Table 27). Although using somewhat questionable parameters, Muckle and Dickson (1973) also found no thermal enhancement in an in vivo tumor model. When lowering the pH in vitro again no enhancement occurred in HA-1 CH cells heated for 1 h at 43° C (Hahn and Shiu 1983).

But using a rather MTX-resistant cell line (CHO cells, exponentially growing), Herman et al. (1982) was able to overcome the resistance by heating the cells at 43° C. There was no difference between heating the cells abruptly or within 30 min, but the effect was markedly reduced when raising the temperature of the cells over as long a period as 3 h. Although thermotolerance occurred, this was not linked to drug tolerance. The mechanism underlaying the thermal enhancement of the cytotoxicity of MTX in MTX-resistant cells is suggested to be a hyperthermia-induced cessation of dihydrofolate reductase synthesis (Herman et al. 1981).

Table 27. Results of preclinical investigations on thermal enhancement of methotrexate

System/ model	Assay	Dose	Temperature duration	En- hance- ment	Special topic	Reference
VX-2 carcinoma in rabbits	Tumor volume, respiration, glycolysis	0.4 mg/kg (6 daily injections	42° C 1 h	–	Uncertain parameters	Muckle and Dickson 1973
L1210 tumor in mice (hind foot)	Plasma *t*/2 drug concentration in tumor, tumor growth delay	3.0 mg/kg (i. v.)	42° C 1 h (waterbath)	+	Targeting, lipid vesicles, pharmacokin- etic	Weinstein et al. 1980
CHO cells (exponentially growing)	% survival	1.0–20.0 μg/ml	43° C 1 h	+	Heating rate, drug resistance	Herman et al. 1981
L1210 tumors in mice s. c.				+	Targeting, lipid vesicles	Margin and Weinstein 1982
SV 40 transformed hamster lymphoid cells	Drug uptake	10 μ*M*	41° C 10 min	+	Drug uptake	Mikkelsen and Wal- lach 1982
TCC bladder tumors in C3H/Bi mice	Drug uptake		42° C 1 h	+	Targeting, lipid vesicles	Tacker and Anderson 1982
HA-1 CH cells	Surviving fraction	298 μg/ml	43° C 1 h	–	pH	Hahn and Shiu 1983

Another possible explanation of the thermally enhanced MTX effect of tumor cells having reduced or even a lack of active carrier-mediated cell membrane transport was reported by Mikkelsen and Wallach (1982). They found MTX uptake by SV-40-transformed hamster lymphoid cells enhanced by thermally induced changes in membrane ion gradients.

Methotrexate has been used to evaluate the possibility of "drug-targeting" by temperature-sensitive lipid vesicles (Weinstein et al. 1980; Magin and Weinstein 1982; Tacker and Anderson 1982). Similar to what has been found using *cis*-DDP, heated tumors accumulated 11.9-fold more [^{3}H]MTX than nonheated tumors when the drug was administered encapsulated in liposomes. Animals receiving free MTX did not exhibit a temperature-dependent difference (Tacker and Anderson 1982). Heat combined with liposome-encapsulated MTX had an overadditive effect on tumor growth delay (Weinstein et al. 1980). The reasons for this may be multiple since not only is the drug uptake enhanced by heating, but the drug effect itself may be thermally enhanced and in addition the serum half-life of MTX markedly prolonged when the drug is administered in vesicles. The size of the vesicles may play an important role, as large unilamellar ones could release the drug more quickly and more completely than small ones at the liquid crystalline phase transition temperature. The reason for this is thought to be the better ratio of internal volume to lipid vesicle surface (Magin and Weinstein 1982).

Pyrimidine Antagonists

5-Fluorouracil

5-Fluorouracil (5-FU), after conversion to fluorouridine monophosphate (FUMP), exerts its cytotoxicity by two "active" nucleotides. The one, fluorouridine triphosphate (FUTP), is incorporated into RNA; the other (5-FdUMP) binds tightly to thymidylate synthetase, thereby inhibiting formation of the DNA-precursor desoxythymidine triphosphate (dTTP). Thus 5-FU is a cell-cycle-phase-specific antitumor agent.

Temperature dependence of the cytotoxicity of 5-FU has been generally not found. Indeed, Mizuno et al. (1980) in their in vitro experiments and Rose et al. (1979) using a colon cancer cell line in mice were both not able to enhance the cytotoxicity of 5-FU at temperatures of 42° C and 38.9° C, respectively (Table 28).

In a review article published in 1983, Hahn first mentioned 5-FU to be enhanced by heating in vitro and in vivo (details unpublished). In 1984, Lange et al. reported thermal enhancement in two human colon cancer cell lines. In both samples the clinical response to the combined treatment seemed to parallel the in vitro findings.

The relevance of timing of the heat was demonstrated by Joshi and Barendson (1984). They exposed rat ureter carcinoma cells for 24 and 48 h to 5-FU. Heating (43° C) during

Table 28. Results of preclinical investigations on thermal enhancement of 5-fluorouracil (5-FU)

System/ model	Assay	Dose	Temperature duration	Enhancement	Special topic	Reference
L5178Y cells	Surviving fraction	-	42° C 1 h	-		Mizuno et al. 1980
Human colon adenocarcinoma cells	Surviving fraction (%)	-	41.8° C 1 h	+	Predictive assay	Lange et al. 1984
Rat ureter carcinoma RUC-2 cells	Surviving fraction	Up to $2.0 \times 10^{-6} M$ (24 h)	41° C 1 h 43° C 1 h	- +	Timing, "heat dose"	Joshi and Barendsen 1984
CC38 colon carcinoma in BDF1 or B6C3F1 mice	Survivors, median survival time	60 mg/kg i.p.	38.8°-38.9° C 45 min	-		Rose et al. 1979
Ehrlich ascites cells in DDD mice	Tumor weight	40 mg/kg (i.v.)	41° C 1 h (waterbath)	(+)	Systemic cooling	Kamura et al. 1979
P388 cells in BDF1 mice			42° C	+		Adwankar and Chitnis 1984
Hepatic artery infusion in dogs	Serum hepatic enzymes, histology	10 mg/kg (i.v.)	43° C 1 h (whole body hyperthermia)	+	Toxicity	Daly et al. 1982
Isolated perfused rat liver in situ	Survival, liver function, hepatic regeneration	0.125-1.5 g/kg	41° + 43° C 5 or 10 min	NE	Toxicity	Miyazaki et al. 1983

NE, not evaluated.

the 1st h of exposure to 5-FU was hardly additive but heating in the last hour was clearly more than additive. Heating at 41° C did not influence the effectiveness of 5-FU. Enhancement of the antitumor effect of 5-FU in vivo was first reported by Kamura et al. (1979), using the thermo differential heating technique. At 41° C a slightly overadditive antitumor effect of 5-FU was found, taking tumor weight as the parameter.

More recently, Adwankar and Chitnis (1984), using P388 cells in BDF1 mice, demonstrated thermally enhanced cytotoxicity of 5-FU at 42° C.

Hepatic artery infusion of 5-FU or systemic hyperthermia (43° C/1 h) in dogs produced no or minor changes in toxicity parameters (serum hepatic enzymes and histology (Daly et al. 1982). But combined application of both "agents" revealed a significant increase in SGOT levels and marked changes in liver histology.

In isolated perfused rat liver in vivo Miyazaki et al. (1983) found heat alone potentially toxic in a time- and temperature-dependent manner. Survival following hyperthermic perfusion was zero after 10 min at 41° C and was 50% after 5 min at 43° C. Perfusion with 0.25 g/kg or less of 5-FU was well tolerated but induced severe suppressive effects on hepatic regeneration. Combination experiments quite obviously were not performed.

Cytosine Arabinoside

Cytosine arabinoside (Ara-C) is converted enzymatically to its active form, Ara-CTP, acting as an inhibitor of DNA polymerase in competition with deoxycytidine triphosphate (d-CTP). No thermal enhancement of Ara-C was found by Mizuno et al. (1980) and Adwankar and Chitnis (1984) (Table 29). [^{3}H]Ara-C was used for experiments to characterize the dynamics of drug encapsulation into large unilamellar liposomes and drug release at their phase transition temperature. The vesicles were found to be stable up to 41° C in serum and to release the drug completely within seconds at 43° C (Magin and Niesman 1984).

Table 29. Results of preclinical investigations on thermal enhancement of cytosine arabinoside (Ara-C)

System/ model	Assay	Dose	Temperature duration	Enhancement	Special topic	Reference
L5178Y cells	Surviving fraction	-	42° C 1 h	-		Mizuno et al. 1980
P388 cells			42° C 1 h	-		Adwankar and Chitnis 1984

Table 30. Results of preclinical investigations on thermal enhancement of Pala

System/ model	Assay	Dose	Temperature duration	Enhancement	Special topic	Reference
V-79 (exponential growth) CH cells	Cell survival	Up to 10 m*M*	25°–42.5° C 1 h	(+)/–	Cooling	Roize-Towle and Hall 1982

N-(Phosphonacetyl)-L-aspartate

N-(Phosphonacetyl)-L-aspartate (Pala) is an antimetabolite which inhibits an enzyme in the de novo pathway of pyrimidine biosynthesis. Temperature dependence of cytotoxicity of this agent was found between 25° C and 37° C without enhancement at 42° C (Roizin-Towle and Hall 1982) (Table 30).

Plant Alkaloids

Vinca Alkaloids

The vinca alkaloids possess cytotoxic activity by virtue of their binding to tubulin and, therefore, they are cell cycle phase specific for mitosis (M-phase). It has been demonstrated that heating 3T3 cells at 41° C reversibly disassembles microtubules and that at 43° C (30 min) may disrupt microtubule organization completely (Lin et al. 1982). So one might have expected thermal enhancement of the vinca alkaloids. But the majority of reports are negative (Hahn 1983).

Vincristine

Mizuno et al. (1980) could not find an enhancement of the cytotoxicity of vincristine (VCR) in L5178Y cells heated to 42° C for 1 h (Table 31). Rose et al. (1979) reported on two out of 18 survivors in normothermic-treated mice, bearing P388 leukemia as compared with 0/19 survivors in hyperthermically (38.8°–38.9° C/45 min) treated mice, using 1.0 mg/kg VCR given intraperitoneally. Perhaps due to the higher temperature (42° C) Adwankar and Chitnis (1984) found markedly enhanced cytotoxicity of VCR also using P388 mouse leukemia cells.

Table 31. Results of preclinical investigations on thermal enhancement of vincristine

System/ model	Assay	Dose	Temperature duration	Enhancement	Special topic	Reference
P388 leukemia cells in BDF1 mice	Survivors	1.0 mg/kg (i.v.)	38.8°–38.9° C 45 min	–		Rose et al. 1979
L5178Y cells	Surviving fraction	Not given	42° C 1 h	–		Mizuno et al. 1980
P388 leukemia cells	Surviving fraction	10 μg/ml	42° C 1 h	+		Adwankar and Chitnis 1984
Human tumor cells	Surviving fraction (clonogenic)	Up to 10^{-2} μg/ml	40.5° C 2 h	+/–	Long-term incubation, therapeutic	Neumann et al. 1985
Human CFU-C in vitro	Surviving fraction (clonogenic)	Up to 10^{-2} μg/ml	40.5° C 2 h	+	ratio	Neumann et al. 1985

In a variety of different human tumors, Neumann et al. (1985) were able to demonstrate enhancement in three out six tumors and at a higher concentration also in normal human bone marrow progenitor cells (40.5° C/2 h). In these experiments, long-term drug exposure was used, what may have contributed to the positive results (Neumann et al. 1985).

Velbe

Exposure to hyperthermia in the first of 24-h drug treatment protected the cells against velbe (VBL)-induced lethality (Joshi and Barendsen 1984) (Table 32). It is supposed that this is not due to drug inactivation by heat, as no protection occurred in 48-h incubation experiments. Exposure to hyperthermia during the last hour of 24-h drug treatment was of no significant influence on the effect at 37° C; 41° C incubations failed to influence the effectiveness of VBL. This is consistent with the findings of Neumann et al. (1985), who could not find thermal enhancement (40.5° C/2 h) either in tumor cells or in human bone marrow progenitor cells.

Vindesine

There was also no thermal enhancement of the cytotoxicity of vindesine (VDS) in exponentially growing CHO cells, as reported by Herman (1983) (Table 33).

Table 32. Results of preclinical investigations on thermal enhancement of velbe

System/ model	Assay	Dose	Temperature duration	Enhancement	Special topic	Reference
RUC-2 cells	Surviving fraction	Up to 40×10^{-8} μg/ml	41° C 1 h 43° C 1 h	–/↙	Timing	Joshi and Barendsen 1984
Human tumor cells and CFU-C	Surviving fraction (clonogenic)	Up to 10^{-3} μg/ml	40.5° C 2 h	–	Therapeutic ratio	Neumann et al. 1985

Table 33. Results of preclinical investigations on thermal enhancement of vindesine

System/ model	Assay	Dose	Temperature duration	Enhancement	Special topic	Reference
CHO cell growth	Survival (%)	Up to 3 μg/ml	42.4° C 1 h	–		Herman 1983

Podophyllotoxins

Although occasionally used clinically in thermochemotherapy (Barlogie et al. 1979, Heppner 1982), there are no publications on in vitro or animal experiments available concerning thermal enhancement of these drugs.

Miscellaneous Agents

Procarbazine

The exact mode of antineoplastic activity of procarbazine (PBZ) is unclear. Overall it seems to act by inhibition of protein, RNA, and DNA synthesis. Thermal enhancement was found by Senapati et al. (1982) when heating at 43 ± 0.5° C for 30 min was performed on LX-1 human lung tumor xenografts in nude mice; 100 mg/kg PBZ was given and tumor growth was the assay. The combined treatment even produced tumor regression. No further enhancement occurred by increasing the number of heat treatments. The drug was given 1 h before heating (Table 34).

Hydroxyurea

The main mode of action of hydroxyurea (HUR) seems to be its immediate inhibition of DNA synthesis. This property is used in experiments designed to elucidate the mechanism of heat injury, as did Palzer and Heidelberger (1973), who found protection against heat injury to HeLa cells (synchronously growing) by HUR-induced inhibition of DNA, and protein synthesis, the latter being additively influenced by cycloheximide (Table 35).

m-AMSA

This derivative of the acridine dye class is chemically methansulfonamide, *N*-((4-(acridinylamino)-3-methoxyphenyl)). Mechanistically, the compound probably intercalates between DNA base pairs (Waring 1976). Herman has studied heat-drug interaction of m-AMSA (1983). Cell killing by m-AMSA was inhibited when drug exposure occurred at 42.4° C as compared with exposure at 37° C. Inactivation of the drug by heating was ruled out by preheating the drug before exposure (Table 36).

Table 34. Results of preclinical investigations on thermal enhancement of procarbazine

System/ model	Assay	Dose	Temperature duration	Enhancement	Special topic	Reference
Human lung tumor in nude mice	Tumor growth	100 mg /kg	43° ± 0.5° C 30 min	+	Human tumors	Senapati et al. 1982

Table 35. Results of preclinical investigations on thermal enhancement of hydroxyurea

System/ model	Assay	Dose	Temperature duration	Enhancement	Special topic	Reference
HeLa cells (synchronously growing)	Survival (%); DNA synthesis	$2.5 \times 10^{-4} M$	42° C 1-2 h	(+)/ ↙	Mechanism	Palzer and Heidelberger 1973

Table 36. Results of preclinical investigations on thermal enhancement of m-AMSA

System/ model	Assay	Dose	Temperature duration	En-hance-ment	Special topic	Reference
CHO cells (exponential growth)	Survival (%)	Up to 10 μg/ml	42.4° C 1 h	(+)/↙	Protection	Herman 1983

Table 37. Results of preclinical investigations on thermal enhancement of alkyl-phospholipids

System/ model	Assay	Dose	Temperature duration	En-hance-ment	Special topic	Reference
Human leukemic blast cells	Viability [^{3}H]ALP absorption	1–40 μg/ ml	25°–41° C	+	Phase I drug	Andreesen et al. 1983

Table 38. Results of preclinical investigations on thermal enhancement of lonidamine

System/ model	Assay	Dose	Temperature duration	En-hance-ment	Special topic	Reference
HA-1 cells	Survival	5–50 μg/ml	43° C 1 h	+	Phase I drug	Silvestrini et al. 1983
S-180 + Ehrlich ascites cells in mice	Survival time	25/50 mg/ kg daily (p.o.)	39° ±0.5° C 7 h × 4 (whole body hyperthermia)	+		
HeLa cells	Surviving fraction	Up to 100 μg/ml	41° C 42° C	+	Mechanism pH	Kim et al. 1984
Fibrosarcoma in BALB-C mice	Tumor control	–	41.6° C 90 min	+		

Alkyl-phospholipids

Alkyl-analogs of 2-lysophosphatidylcholine induced a progressive destruction of neoplastic cells by interfering with the continuous turnover of membrane phospholipids. Using leukemic blast cells from patients with acute leukemias, the cytotoxicity of alkyl-phospholipids (ALPs) was found to be strongly temperature dependent in the range between 25° and 41° C (Andreesen et al. 1983) (Table 37).

Lonidamine

Chemically, lonidamine is dichlorophenyl-methyl-1 *H*-indazole-carboxylic acid. In 1983 Silvestrini et al. reported on thermal enhancement of the cytotoxicity of lonidamine at low drug concentrations (Table 38). This enhancement was much reduced in the absence of

DMSO, suggesting dependence on facilitated drug uptake into the cells. Also in vivo the authors found thermal enhancement of lonidamine. The two modalities applied alone were without any significant effect, using survival time (S-180 Ehrlich Ascites tumors in mice) as the assay. Applied together the two treatments resulted in an increase of survival time of 27% and 35% at doses of 25 and 50 mg/kg lonidamine, respectively.

Kim et al. (1984) in vitro as well as in vivo found significant enhancement of lonidamine by heating. Additionally they reported further potentiation of the effects at lower pH conditions (pH 6.5). Concerning the mechanism of hyperthermic sensitization of lonidamine it is of interest to mention that this compound is an effective inhibitor of lactate transport.

Pilot study application to human cancer patients induced morphological changes in some lesions, supporting the assumption that heat-induced condensation of the mitochondria facilitates the effects of lonidamine (De Martino et al. 1984).

Substances Acting on the Immune System

The broad topic of the role of the interaction of immune processes and hyperthermia is beyond the scope of this review. Two examples of immunoactive substances, however, will be mentioned.

Interferon

Yerushalmi et al. (1982) reported combined treatment of Lewis lung carcinoma in mice, using either endogeneously induced or exogeneously injected mouse interferon and local heating (42.7° C/1 h). Although statistically significant, an absolutely small increase in mouse survival time was observed relative to either treatment alone (Table 39).

Corynebacterium parvum

Urano et al. have published details of thermal enhancement of this unspecific immunostimulant several times (Urano et al. 1978, 1979, 1984). Logically they did not find cytotoxicity or enhancement in vitro. In vivo, however, using a rather immunogenic spontaneous fibrosarcoma (FSa-II) in mice, they observed enhancement of heat effect on tumor growth as well as on foot reaction by systemically applied *Corynebacterium parvum* (Table 40). The mechanism is still obscure, especially as neither total body irradiation nor anti-mouse T-cell serum inhibited the *C. parvum* enhancement of the thermal response.

Table 39. Results of preclinical investigations on thermal enhancement of interferon

System/ model	Assay	Dose	Temperature duration	Enhancement	Special topic	Reference
Lewis lung carcinoma in L57B1/6 mice	Survival time	0.25 ml (titer 3.2×10^{-6} /ml) day 1-8 (i.p.)	42.7° C 1 h (day 2, 5, 7) (warm air)	+		Yerushalmi et al. 1982

Table 40. Results of preclinical investigations on thermal enhancement of Corynebacterium parvum

System/ model	Assay	Dose	Temperature duration	En-hance-ment	Special topic	Reference
CHO cells	Surviving fraction	–	43° C 1 h	–	Thermo-tolerance, mechanism, toxicity	Urano et al. 1984
FSa/II (spontaneous fibrosarcoma)	Tumor growth, foot reaction	350 μg (i.v.) 3 days before heating	42.5°–46.5° C 63–95 min	+		
C3Hf/Sed mice						

Thermosensitizers

In this chapter some of these agents will be discussed, being characterized by an almost complete lack of cytotoxicity of 37° C. At elevated temperatures, however, they act as cell inactivators in a dose-dependent manner.

Ethanol

Ethanol was demonstrated to modify the cytotoxicity and thermal enhancement of some cytotoxic drugs; this has been mentioned in the preceding sections.

Ethanol-induced drug tolerance (Li 1984) may (see section on Adriamycin) or may not (see section on BCNU) be linked to thermotolerance. Ethanol-induced thermotolerance is found to be linked to the appearance of heat shock proteins (Hahn 1983). The role of these proteins in thermobiology is discussed in another chapter of this volume.

In addition, there is an interaction of ethanol and heat insofar as ethanol, being of low toxicity at 35° C, becomes highly effective at elevated temperatures (Hahn and Li 1982). In HA-1 cells the survival curve exhibits a marked threshold when survival fraction is plotted against alcohol concentration as well as against increasing temperature. The two agents are exchangeable insofar as the effect on survival is concerned, thus indicating a cooperative phenomenon responsible for cell death.

In 1981 Massicotte-Nolan et al. investigated sensitization of V-79 hamster lung cells to hyperthermia by ethanol and other alcohols. They found protein denaturation energy, caused by the same chemicals, consistent with hyperthermic cell killing rather than changes in membrane lipid fluidity. The effect of ethanol and heat in normal tissue was investigated by Anderson et al. (1983). They reported an increase in heat-induced damage to the skin of the ear in mice when exposed to ethanol. They also found ethanol-induced protection to heat necrosis, i.e., ethanol-induced thermotolerance.

In vitro experiments using human normal and leukemic (CLL) lymphocytes revealed synergism of heat (42° C) and 0.8% ethanol (Schrek and Stefani 1981); viable lymphocyte counts were used as the assay.

Lidocaine and Procaine

Lidocaine and procaine are clearly membrane-acting local anesthetic drugs. As hyperthermia is thought to exert its cytotoxicity at least in part by membrane injury, these compounds were studied also in respect to their capacity as thermosensitizers (Yatvin 1977).

The first observation on in vitro interaction between lidocaine and hyperthermia was reported by Yatvin et al. (1979). Young BDF1 mice grafted with mammary adenocarcinoma CA755 had lidocaine injected into the tumor before heat treatment at 42° or 43.5° C. At 43.5° C, survival times were increased threefold over the control and even some cures achieved.

Similar results were reported by Robins et al. (1983). These authors even found heat sensitization by lidocaine given systemically at doses within the therapeutic range for control of arrythmias. In addition they had found potentiation of differential hyperthermic sensitivity of AKR leukemia and normal mouse bone marrow cells (Robins et al. 1984) and, therefore, initiated pilot studies in cancer patients (Robins 1984).

Similar to what was reported on alcohols, lidocaine may induce thermotolerance in a time-dependent manner and again production of heat shock proteins seems to be correlated with this phenomenon (Hahn 1983; Li 1984).

Amphotericin-B

This polyene antibiotic is known to bind specifically to cholesterol, which is found almost exclusively in the plasma membrane.

In 1977, Hahn et al. reported that amphotericin-B (am-B), which is of low cytotoxicity at 37° C, shows little drug concentration dependence in its enhanced cytotoxicity at 41° C and exerts high toxicity even at a small dosis at 43° C; in these experiments HA-1 cells were used. When treating EMT-6 cells in situ (up to 20 mg/kg) in combination with 1 h heat treatment (43° C) they found a markedly enhanced antitumor efficiency. In later experiments (Hahn and Li 1982) the authors demonstrated that preheating was necessary for synergism, whereas postheating revealed only additive toxicity. It is concluded that preheating modifies the plasma membrane and thereby renders the cell highly susceptible to the drug. The temperature dependence of the cytotoxicity of am-B was found to be increased when changing the pH to lower as well as to higher values. This symmetrical pattern of increased efficiency is unique to this drug.

Misonidazole

Misonidazole, an electrophilic compound is expected to be preferentially a radiosensitizer. It was shown to be also cytotoxic to hypoxic cells in vitro (Hall et al. 1977). Misonidazole was found to enhance the heat sensitivity of spheroids (EMT-6 cells), and also preheating the spheroids (43° C) increased their response to misonidazole under hypoxic conditions (Morgan and Bleehen 1981).

In CHO cells, Langer et al. (1982) found further thermal enhancement of the cytotoxicity of misonidazole at temperatures of 39° and 41° C when lowering the pH to 6.4–6.7.

An appreciable enhancement of the cytotoxic action of misonidazole against hypoxic cells in vivo was observed by Lehman and Stewart in 1983. But this occurred only when the tumors (mouse mammary carcinoma) were clamped after misonidazole injection in order to provide prolonged presence of the drug as well as hypoxia.

An approximately fourfold increase of intracellular uptake of misonidazole was found when CHO cells were incubated at 44° C compared with the uptake at 37° C (Brown et al. 1983).

Dimethyl Sulfoxide

Hahn reported in 1983 on this compound. The cytotoxicity of dimethyl sulfoxide (DMSO) seemed to be induced by heat in vitro. Additionally he reported on its capability to induce thermotolerance. Thermally induced inhibition of hemoglobin induction by DMSO in a murine erythroleukemia model was reported by Raaphorst et al. (1984). These findings indicate that heat may act on the nuclear level, i.e., on the differentiation of the hemoglobin-producing state in response to DMSO treatment.

Cysteamine

In order to obtain more information on the mechanism of the thermal enhancement of compounds generating activated oxygen species, Issels et al. (1984) investigated the cysteamine toxicity at different temperatures. When drug-treated cells (up to 8 m*M*) were exposed to a 44° C heat treatment (30 min), cytotoxicity was markedly increased. No modifications of cysteamine toxicity occurred when superoxide dismutase was added, but toxicity was completely abolished by addition of catalase. Production of H_2O_2 and subsequent increase in hydroxyl radicals (·OH) is thought to be the cause of the cytotoxicity observed.

AET

AET (*S*-(2-aminoethyl)isothiouronium bromide HCl) was developed as a radiation protector. At 37° C it shows little if any cytotoxicity, whereas at a high temperature it becomes quite effective (Hahn 1979). In vivo (in animals), however, even at 43° C (30 min) no antitumor effect of AET could be demonstrated (Marmor 1979).

Quercetin

This experimentally used compound is chemically a bioflavonoid that produces lactate transport inhibition. Quercetin is not cytotoxic at 37° C (0.1 m*M*/4 h) (Kim et al. 1984). At 41° C and at 42° C, however, hyperthermia-induced cytotoxicity to HeLa cells is markedly enhanced by exposure to quercetin. This effect is even more pronounced at low pH values of the medium. In those tumor cells having an enhanced rate of lactate production, this compound may well reveal a certain degree of selectivity.

Clinical Experience

As shown in the first part of this review, there is a fair amount of preclinical data demonstrating that heat enhances the efficacy of many of the cytotoxic drugs of clinical interest. Some of the so-called milieu factors inside larger tumors, i.e., hypoxia, acidosis, and malnutrition, additionally support the thermal enhancement of the cytotoxicity of some of these drugs. Such tempting rationale initiates widespread optimism regarding the potential of thermochemotherapy in cancer treatment (Magin 1983). This, however, is in sharp contrast to the results of human clinical studies reported so far. This discrepancy is mainly due to the fact that all the drugs available are of limited specificity and hyperthermia is far from being proved to enhance the cytotoxicity of the drugs in a tumor-specific manner.

The majority of the clinical studies performed so far were designed as phase I studies, with equipment evaluation and toxicity assessment as their major goal. Nevertheless, many authors have claimed improved clinical results. But very little reliable information is available dealing with quantitation of a therapeutic gain. As outlined previously, one should not expect heat to induce an increase in the specificity of the drugs in general. Different routes of application of heat and drug, therefore, are under investigation, in order to create selectivity of the thermally enhanced cytotoxicity of the drugs.

There are three main routes for application of heat in thermochemotherapy:

1. Systemic heating (whole body hyperthermia)
2. Local or regional heating by electromagnetic energy supply
3. Hyperthermic isolation perfusion.

Systemic Heat Application

In cancer patients developing high fevers associated with erysipelas, tumor regressions were observed in the late 1800s (Busch 1866; Coley 1893). These observations initiated numerous attempts to treat cancer patients by pyrogenic substances (Nauts 1985 and Westphal et al. 1977). Nauts recently has summarized the clinical data of 896 patients treated throughout the past 2 decades with Coley Toxins (mixed bacterial vaccines). Unfortunately, this compilation of the largest number of case reports, which lacks all the essentials of a clinical trial, is not suitable for proving the clinical efficacy of this approach.

On the other hand, reports on the "Selective Heat Sensitivity of Tumor Cells" (Stehlin Jr. et al. 1975) supported the assumption that the temperature elevation itself could have initiated the tumor responses reported by Busch and Coley. Therefore, externally induced whole body hyperthermia by physical methods was employed, involving heat delivery across the surface of the body or directly into the circulation by means of an extracorporeal heat exchanger. Since 1977 a fairly large number of different methods for inducing whole body hyperthermia have been reported, which are summarized in Table 41.

It has been found that a core temperature of up to 41.5° C is possible without general anesthesia, but above that temperature level general anesthesia has to be employed. Physiological and toxicity studies of these methods in phase I studies revealed a high degree of safety in the temperature range below 42° C if the vital functions are monitored and the procedure is performed by a trained staff (Cronau et al. 1984). Temperatures above 42° C are of markedly increased toxicity, especially to the liver and the brain (Pettigrew and Ludgate 1977). The main side effects (Table 42) seem to be in part related to the specific technique used: Skin burns and peripheral nerve lesions are reported only when epidermal warming is used, whereas extracorporeal heating regularly induces thrombocytope-

Table 41. Methods for induction of whole body hyperthermia and clinical applications published by different groups

Application	Reference
Infrared	Heckel 1979
	Fabricius et al. 1978
Radiant heat	Robins et al. 1983
Diathermia (27 MHz)	Pomp (1978)
+ warm air	Engelhardt et al. 1982
	Wallach et al. 1982
Wax bath	Pettigrew et al. 1978
Water blanket	Larkin 1979
+ warm air	Barlogie et al. 1979
	Herman et al. 1982
	Cole et al. 1979
	Reinhold et al. 1983
Water suit	Bull et al. 1979
Extracorporeal	Krementz 1979
heating	Parks et al. 1979
	Herman et al. 1982
	Yamanaka et al. 1982
	Barlogie et al. 1979
	Lange et al. 1983

Table 42. Toxicity related to whole body hyperthermia (41.5° C)

I	Hyperthermia alone
	Cardiac arrhythmias and failure
	Stroke, seizure
	Peripheral nerve lesions
	Protracted nausea, vomitting, and diarrhea
	Thrombocytopenia and other coagulopathies
	Hypomagnesemia and other electrolyte abnormalities
II	In combination with radiotherapy
	Sudden myelopathy (Douglas et al. 1981)
III	In combination with chemotherapy
	Adriamycin: cardiac arrhythmias (Kim et al. 1979)
	Mitomycin: pulmonary Edema (Bull 1984)
	Cis-Platin: renal failure (Gerad et al. 1983)
	BCNU: bone marrow (Bull et al. 1979)

nia and other coagulopathies. When whole body hyperthermia is combined with chemotherapy the toxicities of the drugs and their potential enhancement by heat also have to be considered. Related to high temperature levels and to the drugs applied, severe, even fatal, complications were observed.

Concerning the results of whole body hyperthermia applied alone, only "anecdotal" tumor responses, lasting briefly, were reported by Pettigrew and Ludgate (1977) and Parks et al. (1979), although a core temperature of up to 42.3° C had been employed by the latter.

An improvement of the effectiveness of both the chemotherapy and the hyperthermia was expected from the systemic combination of heat and drugs, by producing thermal enhancement of the specificity of the drugs in patients. But it was only more recently reported that in selected animal tumors in vivo (Honess and Bleehen 1985) and in some human tumors in vitro (Neumann et al. 1985) a thermal enhancement of the specificity of the drugs could happen. Predominantly patients with far advanced diseases and mostly pretreated were entered into the Phase I trials performed so far, so that little information came from these treatments concerning the question of a possible therapeutic gain. Additionally, conclusions on the efficacy of thermochemotherapy by whole body hyperthermia were hampered by the heterogeneity of the diagnosis and in respect to all those factors influencing the therapeutic effect of chemotherapy alone as well as in combination with hyperthermia (see Table 43). Details of the number of patients treated by whole body hyper-

Table 43. Factors influencing the clinical efficacy of heat-drug combination therapy

Patient - Performance status - Age - Sex - Other diseases - Supportive care	Drug - Type of action - Dosage - Combination - Schedule - Route of application - Pharmacokinetic
Tumor - Type/histology/grade - Burden/stage - Size of lesions - Sensitivity	Temperature - Heating rate - Maximum level - Duration - Timing

Table 44. Results of whole body hyperthermia in combination with chemotherapy

Group, method	Temperature time	Patients (n)	Diagnosis, histology	Drug dosage	Results
Parks (1979, 1980) EC	41.5°-42° C 5 h	102	25 lung cancer (15 squamous, 4 large cell, 3 adenocarcinoma, 2 small cell 1 alveolar)	Cyclo phosphamide 250 mg/m^2 BCNU 50 mg	Incomplete regression in 16/25, duration 4-6 months, 1 CR, 27 months
			23 gastrointestinal cancer	CPDD 100 g/m^2	2 CR, 10/16 months
			7 kidney cancer		
			2 ovary cancer		1 CR, 17 months
			11 breast cancer		1 NC, 1, 2 months
			9 malignant melanoma		1 CR, 1 month
			4 sarcoma		
			9 miscellaneous		
Herman (1982) EC WB	42°-42.4° C 2.5-3 h	11 (5) (6)	3 malignant melanoma	CPDD 60 mg/m^2 or BCNU 50 mg/m^2	2 PR, 6, 12 weeks
			3 large bowel cancer		2 PR, 5, 10 weeks
			3 sarcoma		1 PR, 8 weeks
			2 acute nonmyelogeneous leukemia		1 PR, 24 weeks
Yamanaka and Kato (1982, 1982)	41.5°-42° C 4-8 h	34	9 stomach cancer	Adriamycin n=13 0.5-1.5 mg/kg	7 "more than PR"
			8 large bowel cancer		10 patients alive in NC or PR
			4 kidney cancer		

Table 44 *(continued)*

Group, method	Temperature time	Patients *(n)*	Diagnosis, histology	Drug dosage	Results
EC			2 liver cancer 2 lung cancer 9 miscellaneous	or mitomycin ACNU VCR VBL 5-FU cyclophosphamide CPDD	2–8 months
Lange (1983) EC	41.8° C 5–6 h	14	6 gastrointestinal cancer 4 malignant melanoma 4 miscellaneous	5-FU 1.0 g DTIC 200 mg/m^2 cyclophosphamide 300 mg/m^2	5 NC survival 2–17 months 1 NC survival 3–8 months 1 PR 0.3–13 months
Barlogie (1979, 1979) WB	41.9°– 42° C 4 h	17	7 malignant melanoma 1 rectum cancer 1 breast cancer 1 sarcoma 1 malignant melanoma 1 non-Hodgkin lymphoma	L-PAM* *(n=5) 10–15 mg/m^2 VP-16* *(n=1) 150 mg/m^2	9 NC 3 PR 1 PR 1 PR
Larkin (1979) WB	42° C >2 h	29 5 adjuvant	lung cancer malignant melanoma kidney cancer gastrointestinal cancer stomach cancer sarcoma miscellaneous	5-FU 9 mg/kg CTX 200 mg/m^2 DTIC 200 mg/m^2	17/29 ("objective + subjective") transient
Bull (1979, 1979) WS	41.8° C 4 h	14 13	6 gastrointestinal cancer 3 malignant melanoma 2 sarcoma 1 ovary cancer 1 liver cancer 1 adrenal gland cancer malignant melanoma	Adriamycin 60 mg/m^2 BCNU 225 mg/m^2 Me-CCNU orally BCNU 225 mg/m^2	PR 2/14 "Objective respiration" 5/13 CR 1/13
Pettigrew (1977) Wax	40°–41° C 2 h	15	2 sarcoma 2 gastrointestinal cancer 2 malignant melanoma 1 breast cancer 3 ovary cancer 3 neurogenic tumors 6 miscellaneous	"Cytostatic drugs"	Some "objective + subjective responses"
(1978)	41.8° C 2 h	15	gastrointestinal cancer* malignant melanoma* breast cancer* 1 malignant melanoma	CTX 200 mg "and others" melphalan 1 mg/kg	"Remarkable regression"
Engelhardt (1979,	40.5°– 41° C	27	5 small cell lung cancer (limited disease)	Adriamycin 60 mg/m^2	CR 3/5 PR 2/5

Table 44 *(continued)*

Group, method	Temperature time	Patients (*n*)	Diagnosis, histology	Drug dosage	Results
1982) Dia	1 h		15 small cell lung cancer (extensive disease) 4 Hodgkin's disease 3 miscellaneous	VCR/2 mg Cyclophosphamide 250 mg/m^2 day 2–5	CR 8/15 PR 5/15 NC 1/15 PD 1/15 SR = 50% after 53 weeks

EC, extracorporeal heating; *WB*, water blanket; *WS*, water suit; *Wax*, wax bath; *Dia*, diathermia; *, "all influenced".

thermia-thermochemotherapy including their diagnosis, drug and dosages, heating method applied, and heating criteria are given in Table 44, which was adopted from an earlier review of the author (Engelhardt et al. 1984).

As can easily be seen from these data, it is difficult to draw any final conclusion on the therapeutic value of the methods applied. Only two reports judged their own results to be negative, due to an enormous amount of toxicity (Herman et al. 1982) and absence of any detectable therapeutic gain (van der Zee et al. 1983). The other investigators classified their results as being "promising" although nobody claimed to have found the "magic bullet" in cancer therapy. This indeed seems to be unlikely, bearing in mind the multitude of possible modifications of drug effects by heat as described in the preclinical data part of this review and the long list of factors influencing the drug-heat-effect in patients (Table 43).

Only a few investigators started typical pilot studies, the results of which could be used for comparison to those of standard normothermic chemotherapy:

Bull et al. (1979) treated 13 patients all having malignant melanoma with a standard protocol consisting of 225 mg BCNU/m^2 in combination with 4 h whole body hyperthermia at 41.8° C. Out of five objective responses there was only one complete remission; hematological toxicity was markedly enhanced.

Engelhardt et al. (1982) reported 15 patients having small cell lung cancer, extensive disease, treated first with the standard protocol ACO (Adriamycin, cyclophosphamide, Oncovin), and combined with 40.5°–41° C whole body hyperthermia for 1 h in 3 out of 6 multidrug application cycles.

Response rates were: complete remission (CR) 8/15 (not histologically affirmed!), partial remission (PR) 5/15. Fifty percent survival time was 53 weeks.

More phase II studies are underway (Lange et al. 1984) to answer specific biological questions and to find tumors and protocols on which it would be worthwhile commencing a phase III trial. Preliminary results of such a phase III trial were reported by Engelhardt et al. in 1985. Although the drugs used at an elevated temperature (Adriamycin and vincristine) may be crucial in terms of preclinically explored thermal enhancement, the antitumor effect seemed to be enhanced: The response rate (CR + PR) was 77% in the hyperthermically treated arm ($n = 17$) vs. 50% in the normothermically treated arm ($n = 15$). The 1-year survival rate was also increased from 0% to 29%. Bone marrow toxicity in terms of thrombocytes and white blood cell nadir values seemed to be enhanced too, thus demonstrating the possibility of the thermal enhancement of the drugs used, but concurrently reducing the hope of a therapeutic gain. The final judgment on this trial has to be postponed, however, until the final data are available.

Although of much interest, so far there are few data available on thermally induced changes in pharmacokinetics. The possibilities of thermally induced alterations of drug actions in man as well as thermally induced alterations of the response of the organism have been pointed out by Ballard (1974).

Preclinical studies in animals concerning pharmacokinetics of cytostatic drugs have been mentioned above (see Tables 3, 11–13). Clinical data have been reported on Adriamycin, CTX, and 5-FU. In patients receiving 45 mg/m^2 Adriamycin in combination with whole body hyperthermia (41.8° C for 2 h) excretion of Adriamycin seemed to be impaired, as total fluorescence in the plasma was increased with no change in Adriamycin concentrations, suggesting an increase in Adriamycinol levels (Brenner et al. 1981).

No significant changes in serum t½ and AUC values were found in patients receiving 50 mg/m^2 Adriamycin in combination with 40.5–41° C whole body hyperthermia for 1 h (Maier-Lenz et al. 1983). Cyclophosphamide pharmacokinetics were studied by Van Echo et al. (1982). The drug was given 4 h after or together with a 2-h course of whole body hyperthermia (41.8° C). While peak concentrations of CTX were unchanged as compared with drug administration at 37° C, urinary excretion of CTX was increased and urinary excretion of total alkylating activity was decreased. These data may indicate a decrease in total availability of active alkylating metabolites, thus reducing the cytotoxic activity of the drug. The plasma concentration-time course of 5-FU was studied by Lange et al. (1984). Patients received 1000 mg 5-FU by regional intraarterial infusion to the liver combined with whole body hyperthermia at 41.8° C induced by an extracorporeal heat exchanger. The initial half-life of elimination dropped from 17 min (normothermic application) to 8 min on hyperthermia without measurable increase in urinary excretion. Changes in distribution kinetics are discussed as the reason, and in addition the enlarged circulation volume by the heating technique applied may have contributed to these findings. Another important aspect of thermally induced changes of heat-drug interaction is the pH value of the tumors. A detailed review on this topic was given by Wike-Hooley et al. (1984). The authors reported in another communication (Wike-Hooley et al. 1984) that no substantial pH decrease was found during whole body hyperthermia of 41.8° C for 2 h in various tumors in patients. As these findings are in contrast to the majority of measurements in animal tumors, further investigations are warranted. This seems to be of particular interest as the pH value could perhaps be lowered by glucose infusions as reported by Urano and Kim (1983) in mice.

Local or Regional Heating

Electromagnetic techniques are employed for heating of bulky, deep, or extensive superficial tumors. If controlled heating could be restricted to the tumor volume and if temperature could be distributed homogeneously within the tumors, then the thermal enhancement of the cytotoxicity of concomitantly administered drugs could also be restricted to the tumor and no enhancement of systemic toxicity would occur.

Storm et al. (1982) treated patients with liver metastases from melanoma by systemic or intraarterial hepatic infusion of DTIC in combination with regional trunk heating by a circular current-carrying coil operating at 13.56 MHz (Magnetrode). The results were classified as being "promising."

Lateron Storm et al. reported on a multicenter trial including most probably the patients mentioned previously (Storm et al. 1985): In 1170 patients, 14807 phase I and II treatments were performed, consisting of regional hyperthermia (Magnetrode) combined with radiotherapy or chemotherapy. The data concerning thermochemotherapy were as follows: All patients had advanced primaries or recurrencies. Intratumoral temperatures were recorded occasionally; these were below 40° C in 20% of the measurements, over 50° C in 2% of the tumors, and between 40° and 42.9° C in the majority, 58%, of the tumors. In 95 patients treated by heat plus drugs given i.v., the CR rate was 4% and PR rate 18%. In patients ($n = 165$) who had failed with the same chemotherapy given previously normothermically, CR rate was 2% and PR rate 10%. In patients ($n = 99$) who received chemotherapy by intraarterial infusion, CR occurred in 5% and partial response in 18%. In patients ($n = 46$) who had failed with the same drug, now given intraarterially, complete remission rate and partial remission rate were both 4%. Although the object of this study was primarily to establish the safety of the Magnetrode heating system and the heterogeneity of the material is immense, very cautiously the following conclusions may be drawn: (1) There is no substantial advantage in regional drug administration together with regional heat as compared with the systemic route and (2) there is no striking evidence for the capacity of heating to overcome drug resistance of tumors, when using the heating method applied in this trial.

Two other reports, dealing with the experience in smaller numbers of patients in regional hyperthermia plus chemotherapy, may be mentioned. Fairman in 1982 reported on combined treatment in 22 cases of carcinoma of the head and neck. He had applied Adriamycin and bleomycin and 2450 MHz hyperthermia for 1 h three times a week. A high response rate (85%) was found but temperature measurement was not performed and no controls were mentioned.

Another report on superficially located head and neck tumors was published by Moffat et al. in 1984. MTX and other chemotherapeutic agents were applied together with capacitive radiofrequency heating (13.56 MHz). Single-point temperature measurements ranged between 41.5 and 43° C, lasting 72–120 min. In 14 patients two CRs occurred and, interestingly, three patients responded to MTX who had previously been unresponsive.

There are many other reports on clinical use of hyperthermia together with a wide variety of combinations including all the methods of conventional cancer therapy. But although they are of some value from the toxicity point of view and although some patients may have been palliated, there is no way of evaluating the treatment results because of the lack of controls, the small number and heterogeneity of the patients, the great variety of tumors and treatments, and last not least the lack of temperature control. These reports are beyond the scope of this review.

Regional Heating by Isolation Perfusion

At first glance this seems to be the route providing the most convincing evidence available for the efficacy of thermochemotherapy in human patients. But although employed since 1959 by Krementz's group (Creech et al. 1958), it is still not clear whether hyperthermia or the high drug concentration within the isolated circulation or the combination of both agents is the active principle.

Body extremities can be heated by isolating major arteries and veins and perfusing this isolated closed circuit with extracorporeally heated blood alone or with heated blood containing chemotherapeutic drugs.

The groups with the longest and most extensive experience in this technique are those of Krementz (1983), Stehlin Jr. et al. (1984), and Cavaliere et al. (1983). Since then quite a number of other groups have adopted this method as well and applied it not only to the isolated perfusion of extremities (Schrafford-Koops et al. 1981; Rege et al. 1983; Janoff et al. 1982; Ghussen et al. 1985; Sugarbaker and McBride 1976; Tonak et al. 1983) but also to other organs, e.g., the liver (Aigner et al. 1985).

Stehlin Jr. (1969) reported on the treatment of 39 patients with melanomas of the extremities, using combinations of hyperthermic perfusion and melphalan. The results are difficult to interpret, due to the lack of detailed information on the tumor temperatures, tumor stages, and other treatment details. In follow-up articles the results of the increasing number of patients is updated (Stehlin Jr. et al. 1984). Using the Berkson-Gage projection method for 392 patients with malignant melanoma of the extremities in stage III-A patients, 5 and 10-year survival rates were "greatly increased" to 81% and 49%, respectively. Control values, however, are not given for comparison. Usually the majority of the patients will die from distant metastases, which was not the case in Stehlins heat-treated patients. The authors supposed that stimulation of the immune system by antigenic material derived from destroyed tumor cells could be the reason for this remarkable change in the course of the disease. Stehlin also reported on his 15-year experience with 65 patients having soft tissue sarcomas of the extremities. They were treated with a multimodality plan: hyperthermic perfusion (38.8–40° C including melphalan and act-D), which was followed by radiotherapy (30 Gy within 3 weeks) and, finally, when feasible a 2- to 3-month delayed local excision was performed. This treatment "reduced enormously the need for radical amputation of the extremities" and yielded a 5-year survival rate of 72.7%, which again is difficult to judge due to the lack of stage-corresponding control data.

Cavaliere's group also has long-term experience in regional perfusion hyperthermia. They found better results in malignant melanoma patients (stage III A, B, A-B) when limb perfusion was performed at an elevated temperature without antineoblastic drugs ($n=18$) compared with hyperthermic antiblastic perfusion ($n=20$), producing a 5-year survival rate of 50% vs. 40%, respectively. These authors also reported multistep therapeutic procedure including hyperthermic perfusion in patients with osteosarcoma in order to reduce the necessity of limb amputation. Indeed local tumor control was improved but distant metastases occurred as usual. In 50 patients with different types of soft tissue sarcoma again no difference was found whether the hyperthermic perfusion was applied with or without Adriamycin in the perfusate.

The greatest number of patients was treated by Krementz, who gave a detailed summary of his results in 1983. In Table 45 his data concerning stage III tumor patients are compiled together with those of Stehlin and Cavaliere. Again there are heterogeneities within the groups in respect to treatment modalities, hampering a conclusive comparison. In addition the treatment modalities, especially in Krementz's patients, have substantially changed over the years in respect to the temperature level, the duration of heating, and also the drugs applied. Single-drug as well as multiple-drug regimens (melphalan, thio-tepa, nitrogen mustard, act-D) were used. Krementz himself acknowledges that a randomized trial comparing optimum surgery with and without perfusion therapy especially in stage I melanoma patients was urgently required.

In a nonrandomized prospective trial, Rege et al. in 1983 had reported a significant improvement on actuarial survival and disease-free survival in 39 clinical stage I perfusion patients compared with 72 stage I patients treated by conventional surgery alone. The difference was highly significant in those patients with a Breslow depth of invasion of more than 1.5 mm. The perfusion temperature was 40° C and melphalan was added at a dose level of 1.5 mg/kg.

It was not until 1984 that the first randomized trial on this question was published by Ghussen et al. (1984). Out of 104 stage I–III malignant melanoma patients having a Clark level of more than IV and a tumor thickness greater than 1.5 mm, 53 patients underwent

Table 45. Survival (%) in malignant melanoma patients treated with hyperthermic limb perfusion

	Tumor stage	Number of patients	5 years	10 years	15 years	20 years
Krementz	III A	71	35.6	28.4	28.4	28.4
(1983)	B	124	45.2	44.0	41.9	41.9
	AB	86	32.1	28.0	28.0	28.0
		281	38.9	35.5	34.5	34.5
Stehlin Jr.	III A	57	81.0	49.0		
(1983)	B	46	54.0	34.0		
	AB	38	45.0	35.0		
Cavaliere	III A					
(1984)	B	38				
	AB					
Hyperthermia =		18	50.0			
Hyperthermia + melphalan =		20	40.0			

hyperthermic (42° C) perfusion using melphalan (1.5 mg/kg in the lower extremities and 1.0 mg/kg in the upper extremities) in addition to conventional surgery, which was the only therapy for the 54 patients in the control group. Disease-free survival was used as the end point. The study had to be discontinued since an intermediate evaluation revealed a highly significant advantage for the regionally treated patients. It will be of great interest to see the survival rates within the next few years in order to obtain data comparable to those of the other groups mentioned previously. From the Kaplan-Meier plots of disease-free survival it seems advisable to be cautious in regard to the long-term results as the curves are already narrowing after 3 years, suggesting that the clinical manifestation of recurrencies may be postponed rather than definitely prevented. As the aim of this study was to evaluate the effectiveness of hyperthermic regional cytotoxic perfusion, the question remains to be answered whether the heat or the high drug concentration in the isolated perfusate or indeed the combination of both agents is the effective principle.

Regional isolation perfusion of the liver has been performed by Aigner et al. (1985) in patients having liver metastases of large bowel cancer. The drug used was 5-FU; the temperature applied was 39°-39.5° C. The toxicity of the procedere seems to be considerable and as the results were of relatively short-term palliative value only, it was felt that the therapeutic gain was limited to a highly selected small group of patients.

Another example of isolated application of both heat and drug was reported by Kubota et al. in 1984. Advanced-stage as well as superficial tumors of the urinary bladder were treated by a combination of daily courses of irradiation (150-200 rad, total exposure of 3500-4000 rad in 4 weeks) followed by intracavital irrigation with saline (43°-45° C) containing bleomycin (30 μg/ml). Although only 33 patients were studied, the authors felt that the reponse rate (CR + PR) in T_2-T_3-stage patients of 68% (13/19) is remarkable and "promising."

There is finally one report on regional hyperthermia plus chemotherapy combined with systemic hypothermia (Shingleton et al. 1961). Twenty-five patients with advanced abdominal or pelvic malignancies were cooled to 31°-32° C core temperature and an incomplete isolation perfusion of the tumor-bearing region was performed. The perfusate was heated to 41°-42° C; perfusion duration was 20-30 min and the drugs administered within the perfusate were several alkylating agents (including CTX!). Although the method was reported to be relatively safe and a fairly good palliation was obtained, no further reports were found in the literature.

Conclusions and Outlook

The fascinating field of heat-drug interaction is just beginning to be explored. Due to the intensive effort of a few groups, a fair amount of knowledge about the thermal enhancement of the cytotoxicity of drugs has been gathered within the past years (Hahn 1982). The data available not only include the basic phenomenon of thermally improved ability of many drugs to kill tumor cells in vitro but also confirm many of these results in vivo. In addition an increasing number of factors have been found, modifying the amount of thermally induced changes of the cytotoxicity of the drugs: Drug dosage, temperature level, heating time and heating rate, sequence of heat and drug application, and the so-called milieu factors (pH value, O_2 pressure, tumor size, and malnutrition) as well as interactions with noncytotoxic drugs have also been found to act as modifiers of the heat-drug interactions. In addition data are accumulating which are demonstrating the heterogeneous reaction of different tumors to the same heat-drug exposure. And even heterogeneity within the same tumor obviously occurs also in respect to thermal enhancement of drug toxicity.

The majority of these data were not available when those clinical trials were started which have provided our clinical experience of today. The crucial point, however, is, whether we will be successful in translating the preclinical results into the clinical setting. As always, one has to be cautious in predicting clinical results from experimental work. This is of particular importance in thermochemotherapy, as one must realize that thermal enhancement of the cytotoxicity of a drug of course does not mean thermal enhancement of the therapeutic efficacy of this drug. Although it has been shown that in some cases a therapeutic gain occurs to some extent so to speak intrinsically, this seems to happen more as an exception than as the rule.

The question of increasing the specificity of a drug by heating is of course of special relevance in whole body hyperthermia combined with systemic drug administration. In order to increase the probability of selectively enhanced cytotoxicity, attempts were made to restrict either heat or drug or even both to the tumor region.

Concerning whole body hyperthermia the methods which have been developed are safe and reproducible, the side effects seem to be manageable, but the costs are remarkably high. Clinical results are so far inconclusive, because there are not enough phase II or even phase III trials. Preclinical testing of freshly collected individual tumor specimens may be one way to confine this approach to patients with tumors showing greater sensitivity to the thermal enhancement of the cytotoxicity of the drugs, than compared with the normal tissues at risk, which is predominately the bone marrow. The latter seems to be of rather uniform sensitivity to heat and drugs, without great interindividual variations.

Local or regional heating has so far only been manageable in superficial tumors 3–4 cm in depth. Reports on thermochemotherapy of head and neck tumors are reported to be encouraging. The technical limitation of heating deep-seated tumors is the main reason for the lack of reliable data on clinical results concerning this approach of thermochemotherapy (Engelhardt 1986). Logically there is no advantage in regional administration of the drug compared with systemic administration, when combined with uncontrolled, heterogeneous heating, because parts of the tumor stay heated inefficiently. This might be overcome by permanent (Ethiblock) or intermittent (Spherex) stoppage of the circulation, thereby ruling out the most important factor in creating heterogeneity in heat distribution. In addition the intratumoral milieu could become more hypoxic and acidic, thus enhancing the heat effect and the cytotoxicity of some drugs.

The most reliable data on thermochemotherapy indeed exist concerning the isolation perfusion approach. The methods used are highly developed, provide safety, and confine the drug and the heat to the perfused region of the body, being mostly an extremity. After nearly 30 years of application, a randomized trial on malignant melanoma patients was finally published, demonstrating the significant superiority in patients having stage I–III disease, a Clark level of more than IV, and a tumor thickness greater than 1.5 mm. Whether the superiority will also turn out in other end points (e.g., survival time) remains to be seen; but in the adjuvant setting used, the disease-free survival is a meaningful parameter as well. Nevertheless, it remains to be established whether the elevated temperature, the high drug concentration, or the combination of both is the active principle. So the value of the thermochemotherapy approach to cancer therapy even in isolation perfusion remains to be explored in a very basic sense.

As the detailed knowledge on heat-drug interactions is improving by in vitro and animal experiments, a new generation of clinical trials could be initiated, using the increased amount of preclinical data as a guideline for designing clinical protocols.

Acknowledgements. The contribution is supported by Bundesministerium für Jugend, Familie, Frauen und Gesundheit, Bonn; Project Nr. 341-4719-2/30 (81).

References

Adwankar MK, Chitnis MP (1984) Effect of hyperthermia alone and in combination with anticancer drugs on the viability of P388 leukemic cells. Tumori 70: 231–234

Aigner KR, Walther H, Helling HJ, Link KH (1985) Die isolierte Leberperfusion. Beitr Oncol 21: 43–83, Karger, Basel

Alberts DS, Peng YM, Chen HSG, Moon TE, Cetas TC, Höschele JD (1980) Therapeutic synergism of hyperthermia-*cis*-platinum in a mouse tumor model. J Natl Cancer Inst 65: 455–461

Anderson RL, Ahier RG, Littleton JM (1983) Observations on the cellular effects of ethanol and hyperthermia in vivo. Radiat Res 94: 318–325

Andreesen R, Modolell M, Oepke GHF, Munder PG (1983) Temperature dependence of leukemic cell destruction by alkyl-lysophospholipids (NSC 324368). Exp Hematol 11: 564–570

Ballard BE (1974) Pharmacokinetics and temperature. J Pharmaceut Sci 63: 1345–1358

Barlogie B, Corry PM, Yip E, Lippmann L, Johnston DA, Khalil K, Tenczynski TF, Reilly E, Lawson R, Dosik G, Rigor B, Hankenson R, Freireich EJ (1979) Total-body hyperthermia with and without chemotherapy for advanced human neoplasms. Cancer Res 39: 1481–1489

Barlogie B, Corry PM, Drewinko B (1980) In vitro thermochemotherapy of human colon cancer cells with *cis*-dichlorodiammineplatinum (II) and mitomycin C. Cancer Res 40: 1165–1166

Barranco SC, Novak JK, Humphrey RM (1975) Studies on recovery from chemically induced damage in mammalian cells. Cancer Res 35: 1194–1204

Bistoni F, Lees DE, Bull J, Baskies AM, Chretien PB, Smith R, Whang-Peng J, Schuette W, De Vita VT (1979) Correlation of serum glycoproteins and blood T-cell levels with clinical course during whole body hyperthermia (WBHT) alone with chemotherapy. Seventieth annual meeting of the American Association for Cancer Research, 16–19 May 1979, New Orleans, Louisiana, Proc Am Assoc Cancer Res 20: 194

Bowden GT, Kasunic M, Sim D (1983) Sequence dependence for the hyperthermic potentiation of *cis*-diamine diechloroplatinum (II) induced cytotoxicity and DNA damage. Proc. 3rd annual meeting North American Hyperthermia Group (NAHG), 26 Feb–3 March 1983, San Antonio, Texas, pp 161–162

Braun J, Hahn GM (1975) Enhanced cell killing by bleomycin and 43° C hyperthermia and the inhibition of recovery from potentially lethal damage. Cancer Res 35: 2921–2927

Brenner DE, Van Echo D, Riggs CE, Wesley M, Whitacre M, Aisner J, Wienrik PH, Bachur NR (1981) Hyperthermia induced changes in Adriamycin pharmacokinetics in sarcoma patients receiving whole-body hyperthermia (wbh), adriamycin (adr) and cyclophosphamide (ctx). Proceedings 17th annual meeting, American Society of Clinical Oncology, Washington 22: 260

Bronk BV, Wilkins RJ, Regan JD (1973) Thermal enhancement of DNA damage by an alkylating agent in human cells. Biochem Biophys Res Commun 52: 1064-1070
Brown DM, Cohen MS, Sayerman RH (1983) Influence of heat on the intracellular uptake and radiosensitization of 2-nitromidazole hypoxic cell sensitizers in vitro. Cancer Res 43: 3138-3142
Bull JM, Lees D, Schütte W, Whang-Peng J, Smith R, Bynum G, Atkinson ER, Gottdiener JS, Gralnick HR, Shawker TH, De Vita V Jr (1979) Whole-body hyperthermia: a phase I trial of a potential adjuvant to chemotherapy. Ann Intern Med 90: 317-323
Bull JMC (1984) A review for systemic hyperthermia. In: Veath JM (ed) Hyperthermia and radiation therapy/chemotherapy in the treatment of cancer. Frontiers of radiation therapy and oncology, vol 18. Karger, Basel, pp 171-176
Busch W (1866) Über den Einfluß welche heftige Erysipeln zuweilen auf organisierte Neubildungen ausüben. Verh Naturhist Ges Rhein Westph 23: 28-30
Cavaliere R, Mondovi B, Moricca G, Monticelli G, Natali PG, Santori FS, Di Filippo F, Varanese A, Aloe L, Rossi-Fanelli A (1983) Regional perfusion hyperthermia. In: Storm FK (ed) Hyperthermia in cancer therapy. Hall Medical, Boston, pp 369-399
Chlebowski RT, Block JB, Cundiff D, Dietrich MF (1982) Doxorubicin cytotoxicity enhanced by local anesthetics in a human melanoma cell line. Cancer Treat Rep 66: 121-125
Clawson RE, Egorin MJ, Fox BM, Ross LA, Bachur NR (1981) Hyperthermic modification of cyclophosphamide metabolism in rat hepatic microsomes and liver slices. Life Sci 28: 1133-1137
Cohen GL, Bauer WR, Barton JK et al (1979) Binding of *cis*- and *trans*-dichlorodiammineplatinum (II) to DNA: evidence for unwinding and shortening of the double helix. Science 203: 1014-1016
Cole DR, Pung J, Kim JD, Berman RA, Cole DF (1979) Systemic thermotherapy (whole body hyperthermia). Int J Clin Pharmacol Biopharm 17: 329-333
Coley WB (1893) The treatment of malignant tumors by repeated inoculations of erysipelas: with a report of ten original cases. Am J Med Sci 105: 487-511
Collins FG, Skibba JL (1979) Effect of hyperthermia and mechlorethamine on hepatic function in isolated perfused liver (meeting abstract). Proc Am Assoc Cancer Res 20: 125
Collins FG, Skibba JL (1983) Altered hepatic functions and microsomal activity in perfused rat liver by hyperthermia combined with alkylating agents. Cancer Biochem Biophys 6: 205-211
Creech Jr O, Krementz ET, Ryan RF, Winblad JN (1958) Chemotherapy of cancer: regional perfusion utilizing an extracorporeal circuit. Ann Surg 148: 616-632
Cronau Jr LH, Bourke DL, Bull JM (1984) General anesthesia for wholebody hyperthermia. Cancer Res (Suppl) 44: 4873 s-4877 s
Dahl O (1982) Interaction of hyperthermia and Doxorubicin on malignant neurogenic rat cell line in (3T4C) in culture. In: Dethlefsen LA, Dewey WC (eds) 3rd international symposium: cancer therapy by hyperthermia, drugs and radiation, Ft Collins, Colorado 1980. Natl Cancer Inst Monogr 61: 251-253
Dahl O (1983) Hyperthermic potentiation of doxorubicin and 4-epi-doxorubicin in a transplantable neurogenic rat tumor (BT4A) in BD IX rats. Int J Radiat Oncol Biol Phys 9: 203-207
Dahl O, Mella O (1982) Enhanced effect of combined hyperthermia and chemotherapy (Bleomycin, BCNU) in a neurogenic rat tumor (BT4A) in vivo. Anticancer Res 2: 359-364
Dahl O, Mella O (1983) Effect of timing and sequence of hyperthermia and cyclophosphamide on a neurogenic rat tumor (BT4A) in vivo. Cancer 52: 983-987
Dahl O, Mella O (1984) Timing and sequence of hyperthermia and drugs. Hyperthermic oncology 1984, vol 1. Proc 4th int symp on hyperthermic oncology, Aarhus, Denmark, 2-6 July 1984, pp 425-428
Daly JM, Smith G, Frazier H, Dudrick SJ, Copeland EM (1982) Effects of systemic hyperthermia and intrahepatic infusion with 5-fluorouracil. Cancer 49: 1112-1115
De Martino C, Battelli T, Cavaliere R, Curcio CG, Bellocci M, Manocchi D, Giustini L, Mattioli R, Rinaldi M, Di Filippo F (1984) Morphological damage induced in vivo by Lonidamine on human metastatic cancer cells. Oncology (Suppl 1) 41: 94-103
De Silva V, Tofilon PJ, Gutin PH, Dewey WC, Buckley N, Deen DF (1985) Comparative study of the effects of hyperthermia and BCNU on BCNU-sensitive and BCNU-resistant 9L rat brain tumor cells. Radiat Res 103: 363-372
Dickson JA, Suzangar M (1974) In vitro-in vivo studies on the susceptibility of the solid Yoshida sarcoma to drugs and hyperthermia (42). Cancer Res 34: 1263-1274

Donaldson SS, Gordon LF, Hahn GM (1978) Protective effect of hyperthermia against the cytotoxicity of actinomycin D on Chinese hamster cells. Cancer Treat Rep 62: 1489-1495
Douglass EC, Bull JM, Smith R, Whang-Peng J (1979) Cytogenetic studies in clinical trials of hyperthermia. Seventieth Annual Meeting of the American Association for Cancer Research, May 16-19, 1979, New Orleans, Louisiana. Proc Am Assoc Cancer Res 20: 22
Douglas MA, Parks LC, Bebin J (1981) Sudden myelopathy secondary to therapeutic total-body hyperthermia after spinal-cord irradiation. N Engl J Med 304: 583-585
Douple EB, Strohbehn JW, de Sieyes DC, Alborough DP, Trembly BS (1982) Therapeutic potentiation of *cis*-diammineplatinum (II) and radiation by interstitial microwave hyperthermia in a mouse tumor. In: Dethlefsen LA, Dewey WC (eds): 3rd intern symposium: cancer therapy by hyperthermia, drugs and radiation. Natl Cancer Inst Monogr 61: 259-262
Engelhardt R (1985) Whole-body-hyperthermia. Methods and clinical results. Proceedings 4th international symposium hyperthermic oncology, Aarhus, Denmark, 2-6 July 1984, vol 2. Taylor and Francis, London, pp 263-276
Engelhardt R (1986) Clinical requirements of local and regional hyperthermia application. In: Bruggmoser G, Hinkelbein W, Engelhardt R, Wannenmacher M (eds) Locoregional high-frequency hyperthermia and temperature measurement. Springer, Berlin Heidelberg New York Tokyo, pp 1-6 (Recent Results Cancer Res vol 101)
Engelhardt R, Neumann H, von der Tann M, Löhr GW (1982) Preliminary results in the treatment of oat cell carcinoma of the lung by combined application of chemotherapy (CT) and whole-body hyperthermia. In: Gautherie M, Albert E (eds) Biomedical thermology. International symposium, Straßburg 1981. Liss, New York, pp 761-765
Engelhardt R, Neumann H, Hinkelbein W, Adam G, Weth R, Löhr GW (1984) Clinical studies in thermo-chemotherapy. In: Engelhardt R, Wallach DH (eds) Hyperthermia. In: Spitzy KH, Karrer K (eds) Proceedings of the 13th international congress of chemotherapy, Vienna 1983, vol 18. Egermann, Vienna, pp 273/41-273/46
Engelhardt R, Weth-Simon R, Neumann H, Maier-Lenz H (1985) Thermo-Chemotherapie kolo-rektaler Karzinome mit 4′-Epiadriamycin. In: Nagel GA, Wannenmacher M (eds) Farmorubicin: klinische Erfahrungen; gemeinsames Symposium der Arbeitsgemeinschaft für Internisten. Zuckschwerdt, Munich, pp 204-208 (Aktuelle Onkologie, Vol 15)
Fabricius HA, Neumann H, Stahn R, Engelhardt R, Löhr GW (1978) Klinisch-chemische und immunologische Veränderungen bei Gesunden nach einer einstündigen 40° C-Ganzkörperhyperthermie. Klin Wschr 56: 1049-1056
Fairman HD (1982) The treatment of head and neck carcinoma with adriamycin and bleomycin using adjuvant hyperthermia. J Laryngol Otol 96: 251-263
Field SB, Bleehen NM (1979) Hyperthermia in the treatment of cancer. Cancer Treat Rep 6: 63-94
Fisher GA, Hahn GM (1982) Enhancement of *cis*-diaminedichloroplatinum (*cis*-DDP) cytotoxicity by hyperthermia. In: Dethlefsen LA, Dewey WC (eds) 3rd international symposium: cancer therapy by hyperthermia, drugs and radiation, Ft Collins, Colorado 1980. Natl Cancer Inst Monogr 61: 255-257
Friedmann CA (1980) Structure-activity relationships of anthraquinones in some pathological conditions. Pharmacology 20: 113-122
Fujiwara K, Kohno I, Miyao J, Sekiba K (1984) The effect of heat on cell proliferation and the uptake of anti-cancer drugs into tumour. Hyperthermic oncology, 1984, vol 1, summary papers. Proceedings of the 4th intern symp on hyperthermic oncology 2-6 July 1984, pp 405-408
Gerad H, Egorin MJ, Whitacre M, Van Echo DA, Aisner J (1983) Renal failure and platinum pharmacokinetics in three patients treated with *cis*-diamminedichloroplatinum (II) and whole-body hyperthermia. Cancer Chemother Pharmacol 11: 162-166
Ghussen F, Nagel K, Groth W, Muller JM, Stutzer H (1984) A prospective randomized study of regional extremity perfusion in patients with malignant melanoma. Ann Surg 200: 764-768
Ghussen F, Nagel K, Müller JM, Groth W, Stützer H (1985) Die regionale Extremitätenperfusion bei Patienten mit malignen Melanomen. Tumor Diagnos Ther 6: 74-78
Goss P, Parsons PG (1977) The effect of hyperthermia and melphalan on survival of human fibroblast strains and melanoma cell lines. Cancer Res 37: 152-156
Haas GP, Klugo RC, Hetzel FW, Barton EE, Cerny IC (1984) The synergistic effect of hyperthermia and chemotherapy on murine transitional cell carcinoma. J Urol 132: 828-833
Hahn GM (1974) Metabolic aspects of the role of hyperthermia in mammalian cell inactivation and their possible relevance to cancer treatment. Cancer Res 34: 3117-3123

Hahn GM (1978) Interactions of drugs and hyperthermia in vitro and in vivo. In: Streffer C, Van Beuningen D, Dietzel F, Roettinger E, Robinson JE, Scherer E, Seeber S, Trott KR (eds) Cancer therapy by hyperthermia and radiation. Proceedings of the 2nd international symposium, Essen, 2-4 June 1977. Urban and Schwarzenberg, Baltimore, 1978 pp 72-79
Hahn GM (1979) Potential for therapy of drugs and hyperthermia. Cancer Res 39: 2264-2268
Hahn GM (1982) Hyperthermia and cancer. Plenum, New York
Hahn GM (1983) Hyperthermia to enhance drug delivery. In: Chabner BA (ed) Rational basis for chemotherapy UCLA symposia on molecular and cellular biology - New Series, vol 4. UCLA symp on the rational basis for chemotherapy, Keystone, CO, 1982. Liss, New York, pp 427-436
Hahn GM, Strande DP (1976) Cytotoxic effects of hyperthermia and adriamycin on Chinese hamster cells. J Natl Cancer Inst 57: 1063-1067
Hahn GM, Li GC (1982) Interactions of hyperthermia and drugs: treatment and probes. In: Dethlefsen LA, Dewey WC (eds) 3rd international symposium: cancer therapy by hyperthermia, drugs and radiation, Ft Collins, Colorado 1980. Natl Cancer Inst Monogr 61: 317-323
Hahn GM, Shiu EC (1983) Effect of pH and elevated temperature on the cytotoxicity of some Chemotherapeutic agents on Chinese hamster cells in vitro. Cancer Res 43: 5789-5791
Hahn GM, Braun J, Har-Kedar I (1975) Thermochemotherapy: synergism between hyperthermia (42-43°) and adriamycin (or bleomycin) in mammalian cell inactivation (cancer chemotherapy/ cell membranes). Proc Natl Acad Sci USA 72: 937-940
Hahn GM, Li GC, Shue E (1977) Interaction of amphotericin B and 43° C hyperthermia. Cancer Res 37: 761-764
Hall EJ, Astor M, Geard C, Biaglow J (1977) Cytoxicity of Ro-07-0582; enhancement by hyperthermia and protection by cysteamine. Br J Cancer 35: 809-815
Har-Kedar I, Bleehen NM (1976) Experimental and clinical aspects of hyperthermia applied to the treatment of cancer with special reference to the role of ultrasonic and microwave heating. Adv Radiat Biol 6: 229-266
Hassanzadeh M, Chapman IV (1983) Thermal enhancement of bleomycin-induced tumor growth delay: the effect of dose fractionation. Eur J Cancer Clin Oncol 19: 1517-1519
Hazan G, Ben-Hur E, Yerushalmi A (1981) Synergism between hyperthermia and cyclophosphamide in vivo: the effect of dose fractionation. Eur J Cancer 17: 681-684
Hazan G, Lurie H, Yerushalmi A (1984) Sensitization of combined *cis*-platinum and cyclophosphamide by local hyperthermia in mice bearing the Lewis lung carcinoma. Oncology 41: 68-69
Heckel M, Heckel I (1979) Beobachtungen an 479 Infrarothyperthermiebehandlungen. Beitrag zur Methodik der Ganzkörperüberwärmung. Med Welt 30: 971-975
Heppner F (1982) New technologies to combat malignant tumors of the brain. Anticancer Res 2: 101-109
Herman TS (1983) Temperature dependence of adriamycin, *cis*-diamminedichloroplatinum, bleomycin, and 1,3- bis (2-chloroethyl)-1-nitrosourea cytotoxity in vitro. Cancer Res 43: 517-520
Herman TS (1983) Effect of temperature on the cytotoxicity of vindesine, amsacrine, and mitoxantrone. Cancer Treat Rep 67: 1019-1022
Hermann TS, Cress AE, Sweets C, Gerner EW (1981) Reversal of resistance to methotrexate by hyperthermia in Chinese hamster ovary cells. Cancer Res 41: 3840-3843
Herman TS, Sweets CC, White DM, Gerner EW (1982) Effect of rate of heating on lethality due to hyperthermia and selected chemotherapeutic drugs. J Natl Cancer Inst 68: 487-492
Herman TS, Zukoski CF, Anderson RM, Hutter JJ, Blitt CD, Malone JM, Larson DF, Dean FC, Roth HB (1982) Whole-body hyperthermia and chemotherapy for treatment of patients. Cancer Treatm Rep 66: 259-265
Hinkelbein W, Menger D, Birmelin M, Engelhardt R (1984) The influence of whole-body hyperthermia on myelotoxicity of doxorubicin and irradiation in rats. In: Proceedings of the 4th international symposium on hyperthermic oncology, vol 1, Aarhus, Denmark, 2-6 July 1984, pp 281-283
Hiramoto R, Ghanta VK, Lilly MB (1984) Reduction of tumor burden in a murine osteosarcoma following hyperthermia combined with cyclophosphamide. Cancer Res 44: 1405-1408
Honess DJ (1983) Animal models in the evaluation of therapeutic gain of thermochemotherapy. In: Spitzy KH et al (eds) Proc of the 13th int congress of chemotherapy, Vienna, August 28-September 2, 1983. pp 273/15-273/18

Honess DJ, Bleehen NM (1982) Sensitivity of normal mouse marrow and RIF-1 tumor to hyperthermia combined with cyclophosphamide or BCNU: a lack of therapeutic gain. Br J Cancer 46: 236-248

Honess DJ, Bleehen NM (1985) Potentiation of melphalan by systemic hyperthermia in mice: therapeutic gain for mouse lung microtumours. Int J Hyperthermia 1: 57-68

Honess DJ, Donaldson J, Workman P, Bleehen NM (1985) The effect of systematic hyperthermia on melphalan pharmacokinetics in mice. Br J Cancer 51: 77-84

Janoff KA, Moseson D, Nohigren J, Davenport C, Richards C, Fletcher WS (1982) The treatment of stage I melanoma of the extremities with regional hyperthermia isolation perfusion. Ann Surg 196: 316-323

Ishida A, Mizuno S (1981) Synergistic enhancement of bleomycin cytotoxicity toward tumor cells in culture by a combination of ethanol and moderate hyperthermia. Gann 72: 455-458

Ishida A, Mizuno S (1982) Effects of hyperthermia and ethanol on the cytotoxicity of bleomycin, adriamycin, *cis*-diamminedichloroplatinum (II) and macromomycin toward Hela cells. Gann 73: 129-131

Issels RD, Biaglow JE, Epstein L, Gerweck LE (1984) Enhancement of cysteamine cytotoxicity by hyperthermia and its modification by catalase and superoxide dismutase in Chinese hamster ovary cells. Cancer Res 44: 3911-3915

Johnson HA, Pavelec M (1973) Thermal enhancement of thio-TEPA cytotoxicity. J Natl Cancer Inst 50: 903-908

Joiner MC, Steel GG, Stephens TC (1982) Response of two mouse tumors to hyperthermia with CCNU or Melphalan. Br J Cancer 45: 17-27

Joshi DS, Barendsen GW (1984) Hyperthermic modification of drug effectiveness for reproductive death of cultured mammalian cells. Indian J Exp Biol 22: 251-254

Kamura T, Aoki K, Nishikawa K, Baba T (1979) Antitumor effect of thermodifferential chemotherapy with carboquone on Ehrlich carcinoma. Gann 70: 783-790

Kano E, Furukawa M, Yoshikawa S, Tsubouchi S, Kondo T, Sugahara T (1984) Hyperthermic chemopotentiation and chemical thermosensitisation. Hyperthermic oncology 1984, vol 1, summary papers. Proceedings of the 4th int symp on hyperthermic oncology, Aarhus, Denmark, 2-6 July 1984, pp 437-440

Kato N, Hosoi M, Ohta K, Yamanaka N, Morita K, Ota K (1982) Clinical study of extracorporeal total body hyperthermia. IVth meeting of the European Co-operative Hyperthermia Group, London, 1-2 July 1982. Strahlentherapie 158: 384

Kidera Y, Baba T (1978) Blood flow-interrupting hyperthermic chemotherapy on established autochtonous mouse sarcoma induced by 3-methycholanthrene. Cancer Res 38: 556-559

Kim YD, Lees DE, Lake CR, Whang-Peng J, Schütte W, Smith R, Bull J (1979) Hyperthermia potentiates doxorubicin-related cardiotoxic effects. JAMA 241: 1816-1817

Kim JH, Kim SH, Alfieri AA, Young CW (1984) Quercetin, an inhibitor of lactate transport and a hyperthermic sensitizer of Hela cells. Cancer Res 44: 102-106

Kim JH, Kim SH, Alfieri A, Young CW, Silvestrini B (1984) Lonidamine: a hyperthermic sensitizer of HeLa cells in culture and of the meth-A tumor in vivo. Oncology 41: 30-35

Klein ME, Frayer K, Bachur NR (1977) Hyperthermic enhancement of chemotherapeutic agents in L1210 leukemia. Blood 50 5: 223

Klein ME, Frayer K, Gangji D (1977) Hyperthermic potentiation of daunorubicin in L1210 leukemia. Proceedings Sixty-eighth annual meeting of the American Association of Cancer Research, 18-21 May 1977, Denver, Colorado, AACR Abstracts 209

Koga S, Hamazoe R, Maeta M, Shimizu N, Kanayama H, Osaki Y (1984) Treatment of implanted peritoneal cancer in rats by continuous hyperthermic peritoneal perfusion in combination with an anticancer drug. Cancer Res 44: 1840-1842

Kostic L, Djordjevic O, Brkic G (1978) Effect of hyperthermia and chemotherapeutic agents on the survival of isolated mammalian cells. In: Streffer C, Van Beuningen D, Dietzel F, Roettinger E, Robinson JE, Scherer E, Seeber S, Trott KR (eds) Cancer therapy by hyperthermia and radiation. Proceedings of the 2nd international symposium, Essen, 2-4 June 1977. Urban and Schwarzenberg, Baltimore, pp 281-282

Krementz ET (1983) Chemotherapy by isolated regional perfusion for melanoma of the limbs. In: Schwemmle K, Aigner K (eds) Vascular perfusion in cancer therapy. Springer, Berlin Heidelberg New York, pp 193-203 (Recent results in cancer research, vol 86)

Kremkau FW, Kaufmann JS, Walker MM, Burch PG, Spurr CL (1976) Ultrasonic enhancement of nitrogen mustard cytotoxicity in mouse leukemia. Cancer 37: 1643-1647

Kubota Y, Nishimura R, Takai S, Umeda M (1979) Effect of hyperthermia on DNA single-strand breaks induced by bleomycin in HeLa cells. Gann 70: 681-685

Kubota Y, Shuin T, Miura T, Nishimura R, Fukushima S, Takai S (1984) Treatment of bladder cancer with a combination of hyperthermia, radiation and bleomycin. Cancer 53: 199-202

Lange J, Zänker KS, Siewert JR, Eisler K, Landauer B, Kolb E, Blümel G, Remy W (1983) Extrakorporal induzierte Ganzkörperhyperthermie bei konventionell inkurablen Malignompatienten. Dtsch Med Wschr 108: 504-509

Lange I, Zänker KS, Siewert JR, Blümel G, Eisler K, Kolb E (1984) The effect of whole body hyperthermia on 5-fluorouracil pharmakokinetics in vivo and clonogenicity of mammalian colon cancer cells. Anticancer Res 4: 27-32

Langer M, Weidenmaier W, Röttinger EM (1982) Increased cytotoxicity of misonidazole by pH reduction and 41° C hyperthermia in Chinese hamster cells. Strahlentherapie 158: 688-691

Larkin JM (1979) A clinical investigation of total-body hyperthermia as a cancer therapy. Cancer Res 39: 2252-2254

Latreille J, Barlogie B, Yip E, Reilly E, Freireich EJ (1979) Total body hyperthermia (TBH) with and without chemotherapy (CT) for advanced malignancies. Seventieth annual meeting of the American Association for Cancer Research, 16-19 May 1979, New Orleans, Louisiana. Proc Am Assoc Cancer Res 20: 58

Lehman CM, Stewart JR (1983) In vivo cytotoxicity of misonidazole and hyperthermia in a transplanted mouse mammary tumor. Radiat Res 96: 628-634

Li GC (1984) Thermal biology and physiology in clinical hyperthermia: current status and future needs. Cancer Res (Suppl) 44: 4886s-4893s

Li GC, Hahn GM (1978) Ethanol-induces tolerance to heat and to adriamycin. Nature 274: 699-701

Li DJ, Hahn GM (1984) Responses of RIF tumors to heat and drugs: dependence on tumor size. Cancer Treat Rep 68: 1149-1151

Lin PS, Turi A, Kwock E, Lu RC (1982) Hyperthermic effect on microtubule organization. Natl Cancer Int Monogr 61: 57-60

Lin PS, Cariani PA, Jones M, Kahn PC (1983) Work in progress: the effect of heat on bleomycin cytotoxicity in vitro and on the accumulation of co-bleomycin in heat-treated rat tumors. Radiology 146: 213-217

Lin PS, Hefter K, Jones M (1983) Hyperthermia and bleomycin schedules on V79 Chinese hamster cell cytotoxicity in vitro. Cancer Res 43: 4557-4561

Longo FW, Tomashefsky P, Rivin BD, Tannenbaum M (1983) Interaction of ultrasonic hyperthermia with two alkylating agents in a murine bladder tumor. Cancer Res 43: 3231-3235

Lorenz M, Habs M, Schmähl D (1983) Effect of moderate local hyperthermia combined with chemotherapy by administration of cyclophosphamide or *N*-nitroso-1,3-bis (2-chloroethyl) urea (BCNU) on Yoshida sarcoma implanted in the descending colon of rats. Langenbecks Arch Chir 359: 205-213

Lorenz M, Biwer E, Habs M, Schmähl D (1984) Wirkung der lokalen moderaten Hyperthermie in Kombination mit einer Chemotherapie durch *N*-nitrose-1,3-bis-(2-chloroethyl)-harnstoff (BCNU) auf das in das Colon descendens der Ratte transplantierte Yoshida-Sarkom. 2. Mitteilung: Monochemotherapie in Kombination mit nachfolgender Hyperthermie in unterschiedlichen Zeitintervallen. Langenbecks Arch Chir 362: 253-261

Ludlum DB (1977) Alkylating agents and nitrosoureas. In: Becker SF (ed) Cancer: a comprehensive treatise, vol 5. Plenum, New York, pp 285-307

Maata M, Koga S, Shimizu N, Kanayama H, Hamazoe R, Karino T, Yamane T, Oda M (1984) Effect of extracorporeally induced total body hyperthermia for cancer on cardiovascular function. Jpn Heart J 25: 993-1000

Magin RL (1983) Hyperthermia and chemotherapy: when will they be used in clinical treatment of cancer? Eur J Cancer Clin Oncol 19: 1655-1658

Magin RL, Niesman MR (1984) Temperature-dependent permeability of large unilamellar liposomes. Chem Phys Lipids 34: 245-256

Magin RL, Weinstein JN (1982) Delivery of drugs in temperature-sensitive liposomes. In: Gregoriadis, Senior, Trouet (eds) Targeting of drugs. Plenum, New York, pp 203-221

Magin RL, Sikic BI, Cysyk RL (1979) Enhancement of bleomycin activity against Lewis lung tumors in mice by local hyperthermia. Cancer Res 39: 3792-3795

Magin RL, Cysyk RL, Litterst CL (1980) Distribution of adriamycin in mice under conditions of local hyperthermia which improve systemic drug therapy. Cancer Treat Rep 64: 203-210

Mahaley MS Jr, Woodhall B (1961) Effect of temperature upon the in vitro action of anticancer agents on VX2 carcinoma. J Neurosurg 269-272

Maier-Lenz H, Weth R, Engelhardt R (1983) Pharmakokinetics of cytostatic drugs in men under normothermic and hyperthermic conditions with special reference to antracyclines. In: Engelhardt R, Wallach DH (eds) Proceedings of the 13th international congress of chemotherapy, Vienna. Session 12.10., part 273, Vienna 28 August-2 Sept 1983, pp 23-26

Malangoni MA, Grosfeld JL, Cakmak O, Ballantine TVN (1978) The effect of hyperthermia on survival in transplanted lymphosarcoma. J Pediat Surg 13: 740-745

Mann BD, Storm FK, Morton DL (1983) Predictability of response to clinical thermochemotherapy by the clonogenic assay. Cancer 52: 1389-1394

Marmor JB (1979) Interactions of hyperthermia and chemotherapy in animals. Cancer Res 39: 2269-2276

Marmor JB, Kozak D, Hahn GM (1979) Effects of systematically administered bleomycin or adriamycin with local hyperthermia on primary tumor and lung metastases. Cancer Treat Rep 63: 1311-1325

Massicotte-Nolan P, Glofcneski DJ, Kruuv J, Lepock JR (1981) Relationship between hyperthermic cell killing and protein denaturation by alcohols. Radiat Res 87: 284

Mella O (1985) Combined hyperthermia and *cis*-diamminedichloroplatinum in BD IX rats with transplanted BT4A tumors. Int J Hyperthermia 1: 171-183

Metelmann HR, von Hoff DD (1979) In vitro activation of dacarbazine (DTIC) for a human clonic system. Int J Cell Cloning 1: 24-32

Meyn RE, Corry PM, Fletcher SE, Demetriades M (1980) Thermal enhancement of DNA damage in mammalian cells treated with *cis*-diamminedichloroplatinum (II). Cancer Res 40: 1136-1139

Meyskens FL Jr (1980) Human melanoma colony formation in soft agar. In: Salmon SE (ed) Cloning of human tumor stem cells. Liss, New York, pp 85-99

Mikkelsen RB, Wallach DFH (1982) Transmembrane ion gradients and thermochemotherapy. In: Gautherie M, Albert (eds) Biomedical thermology. International symposium, Straßburg 1982. Liss, New York, pp 103-107

Mikkelsen RB, Lin PS, Wallach DF (1977) Interaction of adriamycin with human red blood cells: a biochemical and morphological study. J Mol Med 2: 33-40

Miller AB, Hoogstraten B, Staquet M, Winkler A (1981) Reporting results of cancer treatment. Cancer 47: 207-214

Mimnaugh EG, Waring RW, Sikic BI, Magin RL, Drew R, Litterst CL, Gram TE, Guarino AM (1978) Effect of whole-body hyperthermia on the disposition and metabolism of adriamycin in rabbits. Cancer Res 38: 1420-1425

Miyazaki M, Makowka L, Falk RE, Falk W, Venturi D, Ambus U, Falk JA (1983) Hyperthermochemotherapeutic in vivo isolated perfusion of the rat liver. Cancer 51: 1254-1260

Mizuno S, Ishida A (1981) Potentiation of bleomycin cytotoxicity toward cultured mouse cells by hyperthermia and ethanol. Gann 72: 395-402

Mizuno S, Ishida A (1982) Selective enhancement of the cytotoxicity of the bleomycin derivate, peplomycin by local anesthetics alone and combined with hyperthermia. Cancer Res 42: 4726-4729

Mizuno S, Amagai M, Ishida A (1980) Synergistic cell killing by antitumor agents and hyperthermia in cultured cells. Gann 71: 471-478

Mizuno S, Ishida A, Amagai M (1981) Potentiation of the action of antitumor agents by hyperthermia. Gan Kagakuryoho 8 (Suppl): 147-153

Moffat FL, Rotstein LE, Calhoun K, Langer JC, Makowka L, Ambus U, Palmer JA, Campbell A, Howard V, Mikkelsaar R, Venturi D, Laing D, Falk JA, Falk RE (1984) Palliation of advanced head and neck cancer with radiofrequency hyperthermia and cytotoxic chemotherapy. Can J Surg 27: 38-41

Morgan JE, Bleehen NM (1981) Response of EMT6 multicellular tumor spheroids to hyperthermia. Br J Cancer 43: 384-391

Morgan JE, Bleehen NM (1981) Interactions between misonidazole and hyperthermia in EMT6 spheroids. Br J Cancer 44: 810-818

Morgan J, Honess D, Bleehen N (1979) The interaction of thermal tolerance with drug cytotoxicity in vitro. Br J Cancer 39: 422-428
Muckle DS, Dickson JA (1983) Hyperthermia (42° C) as an adjuvant to radiotherapy and chemotherapy in the treatment of the allogeneic VX2 carcinoma in the rabbit. Br J Cancer 27: 307-315
Murphree SA, Cunningham LS, Hwang KM et al (1976) Effects of adriamycin on surface properties of sarcoma 180 ascites cells. Biochem Pharmacol 25: 1227-1231
Murray D, Milas L, Meyn RE (1984) DNA damage produced by combined hyperglycemia and hyperthermia in two mouse fibrosarcoma tumors in vivo. Int J Radiat Oncol Biol Phys 10: 1679-1682
Murthy MS, Khandekar JD, Travis JD, Scanlon EF (1984) Combined effect of hyperthermia (HT) and platinum compounds in vivo and in vitro on murine and human tumor cells. Hyperthermic oncology 1984, vol 1. Proc of the 4th int symp on hyperthermic oncology, Aarhus, Denmark (2-6 July 1984), pp 421-424
Nakajima K, Hisazumi H (1983) An experimental study of enhanced cell killing by hyperthermia and bleomycin. Urol Res 11: 43-46
Nauts HC (1985) Hyperthermic oncology: historic aspects and future trends. In: Overgaard J (ed) Hyperthermic oncology 1984, vol 2. Proceedings of the 4th int symp on hyperthermic oncology, Aarhus, Denmark (2-6 July 1984). Taylor and Francis, London, pp 199-209
Neumann H, Engelhardt R, Fabricius HA, Stahn R, Loehr GW (1979) Klinisch-chemische Untersuchungen an Tumorpatienten unter Zytostatica- und Ganzkörperhyperthermiebehandlung. Klin Wschr 57: 1311-1315
Neumann HA, Fiebig HH, Löhr GW, Engelhardt R (1985) Effects of cytostatic drugs and 40.5° C hyperthermia on human clonogenic tumor cells. Eur J Cancer Clin Oncol 21: 515-523
Neumann HA, Fiebig HH, Löhr GW, Engelhardt R (1985) Effects of cytostatic drugs and 40.5° C hyperthermia on human bone marrow progenitors (CFU-C) and human clonogenic tumor cells implanted into mice. JNCI 75: 1059-1066
Neville AJ, Robins HI, Martin P, Gilchrist KW, Dennis WH, Steeves RA (1984) Effect of whole body hyperthermia and BCNU on the development of radiation myelitis in the rat. Int J Radiat Biol 46: 417-420
O'Donnell JF, Mockoy WS, Makuch RW, Bull JM (1979) Increased in vitro toxicity to mouse bone marrow with 1,3-bis(2-chloroethyl)-1-nitrosourea and hyperthermia. Cancer Res 39: 2547-2549
Ohnoshi T, Ohnuma T, Beranek JT, Holland JF (1985) Combined cytotoxicity effect of hyperthermia and anthracycline antibiotics on human tumor cells. JCNI 71: 275-281
Orth PE, Swidler HJ, Zarakov MS (1977) Survival of pleomorphic sarcoma-37 transplanted virgin female DBA/2J mice: hyperthermia and hyperglycemia, alone and in combination with drugs. J Pharmaceut Sci 66: 437-438
Osieka R, Magin RL, Atkinson ER (1978) The effect of hyperthermia on human colon cancer xemografts in nude mice. In: Streffer C, Van Beuningen D, Dietzel F, Roettinger E, Robinson JE, Scherer E, Seeber S, Trott KR (eds) Cancer therapy by hyperthermia and radiation. Proceedings of the 2nd international symposium, Essen, 2-4 June 1977. Urban and Schwarzenberg, Baltimore, pp 287-290
Ostrow S, Van Echo D, Egorin M, Whitacre M, Grochow L, Aisner J, Colvin M, Bachur M, Bachur N, Wiernik PH (1982) Cyclophosphamide pharmacokinetics in patients receiving whole-body hyperthermia. In: Dethlefsen LA, Dewey WC (eds) 3rd international symposium: cancer therapy by hyperthermia, drugs and radiation, Ft Collins, Colorado 1980. Natl Cancer Inst Monogr 61: 401-403
Overgaard J (1976) Combined adriamycin and hyperthermia treatment of a murine mammary carcinoma in vivo. Cancer Res 36: 3077-3081
Palzer RJ, Heidelberger C (1973) Influence of drugs and synchrony on the hyperthermic killing of HeLa cells. Cancer Res 33: 422-427
Parks LC, Smith GV (1980) Extracorporeal induction of systemic hyperthermia techniques; effects on man and malignancy. Department of Surgery, University of Mississippi Medical Center, Jackson, Mississippi
Parks LC, Minaberry D, Smith DP, Neely WA (1979) Treatment of far-advanced bronchogenic carcinoma by extracorporeally induced systemic hyperthermia. 59th Annual Meeting of the American Association for Thoracic Surgery, Boston, Mass, April 30 to May 2, 1979. J Thorax Cardiovasc Surg 78: 883-892

Pettigrew RT, Ludgate CW (1977) Whole-body hyperthermia. A systemic treatment for disseminated cancer. In: Rossi-Fanelli A, Cavaliere R, Mondovi B, Moricca G (eds) Selective heat sensitivity of cancer cells. Springer, Berlin Heidelberg New York, pp 153-170

Pettigrew RT, Ludgate CM, Gee AP, Smith AN (1978) Whole-body hyperthermia combined with chemotherapy in the treatment of advanced human cancer. In: Streffer C, Van Beuningen D, Dietzel F, Roettinger E, Robinson JE, Scherer E, Seeber S, Trott KR (eds) Cancer therapy by hyperthermia and radiation. Proceedings of the 2nd international symposium, Essen 1977. Urban and Schwarzenberg, Baltimore, pp 337-339

Pigram WJ, Fuller W, Amilton LDH (1972) Stereochemistry of intercalation: interaction of daunomycin with DNA. Nature (New Biol) 235: 17-19

Pomp H (1978) Clinical application of hyperthermia in gynecological malignant tumors. In: Streffer C, Van Beuningen D, Dietzel F, Roettinger E, Robinson JE, Scherer E, Seeber S, Trott KR (eds) Cancer therapy by hyperthermia and radiation. Proceedings of the 2nd international symposium, Essen, 2-4 June 1977. Urban and Schwarzenberg, Baltimore, pp 326-327

Raaphorst GP, Azzam EI, Einspenner M, Vadasz JA (1984) Inhibition of DMSO-induced differentiation by hyperthermia in a murine erythroleukemia cell system. Can J Biochem Cell Biol 62: 1091-1096

Rabbani B, Sondhaus CA, Swingle KF (1978) Cellular response to hyperthermia and bleomycin: effect of time sequencing and possible mechanisms. In: Streffer C, Van Beuningen D, Dietzel F, Roettinger E, Robinson JE, Scherer E, Seeber S, Trott KR (eds) Cancer therapy by hyperthermia and radiation. Proceedings of the 2nd international symposium, Essen, 2-4 June 1977. Urban and Schwarzenberg, Baltimore, pp 291-293

Rege VB, Leone LA, Soderberg CH, Coleman GV, Robidoux HJ, Fijman R, Brown J (1983) Hyperthermic adjuvant perfusion chemotherapy for stage I malignant melanoma of the extremity with literature review. Cancer 52: 2033-2039

Robins HI (1984) Role of whole body hyperthermia in the treatment of neoplastic disease: its current status and future prospects. Cancer Res (Suppl) 44: 4878s-4883s

Robins HI, Dennis WH, Slattery IS, Lange TA, Yatvin MB (1983) Systemic lidocaine enhancement of hyperthermia-induced tumor regression in transplantable murine tumor models. Cancer Res 43: 3187-3191

Robins HI, Grossman J, Davis TE, Aubuchon JP, Dennis W (1983) Preclinical trial of a radiant heat device for whole-body hyperthermia using a porcine model. Cancer Res 43: 2018-2022

Robins HI, Dennis WH, Martin PA, Sondel PM, Yatvin MB, Steeves RA (1984) Potentiation of differential hyperthermic sensitivity of AKR leukemia and normal bone marrow cells by lidocaine or thiopental. Cancer 54: 2831-2835

Rochlin DB, Thaxter TH, Dickerson AG, Shiner J (1961) The effect of tissue temperature on the binding of alkylating agents in the isolation perfusion treatment of cancer. Surg Gyn Obstet 113: 555-561

Roizin-Towle L, Hall EJ (1982) Interaction hyperthermia and cytotoxic agents. In: Dethlefsen LA, Dewey WC (eds) 3rd international symposium: cancer therapy by hyperthermia, drugs and radiation, Ft Collins, Colorado 1980. Natl Cancer Inst Monogr 61: 149-151

Rose WC, Veras GH, Laster WR Jr, Schabel FM Jr (1979) Evaluation of whole-body hyperthermia as an adjunct to chemotherapy in murine tumors. Cancer Treat Rep 63: 1311-1325

Rotstein LE, Daly J, Rozsa P (1983) Systemic thermochemotherapy in a rat model. Can J Surg 26: 113-116

Senapati N, Houchens D, Ovejera A, Beard R, Nines R (1982) Ultrasonic hyperthermia and drugs as therapy for human tumor xenografts. Cancer Treatm Rep 66: 1635-1639

Shingelton WW, Parker RT, Mahaley S (1961) Abdominal perfusion for cancer chemotherapy with hypothermia and hyperthermia. Surgery 50: 260-265

Shingleton WW, Bryan FA Jr, O'Quinn WL, Krueger LC (1962) Selective heating and cooling of tissues in cancer chemotherapy. Ann Surg 156: 408-416

Silvestrini B, Hahn GM, Cioli V, Demartino C (1983) Effects of lonidamine alone or combined with hyperthermia in some experimental cell and tumor systems. Br J Cancer 47: 221-231

Skibba JL, Jones FE, Condon RE (1982) Altered hepatic disposition of doxorubicin in the perfused rat liver at hyperthermic temperatures. Cancer Treat Rep 66: 1357

Sugarbaker EV, McBride CM (1976) Survival and regional disease control after isolation perfusion for invasive stage I melanoma of the extremities. Cancer 37: 188-198

Sutherland CM, Krementz ET (1979) Systemic hyperthermia induced by partial cardio-pulmonary bypass for treatment of metastatic tumors. Fifteenth annual meeting of the American Society of Clinical Oncology, 14-15 May 1979, New Orleans. Proc Am Soc Clin Oncol 20: 374

Suzuki K (1967) Application of heat to cancer chemotherapy: experimental studies. Nagoya J Med 30: 1-21

Szczepanski L, Trott KR (1981) The combined effect of bleomycin and hyperthermia on the adenocarcinoma 284 of the C3H mouse. Eur J Cancer Clin Oncol 17: 997-1000

Schraffordt-Koops H, Beekhius H, Oldhoff J, Osterhuis JW, Van der Ploog E, Vermey A (1981) Local recurrence and survival in patients with (Clark Level IV/V and over 1.5 mm thickness) stage I malignant melanoma of the extremities after regional perfusion. Cancer 48: 1952-1957

Schrak R, Stefani SS (1981) Effects of alcohol and hyperthermia on normal and leukemic lymphocytes. Oncology 38: 69-71

Stehlin JS Jr (1969) Hyperthermic perfusion with chemotherapy for cancers of the extremities. Surg Gynecol Obstet 120: 305-308

Stehlin JS Jr, Giovanella BC, Ipolyi de PD, Muenz LR, Anderson RF (1975) Results of hyperthermic perfusion for melanoma of the extremities. Surg Gynec Obstet 140: 339-348

Stehlin JS Jr, Giovanella BC, Gutierrez AE, Ipolyi de PD, Greff PJ (1984) 15 year's experience with hyperthermic perfusion for treatment of soft tissue sarcoma and malignant melanoma of the extremities. In: Veath JM (ed) Hyperthermia and radiation therapy/chemotherapy in the treatment of cancer. Frontiers of radiation therapy and oncology, vol 18. Karger, Basel, pp 177-182

Storm FK, Kaiser LR, Goodnight JE, Harrison WH, Elliott RS, Gomes AS, Morton DL (1982) Thermochemotherapy for melanoma metastases in liver. Cancer 49: 1243-1248

Storm KF, Baker HW, Scanlon EF, Plenk HP, Meadows PM, Cohen SC, Olson CE, Thomson JW, Khandekar JD, Rose D, Nizze A, Morton DL (1985) Magnetic-induced hyperthermia. Results of a 5-year multi-institutional national cooperative trial in advanced cancer patients. Cancer 55: 2677-2687

Tacker JR, Anderson RU (1982) Delivery of antitumor drugs to bladder cancer by use of phase transition liposomes and hyperthermia. J Urol 127: 1211-1214

Takeshita M, Grollmann AP, Ohtsubo E et al (1978) Interaction of bleomycin with DNA. Proc Natl Acad Sci USA 75: 5983-5987

Takiyama W (1984) Experimental studies on combined chemotherapy with hyperthermia and ethanol for advanced esophageal cancer. II. Effects of combined treatments on tumor growth in tumor-bearing mice. Nippon Geka Gakki Zasshi 85: 118 (Abstract)

Teicher BA, Kowal CD, Kennedy KA, Sartorelli AC (1981) Enhancement by hyperthermia of the in vitro cytotoxicity of mitomycin C toward hypoxic tumor cells. Cancer Res 41: 1096-1099

Thuning CA, Bakir NA, Warren J (1980) Synergistic effect of combined hyperthermia and a nitrosourea in treatment of a murine ependymoblastoma. Cancer Res 40: 2726-2729

Tonak J, Hohenberger W, Weidner F, Göhl H (1983) Hyperthermic perfusion in malignant melanoma: 5-years result. Springer, Berlin Heidelberg New York, pp 229-233 (Recent results in cancer research, vol 86)

Twentyman PR, Morgan JE, Donaldson J (1978) Enhancement by hyperthermia of the effect of BCNU against the EMT6 mouse tumor. Cancer Treat Rep 62: 439-443

Umezawa H, Hori S, Sawa T et al (1974) A bleomycin-inactivating enzyme in mouse liver. J Antibiotics 27: 419-424

Urano M, Kim MS (1983) Effect of hyperglycemia in thermochemotherapy of a spontaneous murine fibrosarcoma. Cancer Res 43: 3041-3044

Urano M, Overgaard M, Suit H, Dunn P, Sedlacek R (1978) Enhancement by *Corynebacterium parvum* of the normal and tumor tissue response to hyperthermia. Cancer Res 38: 862-864

Urano M, Suit HD, Dunn P, Landsdale T, Sedlacek RS (1979) Enhancement of the thermal response of animal tumors by *Corynebacterium parvum*. Cancer Res 39: 3454-3457

Urano M, Yamashita T, Suit HD, Gerweck LE (1984) Enhancement of thermal response of normal and malignant tissues by *Corynebacterium parvum*. Cancer Res 44: 2341-2347

Van der Linden PWG, Sapareto SA, Corbett TH, Valeriote FA (1984) Adriamycin and heat treatments in vitro and in vivo. Hyperthermic oncology 1984, vol 1, summary papers. Proceedings of the 4th international symp on hyperthermic oncology, Aarhus, Denmark, 2-6 July 1984, pp 449-452

Vanderzee J, Vanrhoon GC, Wikehooley JL, Faithfull NS, Reinhold HS (1983) Whole body hyperthermia in cancer therapy - a report of a phase I-II study. Eur J Cancer Clin Oncol 19: 1189-1200

Vig BK, Cornforth M, Farook SAF (1982) Hyperthermic potentiation of chromosome aberrations by anticancer antibiotics. Cytogenet Cell Genet 33: 35-41
Wallach DFH (1977) Basic mechanisms in tumor thermotherapy. J Mol Med 2: 381-403
Wallach DFH, Madoc-Jones H, Sternick ES, Santoro JJ, Curran B (1982) Moderate-temperature whole-body hyperthermia in the treatment of malignant disease. In: Gautherie M, Albert E (eds) Biomedical thermology. International symposium, Straßburg 1981. Liss, New York, pp 715-720
Waring MJ (1976) DNA-binding characteristics of acridinil-methanesulphonanilide drugs: comparison with antitumor properties. Eur J Cancer 12: 995-1001
Weinstein JN, Magin RL, Cysyk RL, Zaharko DS (1980) Treatment of solid L 1210 murine tumors with local hyperthermia and temperature-sensitive liposomes containing methotrexate. Cancer Res 40: 1388-1395
West KW, Weber TR, Grosfeld JL (1980) Synergistic effect of hyperthermia, papaverine, and chemotherapy in murine neuroblastoma. J Pediat Surg 15: 913-917
Westphal O, Westphal U, Sommer T (1977) The history of pyrogen research. American Society for Microbiology, pp 221-238
Wike-Hooley JL, Haveman J, Reinhold HS (1984) The review of tumour pH to the treatment of malignant disease. Radiotherapy and Oncology 2: 343-366
Wike-Hooley JL, van der Zee J, van Rhoon GC, van den Berg AP, Reinhold HS (1984) Human tumor pH changes following hyperthermia and radiation therapy. Eur J Cancer Clin Oncol 20: 619-623
Wile AG, Nahabedian MY, Plumley DA, Guilmette JE, Mason GR (1983) Experimental hyperthermic isolation-perfusion using *cis*-diamminedichloroplatinum (II). Cancer Res 43: 3108-3111
Willnow U (1981) The effect of hyperthermia alone or in combination with actinomycin D on the RNA metabolism of solid tumors in children. Neoplasma 28: 721-729
Wüst GP, Norpoth K, Witting U, Oberwittler W (1973) Der in vitro-Effekt der Hyperthermie auf Einbau von Nucleinsäurevorläufern in Tumoren und normale Gewebe. Z Krebsforsch 79: 193-203
Yamada K, Someya F, Shimada S, Ohara K, Kukita A (1984) Thermochemotherapy for malignant melanoma: combination therapy of ACNU and hyperthermia in mice. J Invest Dermatol 82: 180-184
Yamanaka N, Kato N, Hosoi M, Kazuhiro Ota, Morita K, Kazuo O (1982) Total body hyperthermia re-sensitizes doxorubicin against human neo-plasma resistant to doxorubicin. The Bio-Dynamics Res Inst and Shinseikai Hospital, Japan IVth Meeting of the European Co-operative Hyperthermia Group, London, 1-2 July 1982. Strahlentherapie 158: 392
Yamane T, Koga S, Maeta M, Hamazoe R, Karino T, Oda M (1984) Effects of in vitro hyperthermia on concentration of adriamycin in Ehrlich ascites cells. Hyperthermic oncology 1984, vol 1, summary papers. Proceedings of the 4th international symp on hyperthermic oncology, Aarhus, Denmark, 2-6 July 1984, pp 409-412
Yang SJ, Rafla S (1985) Temperature effect on mitoxantrone cytotoxicity in Chinese hamster cells in vitro. Cancer Res 45: 3593-3597
Yatvin M (1977) The influence of membrane lipid composition and procaine on hyperthermic death of mice cells. Int J Radiat Biol 32: 513-523
Yatvin MB, Clifton KH, Dennis WH (1979) Hyperthermia and local anesthetics: potentiation of survival of tumor-bearing mice. Science 205: 195-196
Yatvin MB, Mühlensiepen H, Porschen W, Weinstein JN, Feinendegen LE (1981) Selective delivery of liposome-associated *cis*-dichlorodiammineplatinum (II) by heat and its influence on tumor drug uptake and growth. Cancer Res 41: 1602-1607
Yerushalmi A (1978) Combined treatment of a solid tumour by local hyperthermia and actinomycin-D. Br J Cancer 37: 827-832
Yerushalmi A (1978) Simulation of resistance against local tumour growth of hosts: pretreated by combined local hyperthermia and x-irradiation. Bull Cancer 65: 475-478
Yerushalmi A, Hazan G (1979) Control of Lewis lung carcinoma by combined treatment with local hyperthermia and cyclophosphamide: preliminary results. Isr J Med Sci 15: 462-463
Yerushalmi A, Tovey MG, Gresser I (1982) Antitumor effect of combined interferon and hyperthermia in mice. Proc Soc Exp Biol Med 169: 413
Zupi G, Badaracco G, Cavaliere R, Natali PG, Greco C (1984) Influence of sequence on hyperthermia and drug combination. Hyperthermic oncology 1984, vol 1. Proc of the 4th int symp on hyperthermic oncology, Aarhus, Denmark, 2-6 July 1984, pp 429-432

Subject Index